Biomedical Engineering I
Recent Developments

Biomedical Engineering I
Recent Developments

Proceedings of the First Southern Biomedical Engineering Conference

Edited by
Subrata Saha
Louisiana State University Medical Center

Pergamon Press
New York Oxford Toronto Sydney Paris Frankfurt

Pergamon Press Offices:

U.S.A.	Pergamon Press Inc., Maxwell House, Fairview Park, Elmsford, New York 10523, U.S.A.
U.K.	Pergamon Press Ltd., Headington Hill Hall, Oxford OX3 0BW, England
CANADA	Pergamon Press Canada Ltd., Suite 104, 150 Consumers Road, Willowdale, Ontario M2J 1P9, Canada
AUSTRALIA	Pergamon Press (Aust.) Pty. Ltd., P.O. Box 544, Potts Point, NSW 2011, Australia
FRANCE	Pergamon Press SARL, 24 rue des Ecoles, 75240 Paris, Cedex 05, France
FEDERAL REPUBLIC OF GERMANY	Pergamon Press GmbH, Hammerweg 6 6242 Kronberg/Taunus, Federal Republic of Germany

Library of Congress Cataloging in Publication Data

Southern Biomedical Engineering Conference (1st : 1982:
Louisiana State University Medical Center)
Biomedical engineering I, recent developments.

1. Biomedical engineering--Congresses. I. Saha, Subrata. II. Title. [DNLM: 1. Biomedical engineering--Congresses. W3 S064 1st 1982b / QT 34 S727 1982b]
R856.A2S68 1982 610'.28 82-9008
ISBN 0-08-028826-X AACR2

Printed in the United States of America

CONTENTS
and
TECHNICAL PROGRAM

Opening of the Conference — Monday, June 7, 1982

PLENARY SESSION
Subrata Saha, Chairman, LSU Medical Center

9:00 a.m. — 1st floor auditorium

WELCOMING REMARKS
Perry Rigby, Dean
LSU Medical Center, Shreveport, LA

Martin D. Woodin, President
LSU University System

GENERAL LECTURE
Interfaces
Paul W. Brand, M.D.
Chief, Rehabilitation Branch
National Hansen's Disease Center, Carville, LA
Clinical Professor of Surgery
Professor of Orthopaedic Surgery
LSU School of Medicine

BANQUET ADDRESS
Future Directions of Biomedical Engineering
Samuel F. Hulbert, Ph.D.
President, Rose Hulman Institute of Technology
Terre Haute, IN

Session 1 — Monday, June 7

10:00 - 12:00 a.m. — Room 3-322

BIOMATERIALS

Chairman: L.J. Richelle, M.D., Univ. of New Orleans
New Orleans, LA
Co-Chairman: J.E. Lemons, Ph.D., Univ. of Alabama
Birmingham, AL

Session 15 Tuesday, June 8

COMPUTERS IN MEDICINE

10:20 - 12:00 a.m. Room 8-314

Chairman: G.R. Meneely, M.D., LSU Medical Center
Shreveport, LA
Co-Chairman: W.P. Osborne, M.D., Doctor's Hospital
Shreveport, LA

Session 16 Tuesday, June 8

BONE MECHANICS

1:30 - 3:30 Room 3-322

Chairman: W. Bonfield, Ph.D., Univ. of London
London, England
Co-Chairman: W. C. VanBuskirk, Ph.D., Tulane Univ.,
New Orleans, LA

*Paper unavailable at time of publication.

PREFACE

The First Southern Biomedical Engineering Conference was held at the Louisiana State University Medical Center in Shreveport, Louisiana June 7 and 8, 1982. Our objective in organizing this conference was to provide a forum for publication and timely dissemination of technical information that is being generated in this rapidly developing field of biomedical engineering. Another objective of initiating this new regional conference series in the south was to encourage graduate and undergraduate students of biomedical engineering from neighboring states to participate in a conference and to become familiar with the type of research and development work which is presently being conducted in this field.

This Proceedings contains extended abstracts of most of the 100 papers which were presented at this conference. Short abstracts of all accepted papers have also been published in a special issue of the journal of Biomaterials, Medical Devices and Artificial Organs.

A large and successful conference is the product of dedicated work of many people. I wish to acknowledge the contributions of the authors, session chairmen, reviewers and members of the program and local arrangements committee. Financial support of co-sponsoring organizations is also much appreciated. Finally, many thanks to Mrs. Nadine Lewis, Dr. Subrata Pal and Dr. Sukamal Saha for their technical assistance in finalizing the program.

June 1982
LSU Medical Center
Shreveport, Louisiana 71130

Subrata Saha
Program Chairman

Biomedical Engineering I
Recent Developments

Session 1
BIOMATERIALS

AUGMENTATION OF THE DEFICIENT ALVEOLAR RIDGE
WITH HYDROXYLAPATITE

J. Kent and J. Quinn

Dept. of Oral & Maxillofacial Surgery, Louisiana State University
1100 Florida Avenue, New Orleans, Louisiana 70119

ABSTRACT

A polycrystalline particulate form of hydroxylapatite, was placed subperiosteally to augment deficient alveolar ridges with minor or major resorption, to enable the patients to wear dentures successfully. 55 patients had surgery, the longest follow-up was 42 months with only 10-20% resorption terminating by 6 months postoperatively. Iatrogenic complications only. Patients assessed their ability to wear dentures with stability and comfort as excellent or good using the Cornell Medical Index evaluation.

KEYWORDS

Hydroxylapatite; alloplast; alveolar ridge augmentation; bone graft substitute; preprosthetic surgery.

INTRODUCTION

Approximately 25% of the population of the United States over 30 years of age are edentulous. The very high percentage of these patients have deficient alveolar ridges making it very difficult for them to wear dentures, particularly mandibular dentures. Both autogenous and homogenous bone grafts have been used to augment the deficient alveolar ridges. Problems of extensive resorption of the grafts frequently occurs resulting in a 60-100% loss. Alloplasts of ceramics, porous and solid polymers, and various metals have been also employed. Problems of migration, brittleness, soft tissue pain from denture trauma and occasional infections have resulted.

A dense polycrystalline nonresorbable form of hydroxylapatite (HA)*, free of fine pores and secondary crystalline phases with significantly greater compressive and tensile strength than previously reported sintered (HA) was used. Animal studies have shown HA to be extremely well tolerated by both the bone and soft tissues with an absence of inflammation and immune responses. HA is osteogenetically inductive and forms a strong intimate interface with bone.[1-4] The HA particles were used either alone in minor atrophic defects or with particulate cancellous bone for major atrophic alveolar defects.

*Durapatite, Sterling Winthrop Research Institute Rensselaer, N.Y.

MATERIALS AND METHODS

Preoperative Patient Considerations - Patients received 18-40 mesh HA for augmentation of minor deficiencies or HA in combination with autogenous particulate cancellous bone to correct major deficiencies. This report provides follow-up sugical and dental data on 55 patients with HA augmentation from 12-48 months.

The deficient alveolar ridges were classified and treated as follows:

Class I - adequate height and insufficient width, often with significant undercut areas (Use HA particle alone)

Class II - Ridge deficient in height and width with a knife edge. (Use HA particle alone)

Class III - Complete resorption of alveolus to basilar bone. (Use HA particles alone or with bone)

Class IV - Resorption of some basilar bone (pencil thin mandible). (Use HA particles and bone)

Patients undergoing general anesthesia were given 2 million units of aqueous penicillin intravenously at surgery. Oral phenoxy-methyl penicillin 500 mg. was given 30 minutes preoperatively to outpatients. Penicillin sensitive patients were given a suitable substitute antibiotic.

SURGICAL TECHNIQUE

All procedures using HA alone are performed under local anesthesia as an outpatient. Those requiring HA and bone are obviously performed in a hospital under general anesthesia.

The necessary number of syringes containing the HA particles are filled with patients blood at least 30 minutes before being injected into the subperiosteal pockets. Blood acts as a cohesive vehicle for HA preventing their dislodgment from the syringe during manipulation.

Augmentation of an anterior alveolar atrophy, which most often occurs in the maxilla, involved a single vertical midline incision approximately 2 centimeters in length made sharply down through the periosteum. Mucoperiosteum was then reflected on the lateral and crestal alveolus creating a subperiosteal pocket. The pocket was limited only to the area to be augmented and the opposite side is prepared in a similar fashion.

Complete mandibular or maxillary alveolar supplementation involve bilateral vertical 2 cm. incisions in the cuspid areas again down carried through the periosteum. The dissection of subperiosteal pockets was performed as previously described. Usually 3-4 syringes are necessary to augment the anterior aspect of the alveolar ridge and 6-8 syringes are needed to augment the complete ridge. The filled syringe was placed subperiosteally to the posterior ends of the pockets. In complete ridge augmentation the posterior pockets were filled bilaterally almost to the cuspid area incisions. Then the remaining anterior pocket is filled from both cuspid area incisons. Hydroxylapatite is then ejected from the syringe with the barrel of the syringe being retracted out of the incisions as the plunger maintains a constant relationship with the incision. This provides a smooth uniform placement of the material. The operator by placing his finger on the overlying mucosa molded the material to obtain the proper and contour. Mattress and interrupted sutures provided a water tight closure.

TOTAL RIDGE AUGMENTATION WITH HYDROXYLAPATITE AND BONE

Augmentation with a combination of cancellous bone and hydroxylapatite was indicated in patients with severe mandibular atrophy who required extensive alveolar ridge augmentation beyond 5 mm. The incisions and dissection when using combined material was essentially the same as previously described. Autogenous cancellous bone is harvested from the illiac crest. The bone is reduced to small particles using rongeur forceps or a bonemill and was thoroughly mixed with hydroxylapatite granules in a sterile medicine glass using a ratio of 1 gram of hydroxylapatite to 1 cc. of bone. A custom syringe or a 3 cc. syringe with the end removed is used and the mixture is packed into the syringe until it is about 3/4 full. It is then placed into the pocket and the composite materials ejected into the pocket and is manipulated and molded with fingers and instruments to the desired contour. 10-14 grams of hydroxyapatite is necessary for total ridge augmentation with bone.

HYDROXYLAPATITE - MAXILLA - ALL PATIENTS

1 - 4 years

	Patients	Class I	II	III	IV
Ant. Ridge	5	2		2	1
Post. Ridge	0				
Complete Ridge	4	1		3	·
TOTAL	9	3		5	1

HYDROXYLAPATITE - MANDIBLE - ALL PATIENTS

1 - 4 years

( ) with Staple
[ ] with Visor

	Patients	Class I	II	III	IV
Ant. Ridge	8	8			
Post. Ridge	11	1	1	2(5)	1(1)
Complete Ridge	27	4	7,[1]	11[2]	2
TOTAL	46	13	9	20	4

Figure 1

POSTOPERATIVE CARE

Postoperatively antibiotic coverage is instituted for one week. Hospitalized patients were given 1 million units of penicillin intravenously every 4 hours for 24 hours and then oral penicillin VK 2 grams daily for 6 days. Appropriate antibiotics are substituted for penicillin sensitive patients. Germicidal rinses were begun as soon as possible every 4 hours while awake. Liquid and soft diet is used until dentures are constructed.

PROSTHODONTIC EVALUATION

Denture construction began as early as 4 weeks. In some cases temporary dentures were utilized and worn as soon as the augmentation site achieves stability. Various techniques were used including; definite impression procedures, schemes of occlusion and different types and designs of teeth. Patients were evaluated at 3-6 month intervals up to 24 months for denture retention stability, centric relations, esthetics and phonetics. Subjective patient parameters included; comfort, speech and abilty to chew.

RESULTS:

55 patients, 3:1 female to male, average age 56 (Fig. 1), had Hydroxyalpatite alone or with cancellous bone augmentation for deficient alveolar ridges. The mandibular alveolar ridge was augmented in 46 patients and the maxillary ridge in 9 patients. The areas involved either the entire alveolus or only the anterior or posterior areas. HA was combined with a mandibular staple in 6 patients and a visor osteotomy in 3 patients.

Patients were routinely observed postoperatively at 1, 6, 12, 18, and 24 months. The follow up period extended from 12-48 months with a mean follow up of 32 months. Height dimensional changes were evaluated by:

1. Cephlometric radiographs.
2. Serial panoramic radiographs.

Panoramic radiographs were evaluated by comparing preoperative, immediate postoperative and follow up films. A digital computer was used to measure the height increase and follow up changes at 10 locations both right and left sides relative to base line anatomic standard points. Dimensional width changes were evaluated by comparing preoperative and postoperative models. Caliper measurement were taken 10 mm apart, measuring from the facial to lingual aspect of the alveolus.

The results of these observations indicated less than 10% postoperative demensional reduction of the alveolar ridges with HA alone and less than 20% with HA and bone which occurred primarily due to particle consolidation for a period up to six months. No obvious changes occured after six months postoperatively.

The surgeons evaluated the status of the implant material using postoperative panoramic radiographs. The following parameters were used; implant maintenance, bone opposition, bone infiltration and overall surgical assessment. Figs. 2 & 3.

Denture satisfaction was evaluated by using the Cornell-Medical Index. This evaluates the following paramaters:

1) Denture stability, 2) Ability to chew, 3) Taste, 4) Appearance, 5) General satisfaction, 6) Denture comfort.

The results of the evaluation performed by the prosthodontist indicated that of the 55 patients wearing dentures, there were six excellent and 49 good evaluations for general denture comfort and stability. Fig. 3

CONCLUSION:

The correction of major and minor alveolar ridge deficiencies with and without undercuts has been achieved with the use of hydroxyapatite particles alone or in combination with autogenous cancellous bone with considerable success at this institution. The experiences during the past four years, using the hydroxyapatite particles, have resulted in predictably stable agumentation of the dificient alveolar ridges and satisfied denture patients.

A simple subperiostal tunneling technique and a custom plastic syringe is used to deliver the material uniformly over the alveolar ridge. Additional dental implants such as the mandibular staple or Swiss bone screws can be placed simultaneously with the hydroxyapatite augmentation or following the correction of the deficient ridge.

Denture construction can be started as early as four to six weeks postoperatively. Skin or mucosal grafting if indicated may be performed at two to three months after HA augmentation.

Consolidation producing a loss of ridge height which occurred during the first six months was found to be 10% or less with HA alone and 20% or less with HA and autogenous cancellous bone.

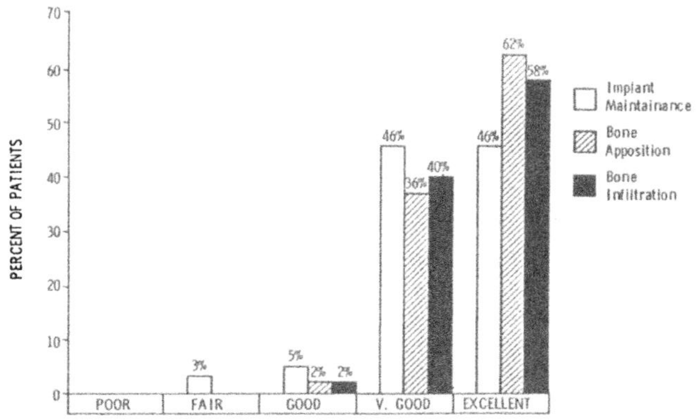

Figure 2

Denture stability is rapidly obtained and relines are much less frequent than with only bone grafting.

The use of hydroxylapatite alone or in combination with cancellous bone for augmentation of atrophic ridges has essentially eliminated the need for onlay rib and iliac crest grafting.

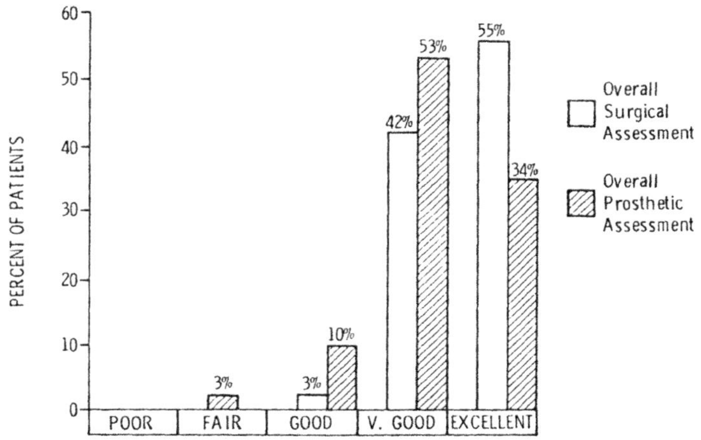

Figure 3

REFERENCES

Jarcho, M., Bolen, C.H., Thomas, M.B., Bobick, J., Kay, J.F. and Doremus, R.H., "Hydroxlapatite Synthesis and Characterization in Dense Polycrystalline Form," Journal of Materials Science 11, 2027-2035, (1976).

Jarcho, M., Kay, J.F., Gumaer, K.I., Doremus, R.H., and Drobeck, H.P., "Tissue, Cellular and Subcellular Events at a Bone Ceramic Hydroxylapatite Interface," Journal of Bioengineering, Vol. 1, No. 2, 79-92, (1977).

Kent, J., James, R., Finger, I., Jarcho, M., Taggart, J., and Cook, S., "Augmentation of Deficient Edentulous Alveolar Ridges with Win 40350 (Durapatite), A Preliminary Report," Presented at the First World Biomaterials Congress, Vienna, Austria, April 8-12, 1980.

Quinn, J.H., "Alveolar Ridge Preservation with Hydroxylapatite Root Implants in Baboons - Surgical Methodology Study," Presented at Conference on "New Concepts in Preprosthetic Surgery and Implant Dentistry" at Louisiana State University School of Dentistry, Jan. 11-12, 1980.

MAJOR TENDON AND LIGAMENT REPAIRS WITH A DACRON PROSTHESIS (GRAFT)

Calvin H. Frazier, M.D.

Verdugo Hills Hospital, Glendale, California

ABSTRACT

The author will present his experience, using Dacron graft material for major tendon and ligament replacements in actual clinical use. This work includes replacement of the sternoclavicular joint ligaments, acromioclavicular joint ligaments, fractures of the patella, and ruptures of the Achilles tendon. It is the author's experience that this is an excellent replacement technique, simple to execute, and permitting early and rapid rehabilitation of the patient.

KEYWORDS

Major tendon; Dacron graft repair.

MATERIALS AND METHODS

Acromioclavicular Dislocations

We first used Dacron for the acromioclavicular dislocations in 1978 and did not keep a record of all of our cases, but estimate we have 12 cases, all of which have been satisfactory both to the patient and the surgeon, using basically the technique of lashing the clavicle to the coracoid.

Postoperative care consisted of a sling for two weeks. Showering was permitted the first postoperative day. Full use of the extremity was delayed until six weeks after surgery.

Achilles Tendon Repairs

See illustrations. (Figs. 1,2,3)

The first patient was a 68-year-old woman whom we first saw three months post-injury. At surgery she had a one centimeter gap at the rupture site with maximum effort at approximation of the ends of the tendon.

Four years post-operation she died, and at autopsy, the Achilles tendon appeared normal to me. The microscopic appearance of the tendon-graft interfaced showed no inflammation or foreign body reaction.

Five other Achilles tendon repairs have been done. Two of these had problems, one in an elderly diabetic patient, had a superficial infection which responded well to soaks, elevations, and antibiotics. The graft was not removed.

Another patient, 73 years old, weighing 240 pounds and being 4'10" tall, had a dehiscence of the incision and the graft was removed. No additional repair was attempted.

Patellar Tendon Repairs

Five patients had fractured patellae repaired. (Figs. 4,5,6,7). Initially, our post-op care was the same as for the traditional patellar fracture repair. In P.E., we permitted immediate motion and weight bearing. His three week range of motion is shown here. He was a paint salesman and, I believe, an alcoholic. He returned to work three weeks post-op and has not been seen since.

Sternoclavicular Dislocations

One patient has had a chronic subluxation of the sternoclavicular joint repaired using a Dacron graft. (Fig. 8,9,10). She was placed on the same regime as the acromioclavicular joint patients and is now seven months post-op, and has no symptoms and an apparently normal functioning sternoclavicular joint.

SURGICAL TECHNIQUE

See illustrations. (Figs. 1,2,3,4,5,6,7,8,9,10).

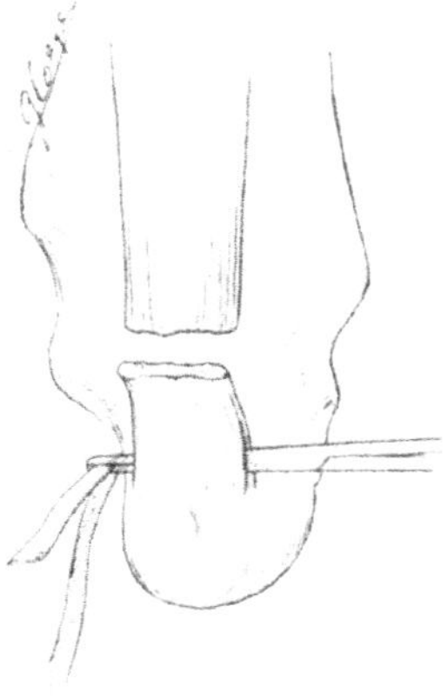

Fig. 1. 1st stage of Achilles tendon repair.

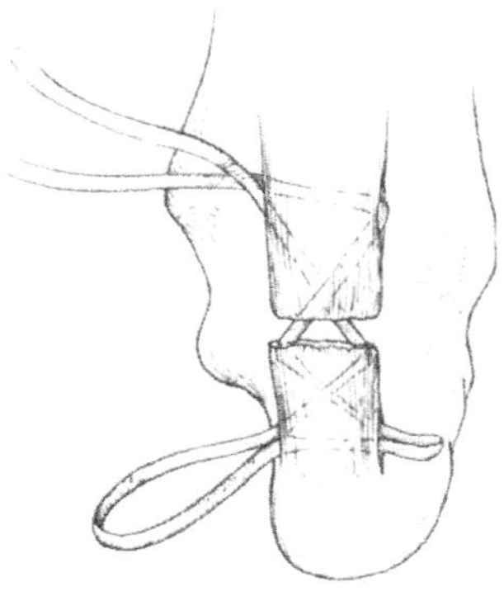

Fig. 2. Prosthesis passed through tendon, ready for pulling up slack and tying.

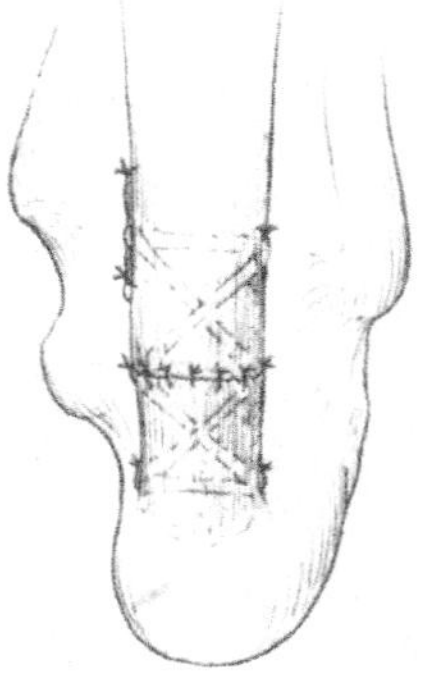

Fig.3. Completion of Achilles tendon repair.

Fig. 4. 1st stage of patellar fracture with or without excision of fragments.

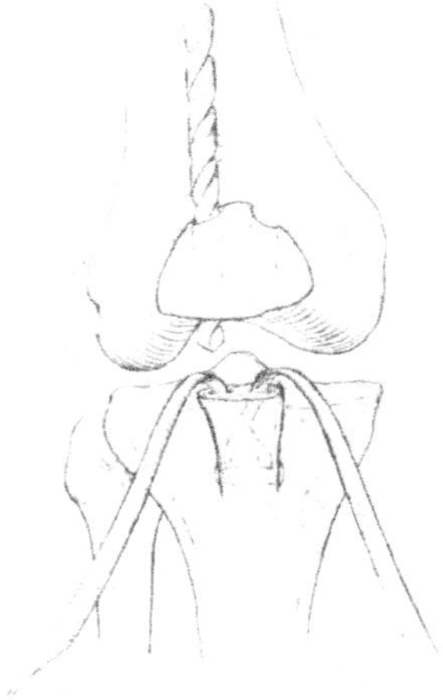

Fig. 5. Dacron prosthesis ready to be passed through proximal patella.

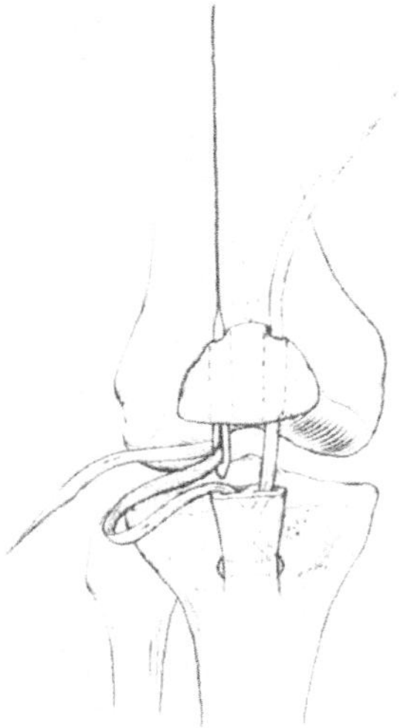

Fig. 6. Passing of Dacron prosthesis through patella.

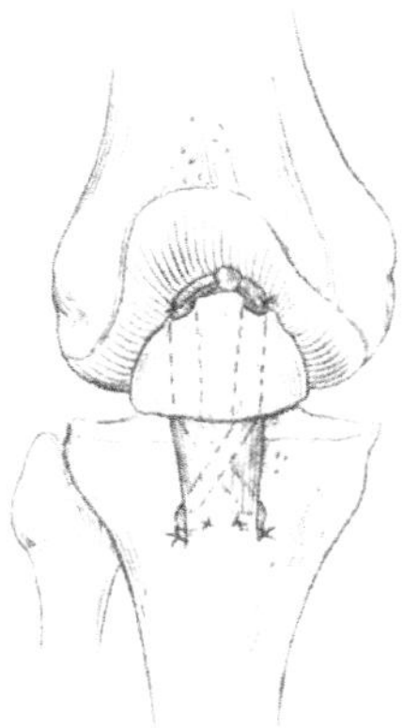

Fig. 7. Completion of patellar repair.

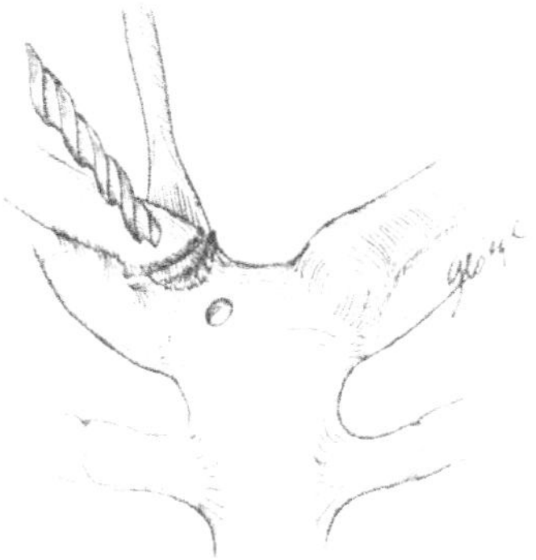

Fig. 8. Holes being made in the sternum and clavicle at sternoclavicular joint.

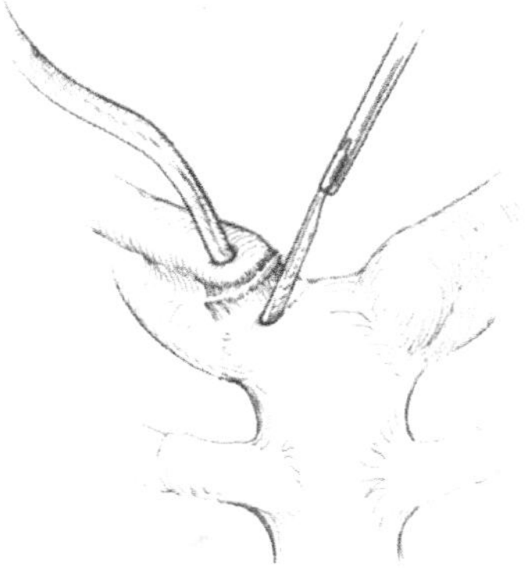

Fig. 9. Passage of Dacron suture through the holes.

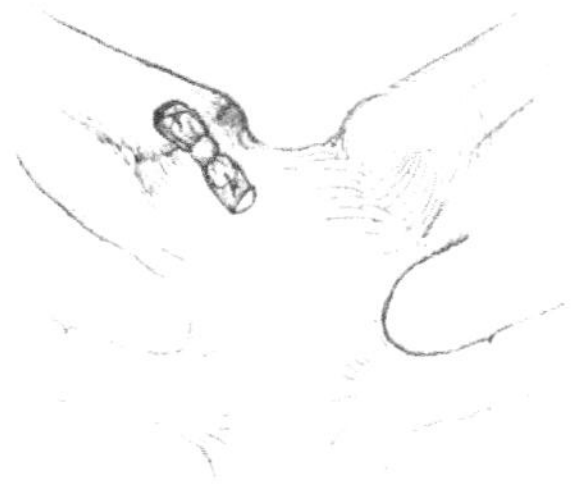

Fig. 10. Fixation of clavicle to sternum with Dacron suture.

THE FATIGUE LIFE OF INORGANIC BONE -- PMMA COMPOSITES

Y. King Liu, D. Stienstra and G. Njus

Center for Materials Research, College of Engineering
University of Iowa, Iowa City, IA 52242

ABSTRACT

Presented in this paper are fatigue crack propagation characteristics of PMMA bone cement with addition of inorganic bone particles. It was found that crack extension rates decreased with increasing bone concentration. At the 30 percent bone level fatigue life was improved by approximately a factor of ten. In consideration of the molding characteristics of this composite, several different cement viscosities were used.

KEYWORDS

Fatigue, PMMA, Bone Cement, Inorganic Bone

INTRODUCTION

One of the major contributory factors to total joint prosthesis failure is a loosening of the cement bone interface. Failure of this interface is primarily due to two reasons: fatigue cracking in the bone cement interface and resorption of bone. Any material contains numerous microscopic loci for crack initiation, and under cyclic loading, crack extension could occur. Clinical studies indicate the cement cracking is intimately involved in premature loosening. A contributing factor to this interface failure is due to the widely different material properties of bone and bone cement.

Polymethylmethacrylate (PMMA) is routinely used as a bone cement in total joint replacements. The addition of particles to PMMA, including glass spheres (Beaumont 1977), carbon (Wright, 1982), sugar and tricalcium phosphate (Rilke, 1972), has been investigated over the years in an effort to improve the properties of the cement, i.e., to narrow the difference between the material properties of the bone and bone cement.

The ideal modifications to the material properties of the current bone cement are: 1) an increase in its fatigue life, tensile strength, and fracture toughness; 2) a decrease in its modulus of elasticity to make the cement more similar to bone; 3) the provision of biocompatibility with the human body to the satisfaction of the FDA; and finally, 4) the creation of a milieu that will allow tissue ingrowth into the PMMA . Concurrent with the above, increases in the creep characteristics or paste viscosity must be avoided. Inorganic human bone was chosen as a likely candidate to satisfy all the above requirements.

MATERIALS AND METHODS

Long bone from human cadavera was cut into manageable pieces and soaked in sodium hydroxide (0.4N) to remove the organic phase. After approximately 24 hours, the inorganic phase was ground into particles. The particles were passed through sieves to yield those in the range of 200-300 μ. The choice of particle size range was made in consideration of the work done by Predecki (1972), who determined the above hole size range to facilitate bone ingrowth. Specimens containing 0, 5, 10, 20 and 30 percent bone by weight were cast and machined.

Two sample mixing techniques were employed. The first method was addition of ground bone to a standard surgical pack of cement, i.e., constant monomer to powder ratio. In the second method, an amount of powder equal to the weight of the added bone was removed. Samples made by the second method will hereafter be referred to as powder removed (PR) samples. For each method bone was first blended with the powder before monomer addition. No attempt was made when mixing or casting to reduce porosity by artificial means such as increased temperature, pressure or ultrasound.

All material tests were conducted on a closed loop servo-hydraulic MTS test machine. Fatigue crack propagation tests were run on conpact tensile specimens according to ASTM standard E647 with sinusoidal tension-tension loading at an R ratio of 0.05. Crack lengths were monitored to .01 mm with a traveling microscope at a testing frequency of 16 cycles per second. Tensile tests were in accordance with ASTM Standard D638.

RESULTS

Data from this study are presented in the form of crack growth rate as a function of stress intensity, da/dN vs ΔK. Results from six fatigue specimens are presented in Fig. 1. A trend toward decreased crack extension rates with increasing bone concentrations may be seen. At the 30% bone level, fatigue is retarded an order of magnitude when compared to that of plain cement. There is evidenced a great deal of scatter in this plot. Each test individually contained significant scatter, and no curve smoothing routines were used for data presentation in this paper. The majority of this variability results from the nonhomogeneous distribution of micro- and macro- porosity of the specimens.

The trend of crack growth retardation is more vividly seen in the samples prepared by removing powder, Fig. 2. This diagram, representing five tests, shows less scatter and demonstrates more clearly the trend toward decrease in crack extension rate with an increase in bone concentration. The 20% (PR) sample shows approximately an order or magnitude improvement over the plain cement.

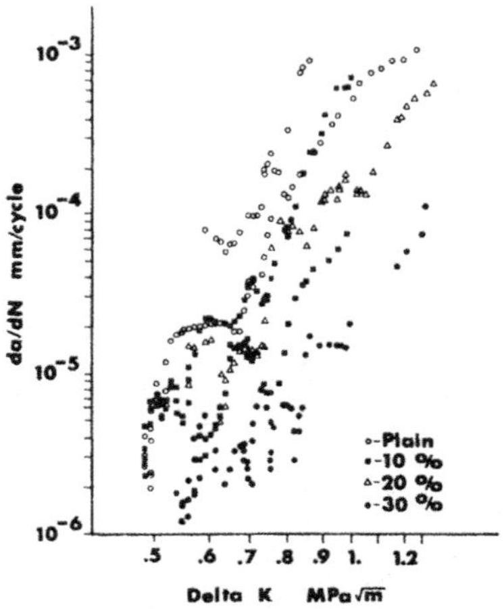

Fig. 1. da/dN vs. ΔK for PMMA with inorganic bone particles.

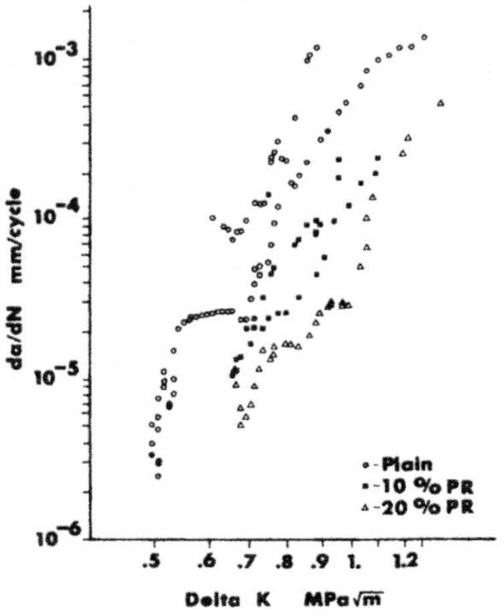

Fig. 2. da/dN vs. ΔK for PMMA with inorganic bone particles (PR).

In Fig. 3, the powder removed samples are compared with samples from the first group at the 10 and 20 percent concentrations. This comparison is evidence that the fatigue characteristics of the powder removed specimens are as good or better than the specimens from the first set.

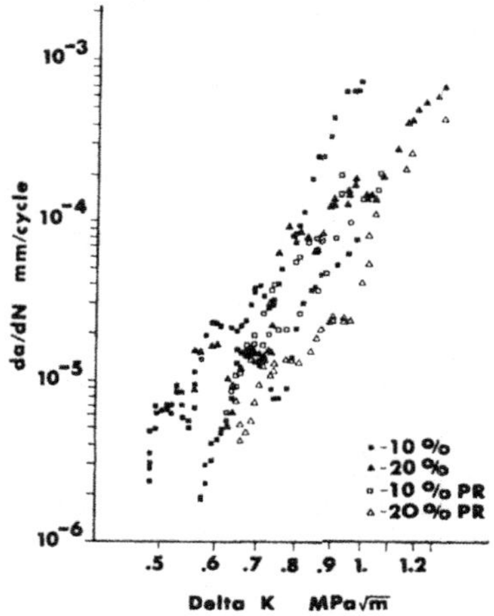

Fig. 3. da/dN vs. ΔK comparison of two mixing methods.

Tensile tests were also conducted for comparison. In general, a decrease in tensile strength and a slight increase in stiffness was observed with increasing bone content.

The viscosity of the samples made by the first method increased with increasing bone content, i.e., the higher concentration samples were more difficult to stir. The viscosity of the powder removed samples, on the other hand, was comparable with or lower than the plain cement.

In all cases the crack front was within the limits for curvature prescribed by the ASTM. The fracture surfaces of the specimens containing bone are rougher and more jagged than the pure PMMA specimens when examined both by eye and with scanning electron microscope. Crack propagation also appeared more irregular for the bone-containing samples, e.g., there were more branching and directional changes.

Besides material properties, biocompatibility as approved by the FDA is an important goal. Since the bone acquired for the composite could come from the patient at the time of surgery, foreign tissue rejection should not be a problem. Hodosh, Povar and Shklar (1968) used up to 20 percent inorganic cowbone in PMMA for tooth implants in baboons without rejection.

CONCLUSION

The result of the present research is that the addition of particles of inorganic bone to PMMA significantly improves its fatigue properties. This decrease in the rate of fatigue crack propagation warrants further research to determine the optimal bone concentration and particle size for fatigue crack inhibition. In light of this study, the candidate for further study is the powder removed mixing method.

REFERENCES

Beaumont, P.W. (1977). The strength of acrylic bone cement and acrylic cement --stainless steel interfaces. J. Mat. Sci., 12, p. 1845.

Hodosh, M, M. Povar, and G. Shklar, (1968). Anatomic inorganic bone-polymethylmethacrylate endosteal dental implant. Oral Surg., 25, 883.

Predecki, Paul, et al. (1972). Kinetics of bone growth into cylindrical channels in aluminum oxide and titanium. J. Biomed. Mater. Res., 6, 375-400.

Rilke, A.M. et al. (1977). Porous acrylic cement. J. Biomed. Mater. Res., 11, 373-397.

Wright, T.M., A.H. Burstein, R.P. Robinson (1982). Fatigue crack propagation in polymethylmethacrylate. Trans. of Orth. Res. Soc., 7, 162.

STRENGTHENING OF CALCIUM HYDROXYAPATITE BY ION EXCHANGE

C. Sombuthawee*, E. A. Monroe* and J. P. Rausch**

*SUNY, College of Ceramics at Alfred Univ., Alfred, NY

**Dept. of Physiology, Alfred Univ., Alfred, NY

ABSTRACT

These experiments attempted to increase the flexural strength of bioceramic sintered rods of hydroxyapatite, $Ca_5(PO_4)_3(OH)$, by means of ion exchange. To do this solid state reaction between the ceramic rods and powders of the reactant salts at elevated temperatures was used. The reactant salts used were $Sr(NO_3)_2$, $CaC\ell_2$ and $SrC\ell_2$. A strength increase occurred only when the reactant salt $SrC\ell_2$ was used. A maximum increase of strength from 65 MN/m^2 to 79 MN/m^2 corresponding to a 22% increase occurred using the conditions of 650°C for 12 hours.

KEYWORDS

Bioceramic; hydroxyapatite; ion-exchange; prosthetics; chemical strengthening.

INTRODUCTION

These experiments used ion exchange in an attempt to increase the flexural strength of apatite ceramic rods. The basis for accomplishing this is to develop a compressive stress in the outer surface layer of the rod by exchanging smaller ions present in the surface by larger ions from an external source. The surface compression is a result of the larger ions expanding the surface which is restrained by the underlying material, thus leading to the compressive surface stress. This technique has been applied to glass in order to achieve increased strength of up to twenty times (Doremus, 1973) and increased strengths after abrasion of greater than ten times have been reported (Nordberg, 1964).

Kistler (1962) reported that when a large ion is exchanged for a smaller one at temperatures below the glass transition, a stress builds up in the exchanged region. For example, Burggaaf (1965) reported a sodium aluminosilicate glass (mole %: SiO_2; 70, $A\ell_2O_3$, 7.5; Na_2O_3, 22.5) which was placed in molten KNO_3 for 24 hrs at 350°C developed a stress of about 100 kg/mm^2 at the outer surface and gradually diminishing to zero at a depth of about 100 μm below the surface. This stress profile corresponded to the concentration of potassium ions which was a maximum at the outer surface and then decreased to zero at a depth of about 100 μm. In practice the compressive layer must be about 50 μm (LaCourse, 1972).

The thermally activated diffusion of the smaller positive ion from the glass is favored by the increase in entropy of mixing. The resulting net negative charge is balanced by the available larger ion diffusing into the glass under the driving force of the chemical gradient or mass action (LaCourse, 1972). The amount of compressive stress at the surface depends upon various factors including the diffusion coefficients of the ions, the relative size of the exchanging ions, the amount and depth of ion exchange which in turn is proportional to the temperature and time of the exchange reaction. Also, any stress relaxation in the surface due to structural rearrangement may be expected to diminish the stress.

The application of ion-exchange strengthening to crystalline ceramic materials has been reported for some solid solution systems such as $A\ell_2O_3$-Cr_2O_3 and TiO_2-SnO_2 (Kirchner, 1979). Here the aim is to induce a solid solution surface layer having a lower thermal expansion by diffusing an appropriate element into the ceramic body at high temperature. Compressive surface stresses are induced during subsequent cooling because the interior contracts more than the surface, thus placing the surface into compression. The diffusing ion into the surface layer must lower the thermal expansion (contraction) in order to achieve this condition. Glasses have also been strengthened by this method of ion exchange to lower the surface expansion. In the case of the polycrystalline $A\ell_2O_3$-Cr_2O_3 solid solution system, bars of alumina ($A\ell_2O_3$) were packed in Cr_2O_3 powder and fired to 1400-1650°C with the result that a thin surface layer of Cr_2O_3-$A\ell_2O_3$ formed. The flexural strength was thereby increased by almost 30%.

No literature reports were found where ion exchange was used to strengthen a crystalline material by expansion of the surface layer analogous to the glass systems where volume expansion rather than thermal expansion is the effective mechanism for inducing the surface stress and strengthening. Our experiments attempt to strengthen a crystalline material in this way.

METHODOLOGY

Sintered polycrystalline ceramic rods of calcium hydroxyapatite (CaHA) about 3.5 mm x 35.0 mm in size were placed in alumina crucibles and packed with powder of the reactant compounds. The crucibles were heated in a furnace at various temperatures and periods of time below the melting point to allow for solid state reactions between the ceramic rods and the powder. The rods were then washed in water, dried and tested. The reactant salts used were $Sr(NO_3)_2$. $CaC\ell_2 \cdot 2H_2O$ and $SrC\ell_2 \cdot 6H_2O$. The latter two salts were dehydrated at 400°C for 2 hrs prior to their use.

In the case of the strontium salts, the hypothesis was that the larger Sr ion would replace the smaller Ca ion in the ceramic surface resulting in a volume increase of the unit cells and a concomittant compressive surface stress and strengthening of the material. In the case of the salts containing the Cℓ ion, the expected exchange was between the larger Cℓ ion and the smaller OH-ion with a concomittant strengthening. X-ray diffraction, X-ray spectroscopy, electron microscopy and mechanical tests were used to evaluate the experimental results. The mechanical tests were performed on ten specimens of similar treatment having a 10:1 aspect ratio, and dried at 100°C for 2 hrs, stored in a desiccator and using three-point bending on the Instron universal machine at a span of 3.4 cm, cross-head load of 500 kg and cross head speed of 0.1 cm/min.

RESULTS AND DISCUSSION

Using $Sr(NO_3)_2$ at the conditions of temperature, °C/time, hrs of 520/24, 550/24 and 560/6, there was no diffusion of Sr ions into the surface of the apatite ceramic (see Table 1).

Using $CaC\ell_2$ as the reactant powder at 550°C for 6, 8, 12 and 24 hrs resulted in $C\ell$ ion penetration of 5 - 125 μm below the ceramic surface depending upon the conditions of time and temperature as shown in Table 1. But instead of the $C\ell$ ions exchanging for the OH ions in the apatite, a new undesirable phase of $Ca_2PO_4C\ell$ was formed. Although at the 550°C/6 hrs condition no $Ca_2PO_4C\ell$ was detected by X-ray diffraction, the detection of $C\ell$ ions at a depth of 5 μm indicated that probably this new crystalline phase was present here too, but below the detection limits of the X-ray diffraction.

Table 1 Result of the Surface Ion Exchange Reactions in Calcium Hydroxyapatite Ceramic Rods

Reactant	Heat Treatment (temp °C/hrs.)	X-ray Diffraction (phases present in surface)	EDS Analysis* elements	EDS Analysis* diffusion depth, μm
$Sr(NO_3)_2$	520/24	CaHA **	Ca, P	-
	550/24	CaHA	Ca, P	-
	560/6	CaHA	Ca, P	-
$CaC\ell_2$	550/6	CaHA	Ca, P, $C\ell$	5
	550/8	CaHA + $Ca_2PO_4C\ell$	Ca, P, $C\ell$	30-40
	550/12	CaHA + $Ca_2PO_4C\ell$	Ca, P, $C\ell$	50-70
	550/24	$Ca_2PO_4C\ell$	Ca, P, $C\ell$	100-125
$SrC\ell_2$	550/6	CaHA	Ca, P	-
	550/12	CaHA	Ca, P	-
	550/24	CaHA	Ca, P	-
	650/6	CaHA	Ca, P, $C\ell$	30
	650/12	CaHA	Ca, P, $C\ell$	50
	650/24	CaHA	Ca, P, $C\ell$	300
	700/6	CaHA + $CaC\ell A$ †	Ca,P,$C\ell$,Sr (trace)	30
	700/10	CaHA + $CaC\ell A$ + CaO	Ca,P,$C\ell$,Sr (trace)	20
	700/15	CaHA + $CaC\ell A$ + CaO	Ca, P, $C\ell$, Sr	10-100
	700/20	CaHA + $CaC\ell A$ + CaO	Ca, P, $C\ell$, Sr	10-150

* EDS = energy dispersive X-ray spectroscopy (electron probe)
** CaHA = calcium hydroxyapatite, $Ca_5(PO_4)_3OH$
† $CaC\ell A$ = calcium chlorapatite, $Ca_5(PO_4)_3C\ell$

The results of the reactions between the CaHA ceramic rods and $SrC\ell_2$ powder at the various conditions are given in Table 1. At 550°C neither Sr nor $C\ell$ ions were detected in the ceramic surface. Raising the temperature of reaction to 650°C, again no Sr ions had moved into the ceramic but $C\ell$ ions were now found to be present in the surface layer. After 6 hrs the depth of $C\ell$ ion diffusion was 30 μm. Extending the time period to 12 hrs and then 24 hrs, the $C\ell$ ion penetration reached 50 μm and 300 μm, respectively. Although X-ray diffraction revealed only the presence of calcium hydroxyapatite and not calcium chlorapatite, it was believed that the $C\ell$ ions were present in the apatite structure as a solid solution $Ca_{10}(PO_4)_6(OH)_{2-x}C\ell_x$. Insufficient replacement of the OH by the $C\ell$ ion had occurred in order to be detected by diffraction.

When the reaction was conducted at 700°C for 6 hrs both Sr and $C\ell$ ions diffused into the apatite to a depth of 30 μm (see Table 1). X-ray diffraction of the ceramic surface showed the presence of both calcium hydroxapatite and calcium chlorapatite. Strontium was probably substituted for Ca in the apatite to form

$(Ca,Sr)_5(PO_4)_3(OH,Cl)$. With longer reaction time periods the additional phase of CaO appeared, indicating that the liberated Ca ions from the Sr for Ca ions exchange resulted in CaO formation. This water soluble CaO phase leached out upon washing the sample resulting in surface deterioration of the ceramic and hindering the aim of strengthening.

Flexural strength of the ceramic rods was tested before and after the ion exchange treatments for two of the exchange reactions giving promising results. These were the reactions with $CaCl_2$ at 550°C for 12 hrs and with $SrCl_2$ at 650°C for 12 hrs (see Table 2).

In the case of the treatment with $CaCl_2$ at 550°C/12 hrs, the Cl ion diffused 50-70 μm below the surface and there occurred the formation ofthe new phase Ca_2PO_4Cl. The average strength of the ten treated ceramic rods was 38 MN/m^2 (5510 psi) whereas the untreated ceramic averaged 65 MN/m^2 (9425 psi). This represented a decrease in the flexural strength of 42% probably due to the formation of the new phase Ca_2PO_4Cl which weakened the apatite ceramic.

In the case of the treatment with $SrCl_2$ at 650°C/12 hrs, the Cl ions migrated into the surface 50 μm without any detectable diffusion of Sr ions at these conditions. The average flexural strength of ten treated apatite rods was 79 MN/m^2 (11,455 psi) which represented a 22% increase in strength as shown in Table 2.

Table 2 Flexural Strength of Surface Treated and Untreated Calcium Hydroxyapatite Ceramic

Reactant	Treatment (temp °C/hrs)	Diffusion Depth, μm	Flexural Strength* (MN/m^2) untreated	treated
$CaCl_2$	550/12	50-70	65 (29)	38 (12)
$SrCl_2$	650/12	50	65 (29)	79 (35)

*Standard deviations given in parenthesis

This strengthening may be explained by Cl ions exchanging for OH ions in the apatite. Since the OH or Cl ions reside in an open tunnel in the structure, easy diffusion is possible (Elliott, 1967). This resulting solid solution, $Ca_{10}(PO_4)_6(OH)_{2-x}Cl_x$, has an increase in the unit cell volume compared to the origin CaHA. The calculated volume of the solid solution end-members, CaHA and CaClA is 520.2 A^{o3} and 545.2 A^{o3}, respectively, which represents a 4.8% volume increase for the chlorapatite. This volume increase may account for the origin of a compressive surface stress and the observed strengthening.

REFERENCES

Burggraaf, A. J. (1965). The mechanical strength of alkali silicate glasses after ion exchange. Thesis, Eindoven.

Doremus, R. H. (1973). Glass Science. J. Wiley and Sons, N.Y., 310-313.

Elliott, J. C. and R. A. Young (1967). Conversion of single crystals of chlorapatite into single crystal hydroxyapatite. Nature 214, 904.

Kirchner, H. P. (1979). Strengthening of Ceramics. Marcel Dekker, Inc., N.Y., 163.

Kistler, S. S. (1962). Stresses in glass produced by nonuniform exchange of monovalent ions. J. Am. Ceram. Soc., 45, 59.

LaCourse, W. C. (1972). Strength of glass. In Intro. Glass Science, Pye, L. D. et al., eds. Plenum Press, N.Y., 501.

Nordberg, M. E. et al. (1964). Strengthening by ion exchange. J. Am. Ceram. Soc., 47, 215.

Session 2

REHABILITATION ENGINEERING

AN EVALUATION SIMULATOR FOR THE SCOTT VAN

A.Y.J. Szeto, Ph.D. and H.A. Hogan, M.S.

Department of Biomedical Engineering
Louisiana Tech University
Ruston, Louisiana 71272

ABSTRACT

A simple driving simulator for evaluating the physical capabilities of severely handicapped prospective drivers of the Scott van was developed and evaluated. The simulator utilized a two-dimensional tracking task using the driving controls of the Scott van, a uni-lever servo controlled vehicle designed especially for the severely disabled population. Twenty-five able-bodied subjects and 9 quadriplegics tested the simulator and then drove the Scott van. Simulator RMS tracking error and the number of traffic cones knocked down during prescribed maneuvers were the respective performance measures for these two tests. Experimental data indicated that the simulator performances correlated very well with the corresponding driving performances. Our simulator, therefore, appears to be a valuable evaluation tool by giving objective, repeatable, and quantitative data on severely handicapped prospective drivers. A person who performs well in our simulator tests is likely to perform well in behind-the-wheel driving tests; he is thus a good candidate for further driver training.

INTRODUCTION

Present driving programs for the handicapped have been adequate, but they rely heavily on subjective judgments of medical professionals or driver evaluators, who can sometimes be inconsistent, contradictory, and perhaps inaccurate (Less, et al., 1978). Since the physical abilities of a handicapped person usually limit his potential for independent mobility, evaluators often measure strength, range of motion, reaction time, and vision of the person requesting driver training. Unfortunately these laboratory tests are typically done separately whereas the driving task requires a simultaneous and coordinated combining of these abilities.

Recognition of the need to more realistically evaluate physical capabilities of driving candidates has led to the use of commercially available driving simulators and actual driving attempts as part of the evaluation process. Conventional driving simulators were originally developed for the able-bodied and thus do not accurately indicate physical abilities (Reger, et al., 1978); the sophisticated, moving-base simulators are too complex, expensive, and not readily available for driver evaluation

To meet the need for better evaluation methods, without resorting to potentially hazardous driving attempts, a fix-based simulator was developed with the following criteria:

1) provide objective, quantitative, and reliable measures of the physical abilities of handicapped driving candidates;

2) measure these abilites interactively on relevant driving control tasks; and

3) be as simple and inexpensive as possible to minimize cost and thus increase its potential clinical availability.

In order to validate the accuracy and usefulness of this simulator, simulator and behind-the- wheel driving performances of able-bodied and handicapped individuals were compared for consistency and similarity. Results of this effort are reported below.

METHODS AND PROCEDURES

The main body of the simulator consists of a replica of the driving controls of the Scott van (Fig. 1), a uni-lever servo controlled van for the severely physically disabled (Scott and Prior, 1976). The van is steered by turning its 9 inch (23 cm) diameter steering wheel 90° clockwise or counterclockwise. The vertical post, to which the wheel is attached, pivots at the base so that pushing the wheel forward depresses the accelerator pedal and pulling it backward applies the brakes. Total travel fore and aft is approximately 6 inches (15 cm). The resistive forces for steering, braking, and accelerating were provided by a system of springs and hydraulic valves inside the simulator. These forces were set at one pound (4.4 N) for steering, 2 pounds (8.9 N) for braking, and 4 to 5 pounds (17.8-22.2 N) for full acceleration -- the same forces needed to drive the actual van. The range of control motions used in the simulator were also matched with those of the van.

FIG. 1. THE DRIVING CONTROLS OF THE SIMULATOR (LEFT) AND THE SCOTT VAN (RIGHT).

A two-dimensional pursuit tracking task was chosen as the simulation method because it has been successfully used in similar applications (Potvin, et al., 1977; Lynn, et al., 1977). The visual display for the tracking task was a 6 in. X 6 in. (15 cm X 15 cm) screen located about 30 inches (76 cm) in front of the test subject. The response and target signals were presented as different dots moving in an X-Y plane. The subject's response signals were generated by two linear displacement potentiometers mounted at the base of the vertical post of the driving controls. One monitored throttling or braking motions, and the other monitored steering motions. Braking or accelerating actions produced vertical movements of the response dot, while steering inputs produced horizontal movements of the same dot. The target signal for the tracking task came from a cassette tape played on a TEAC R-61 data recorder. The signals on the target tape represented the control inputs of a skilled driver performing a series of basic driving maneuvers in the Scott van. The potentiometers that detected these inputs were mounted exactly like they are in the simulator.

Seven basic maneuvers were selected as being representative of various driving situations. Each maneuver began and ended with the van parked and fully braked and its steering wheel centered. Each maneuver lasted 30 to 35 seconds. The distinquishing features of each maneuver were as follows: maneuver 1 -- straight ahead driving; maneuver 2 -- right turn at a constant speed; maneuver 3 -- left turn at a constant speed; maneuver 4 -- right turn while accelerating; maneuver 5 -- left turn while accelerating; maneuver 6 -- serpentine (S-curve) at 10 mph (16 km/h); and maneuver 7 -- serpentine (S-curve) at 15 mph (24 km/h). The serpentine course for maneuvers 6 and 7 was a double S curve that narrowed gradually from 10 feet to 8 feet along a length of 165 feet. The signals recorded for these basic maneuvers were then used to construct the target tape that contained a series of tracking runs. Each run consisted of all seven maneuvers in various orders and lasted about 4½ minutes.

The target and response signals of the simulator were connected to a digital computer for scoring. Upon conversion into digital form, the computer scored the tracking performance in terms of the RMS error between the target and response signals.

Twenty-five able-bodied subjects were tested on the simulator to provide reference data against which the handicapped subjects would be compared. Each subject was tested in one test session that usually lasted one to two hours. After each simulator run, the subject's tracking scores were calculated and shown to him. Each subject performed simulator tracking runs until he reached a learning plateau, as indicated by three consecutive runs of similarly high quality. Scores from these three runs were then averaged to represent the subject's simulator performance.

Because of the limited availability of the Scott van for use in this research, only 13 of the able-bodied subjects were given driving tests and only 2 of the 7 driving maneuvers were used for the driving tests. Maneuvers 6 and 7, the serpentine maneuvers, were selected for the driving tests because they have been successfully used by others (Sivak, et al., 1980) to measure driving ability and because these maneuvers were amenable to objective scoring. Ten to 20 minutes were spent driving the van in an empty parking lot for familiarization before 2 or 3 practice runs along side the serpentine course were made. The subject was then asked to drive through the course at 10 mph (16 km/h), and the number of traffic cones knocked down and their locations on the course were noted for each trial. The driving test at 10 mph was concluded when the driver appeared to have reached his maximum level of performance (i.e. 3 consecutively "good" driving trials). The subject then repeated this maneuver at 15 mph (24 km/h), preceded by 2 or 3 practice runs at this higher speed.

Nine severely physically handicapped subjects, designated A through I, also participated in these simulator and driving tests. All 9 were medically classified as quadriplegics, reflecting a total loss of motor function in their lower limbs plus a partial loss of motor function in their upper limbs.

EXPERIMENTAL FINDINGS

The composite simulator and driving test results for the 25 able-bodied subjects on maneuver 6 were 0.588 $\pm$ 0.092 cm of RMS error and 1.42 $\pm$ 0.63 cones knocked down, respectively. For maneuver 7, the simulator and driving test results were 0.692 $\pm$ 0.095 cm of RMS error and 1.28 $\pm$ 0.77 cones knocked down, respectively.

In order to compare the simulator performances of the handicapped subjects with their driving performances, every individual score was converted into a T-score using the performances of the able-bodied subjects as a common point of reference. The T-score was calculated using the formula: T-score = $\{(x - \bar{x})/S\}\{10\} + 50$, where x is the score being converted, $\bar{x}$ is the mean of the reference data, and S is the standard deviation of the data. A T-score of 50 corresponds to this mean, and each standard deviation that a value is above or below the mean adds or subtracts 10 points from 50. Higher T-scores represent larger RMS tracking errors or more cones knocked down and therefore represent poorer simulator performances or poorer driving performances, respectively.

DISCUSSION OF RESULTS

The results of maneuver 6 (Fig.2) clearly divide the handicapped subjects into two groups -- those who performed nearly as well as did the able-bodied and those who performed much poorer. Subjects F, G, and H composed this second group. Also evident from Fig. 2 is the generally strong agreement between the results from the simulator and driving tests. The Spearman rank correlation coefficient was significant at the $\alpha = 0.05$ level.

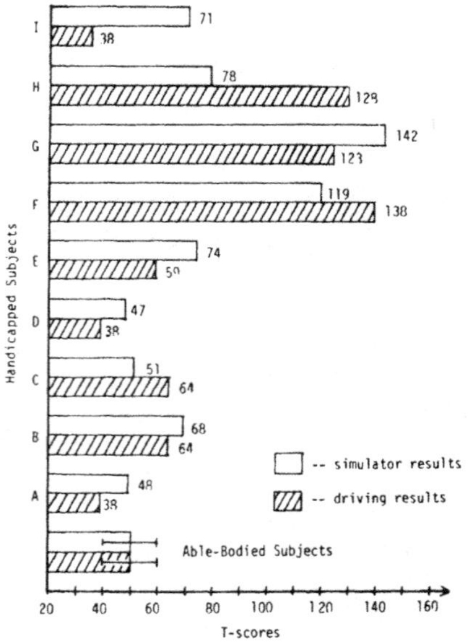

FIG. 2. SIMULATOR AND DRIVING PERFORMANCE T-SCORES OF HANDICAPPED SUBJECTS ON MANEUVER 6.

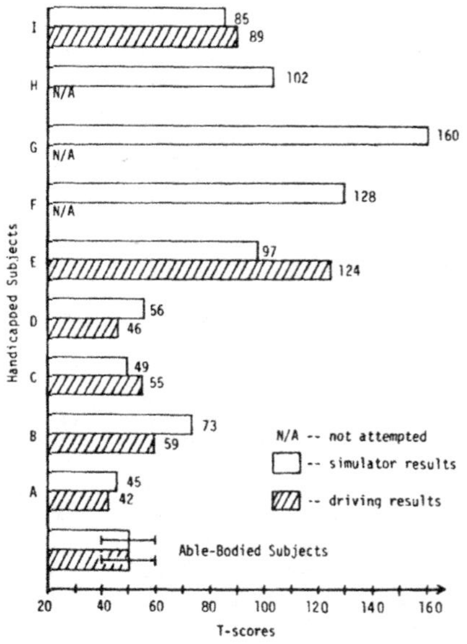

FIG. 3. SIMULATOR AND DRIVING PERFORMANCE T-SCORES OF HANDICAPPED SUBJECTS ON MANEUVER 7.

The results for maneuver 7 (Fig. 3) also showed strong agreement between the simulator and driving T-scores for the 6 subjects who completed both tests. The Spearman rank correlation coefficient was again significant at the $\alpha = 0.05$ level. The higher speed involved with this maneuver caused subjects E and I to do much poorer than they did on maneuver 6. However subjects A,B,C, and D appeared to handle the more difficult driving maneuver without great difficulty. Our simple simulator therefore appears capable of detecting subtle changes in subjects' performances in situation requiring slightly different physical movements. These results provide encouraging support for the usefulness of our simulator as an evaluation tool.

ACKNOWLEDGMENT

This project was supported by a grant from the Division of Vocational Rehabilitation, State of Louisiana.

REFERENCES

Less, M., E.C. Colverd, G.E. De Mauro, and J. Young (1978). Evaluating Driving Potential of Persons with Physical Disabilities. Albertson, New York: Human Resources Center.

Lynn, P.A., G.A.L. Reed, W.R. Parker, R.L. Hewer (1977). Some applications of human operator research to the assessment of disability in stroke. Med. & Biol. Eng. and Comput, 15, 184-188.

Potvin, A.R., J.T. Estes, J.A. Doerr and W.W. Tourtellotte (1977). Portable clinical tracking task instrument. Med. & Biol. Eng. and Comput., 15, 391-397.

Reger, S.I., D.F. Law, A.T. Mc Gloin, and R.E. Spence (1978). Development of a driver evaluator-trainer aid for the handicapped. Pro. 5th Ann. Conf. on Sys. & Devices for the Disabled, Houston, TX., 106-108.

Scott, C.M. and R.E. Prior (1976). Mobility aids for severely handicapped - A status report. Bull. Prosth. Res., BPR 10-26, Fall, 192-213.

Sivak, M., P.L. Olson, D.G. Kewnan, H. Won and D.L. Henson (1980). Perceptual & Cognitive Skills and Driving: Effects of Brain Damage. Ann Arbor, Michigan: The University of Michigan Highway Research Institute.

THE SEVERELY DISABLED DRIVER: GEAR RATIO AND DRIVER TIME DELAY EFFECTS ON THE DYNAMICS OF STEERING

M. Y. Zarrugh and J. D. Miller

The Rehabilitation Engineering Center and the
Department of Mechanical Engineering and Applied
Mechanics, University of Michigan, Ann Arbor, MI 48109

ABSTRACT

Adaptation of standard power-assisted steering systems to the capabilities of the severely disabled driver requires (1) reducing driver effort at the steering wheel and (2) reducing the range of steering wheel rotation by reducing the gear ratio. This reduction, although desirable at low speeds, produces at high speeds unnecessary sensitivity which taxes the attention of the disabled driver and reduces the stability margin of the system. Through a simplified closed loop computer model of the driver/vehicle system (based on McRuer's cross-over model), the effects of steering ratio and driver time delay are investigated. Preliminary results show that a driver time delay greater than 0.4 s or a gear ratio smaller than 12 renders the system unstable.

KEYWORDS

Rehabilitation; steering; severely disabled; vehicle dynamics.

INTRODUCTION

Severely disabled persons can drive an automobile only after modification of the steering system to provide for a lower effort and sometimes for a lower steering ratio. Determining the effects of these modifications requires a careful study of the interactions between driver and vehicle parameters that influence steering. Numerous studies of this interaction have been reported for able-bodied drivers, but little has been reported for disabled drivers.

Dynamics of vehicles have been studied thoroughly, but only relatively recently have driver models received attention. One major advance occurred when McRuer et al. (1975) presented the "cross-over model" which relates the parameters of the vehicle to those of the driver in simple terms. On the basis of this model, Garrot et al. (1982) developed a computer program simulating steering control. The objective of our work is to use Garrot's program to study the stability of the steering system when it is modified to suit the capabilities of the severely disabled driver.

Modification of the steering gear ratio alters driver gains and thus profoundly affects stability of the steering system. Another important parameter that influences stability is the time delay inherent in any neuromuscular action, such as driving. Kochhar et al. (1982) compared the effective time delay of disabled drivers to those of able-bodied drivers using a simulator. Their results indicate that, on the average, disabled drivers have larger time delays (230 ms vs. 165 ms).

DRIVER/VEHICLE MODEL

The simplest driver/vehicle model for studying steering performance is the one offered by McRuer et al. (1975). As depicted in Fig. 1, the model contains two nested loops: an internal loop for controlling the heading angle, ψ and an outer

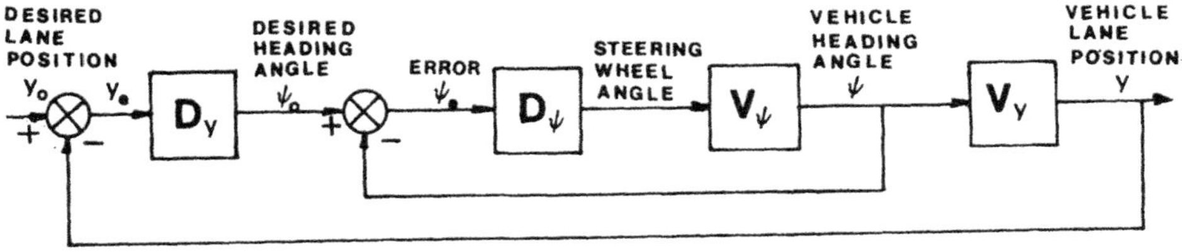

Fig. 1. Block diagram of driver/vehicle control system.

loop for the lateral lane position y. The overall gain in each loop must satisfy the cross-over relationship:

$$D(s)\ V(s) = \omega_c\ e^{-s\tau}/s \tag{1}$$

D(s) represents the driver transfer functions $D_\psi(s)$ or D_y (s) and V(s) represents the vehicle's functions V_ψ (s) or V_y (s). ω_c is the cross-over frequency above which the loop gain rapidly decreases below unity indicating decreased driver sensitivity to changes in the heading or lateral position of the vehicle. This frequency is closely related to the fundamental parameters of the driver and vehicle models:

$$\omega_c = K_\psi\ K_s\ U_o\ /\ell(\ 1 + K\ U_o) \tag{2}$$

Here K_ψ is the driver heading angle gain. K_s is the steering gear ratio. U_o denotes the steady state vehicle speed (m/s) and K, the understeer/oversteer coefficient of the vehicle (s^2/m^2). For disturbances or commands having component frequencies below ω_c , the driver maintains stability by adjusting K_ψ contained both in D_ψ and D_y of Equation (1) to compensate for changes in the dynamics of the vehicle expressed in V_ψ and V_y . Since increased heading gain K_ψ results in a higher cross-over frequency ω_c , stable adjustment can continue only up to the stability limit indicated by the maximum value of ω_c (Allen et al. 1977):

$$\tau\ \omega_u^2\ - \pi/2\ \omega_u + K_y\ U_o + 0.5 = 0 \tag{3}$$

Thus, the frequency ω_u depends on driver time delay τ , driver lateral distance K_y , which is the inverse of the distance of a hypothetical point at which the driver aims the vehicle (m^{-1}), and vehicle speed U_o . Equation (3) shows that increased τ is detrimental to stability since it decreases ω_u which limits the range of frequencies over which stable control can be maintained. It is also evident from the equation that increased vehicle speed will also reduce ω_u.

One of the most important vehicle parameters influencing closed loop stability is the vehicle's heading response time constant Teq. This accounts for the tendency of the vehicle not to respond instantly to sudden changes in steering wheel position. Experiments have shown (McRuer and Klein 1976) that drivers anticipate this sluggishness and compensate for it through a lead time constant T_L negating the effects of Teq. McRuer and Klein (1976) have also shown that able-bodied drivers increase both their heading gain K_ψ and lane position gain Ky as Teq decreases. Thus, the faster the steering dynamics, the greater becomes the driver's sensitivity. The authors also found that drivers decrease their lead time constant T_L and time delay τ when vehicle responsiveness increases (i.e. decreased Teq).

RESULTS

The effects of increased driver time delay τ and reduced gear ratio were investigated at 80 km/hr using a computer simulation based on the vehicle/driver model described above and vehicle parameters of a 1971-Mustang. The response of the vehicle to a desired step change of 50 mm in vehicle lateral position is computed for three different values of driver time delay (0.2, 0.3 and 0.4 s) at a constant steering gear ratio of 17.5 and for three gear ratios (17.5, 14 and 12) at a constant time delay of 0.2 s. Effects of increased are displayed in Fig. 2 which indicates that the vehicle becomes unstable when τ is 0.4 s or greater. Instability here means that the response of the vehicle increases indefinitely in amplitude with time.

Fig. 3 depicts the effects of gear ratio reduction. The response becomes unstable when the ratio is reduced from the normal 17.5 to 12. Attempts to obtain simulated responses at lower ratios that are more suitable for the disabled driver have not been successful. Improvement of the integration algorithm used in the simulation may produce the needed responses. Our initial experimental results indicate that able-bodied drivers can steer a vehicle with a ratio as low as 2:1 which is contrary to what the model indicates. In these preliminary runs, driver gains K_y and K_ψ and driver lead time constant T_L were not changed. Further results will be obtained by varying these important parameters, as well as by lowering the gear ratio to the desired level of 4:1.

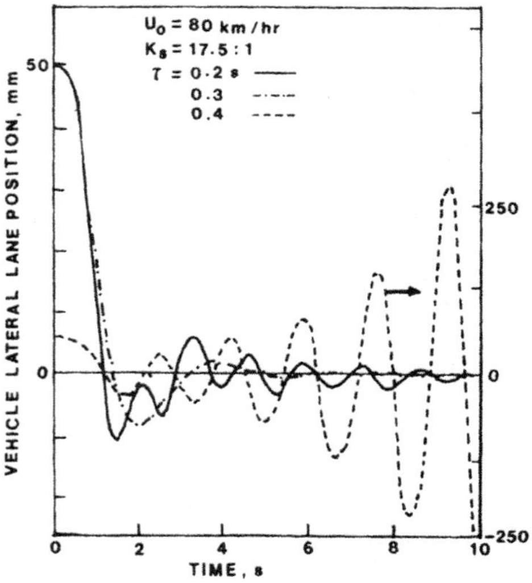

Fig. 2. Effects of driver time delay on vehicle lane position.

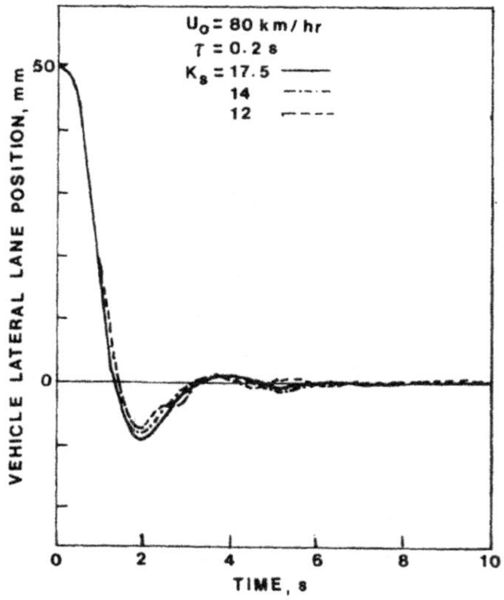

Fig. 3. Effects of steering gear ratio on vehicle lane position.

CONCLUSIONS

Intuitively, the cause of instability is quite clear. A driver who cannot react quickly or is insensitive to the behavior of the vehicle will continue to overshoot the desired path. Similarly, it becomes very difficult to control a vehicle with lower gear ratios at high speeds due to the unnecessary sensitivity. The analogous situation would be trying to steer a toy wagon (1:1 gear ratio) down a steep hill. The literature clearly indicates that able-bodied drivers adjust all of their parameters to maintain stability under a wide range of vehicle parameters. Similarly, it is hypothesized that disabled drivers adapt to these variations with a higher state of alertness to compensate for their relatively larger time delay and reduced ability to change gains. More definitive conclusions must await the completion of further simulation and testing of disabled drivers both with a simulator and on the road.

ACKNOWLEDGEMENT

This research was supported by Research Grant #G008005816 from the National Institute of Handicapped Research, Department of Education, Washington, D.C.

REFERENCES

Allen, R. W. and D. T. McRuer (1977). Driver steering dynamics measured in a car simulator under a range of visibility and road marking conditions. Proc. 13th Ann. Conf. on manual control, NASA report No. CR-158107, 1977, 180-196.

Garrot, W. R., D. L. Wilson and R. F. Scott (1982). Closed loop automobile maneuvers using describing function models. Society of Automotive Engineers paper No. 820306, SAE, Warrendale PA 15096.

Kochhar, D. S., J. M. Miller and L. E. Boydstun (1982). An approach to training the disabled driver. Submitted for publication.

McRuer, D. T. et al. (1975). Automobile controllability -- Driver/vehicle response for steering control. NHTSA Report No. DOT HS-801 407 and DOT HS-801 406, two vols.

McRuer, D. T. and R. H. Klein (1976). Effects of automobile steering characteristics on driver/vehicle performance. Society of Automotive Engineers paper No. 760778, SAE, Warrendale PA 15096.

THE PROSTHETIC PACECOUNTER

A.F. Tencer*#, V.Mooney, # T.W. Carmichael+

* Biomedical Engineering, Univ of Texas at Arlington, Texas
\# Department of Orthopedics, U. Texas Health Science Center, Dallas, TX
\+ Rehabilitation Engineering, VA Medical Center, Dallas, Texas

ABSTRACT

An electromechanical pace counter has been developed to determine the magnitude and number of cycles of axial load transmitted through temporary above/below knee prosthesis. The device was found to have repeatable trigger loading (± 5%), to be insensitive to lateral loads, and to provide reliable pace counts up to 16x 10^3 cycles.

KEYWORDS

Above knee prosthesis, pace counter, load magnitude and cycles.

INTRODUCTION

As a method for rehabilitating patients with above or below knee amputations, a temporary prosthesis has traditionally been fitted to the healing stump. This temporary device is replaced with a permanent one once the stump has healed and formed. Although devices are commercially available which detect the load transmitted through a prosthetic device (Leiper,1981), there appears to be no device available which can identify the duty cycle of temporary prosthetic devices.

OBJECTIVE

The short term objective of this project is to develop a reliable electromechanical device to measure the magnitude and number of load impulses transmitted through a temporary prosthesis. This will allow evaluation of the useage of various types of prostheses. A longer term objective is to study the changes in vivo of healing bone and soft tissue due to compressive load cycling. This situation provides a unique opportunity for such an in vivo study.

PROCEDURE

Requirements: The following performance requirements were established for the device: (a) it should fit into currently used temporary prosthetic systems,

(b) the load magnitude required to trigger the device should be adjustable so that it can be adjusted to the patient's weight, (c) the displacement within the transducer and its weight should be minimal so as not to interfere with the normal functioning of the prosthesis, (d) the device should record at least 1×10^4 load pulses at any particular load setting, (e) it should be entirely self-contained, and rugged enough to withstand normal use outside the clinic.

Method: An electromechanical transducer,Fig.1,has been developed for this purpose. The device is based on the action of a compression spring. The pacecounter pylons fit into the standard socket and shoe. The lock screws allow axial loads transmitted through the socket end pylon to be transferred to the spring cup. The spring tension is preset by rotating the stud so that a particular trigger force can be set. If the axial load transferred to the spring exceeds the preset level,the socket end pylon, spring cup, contact cap, and moving contact displace slightly as a unit. The displacement is limited to the gap width between the socket end pylon and the inner surface of the shoe end pylon. However, the displacement is sufficient to create contact between the moving and leaf contacts. This generates

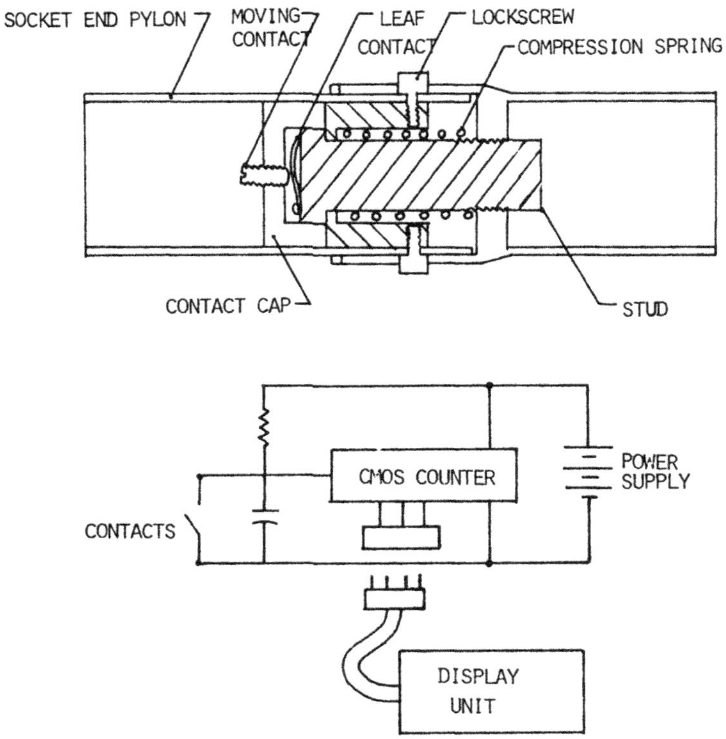

Fig 1. Schematic diagram of the prosthetic pace counter

a pulse which is recorded by a CMOS counter with 16×10^3 capacity. The electronics and battery package are located within the socket end pylon and the counter setting is decoded during a clinic visit.

Calibration: Calibration and sensitivity of the device were investigated, in several ways. A universal materials tester was used to determine repeatability of the trigger setting at various loads. In this test, the device was loaded axially and the load measured using a load cell (Interface No. 500 accuracy ± 2.0 % full scale) and deflection using a linearly variable differential transformer (Shavitz No. 2000 accuracy, ±1%). Both instruments are mounted to the universal tester. In the test, the device is preset for a particular trigger load level and the load deflection curve recorded on a Hewlett-Packard Type 25 X-Y recorder. Twenty load-deflection results are recorded at each of the five trigger load levels. Fig 2 shows typical results. The trigger contact loads are represented as points on the curve. The trigger load level was found to be repeatable at any preset load level. The characteristic curve is that of a preloaded compression spring, with high stiffness, and a sharp change in slope when the preset load level is reached. A small frictional component of axial load is superimposed upon the load-deformation characteristic due to sliding between components. This value is, shown in Fig 1 as the load required to produce deflection in the device with zero preload on the spring.

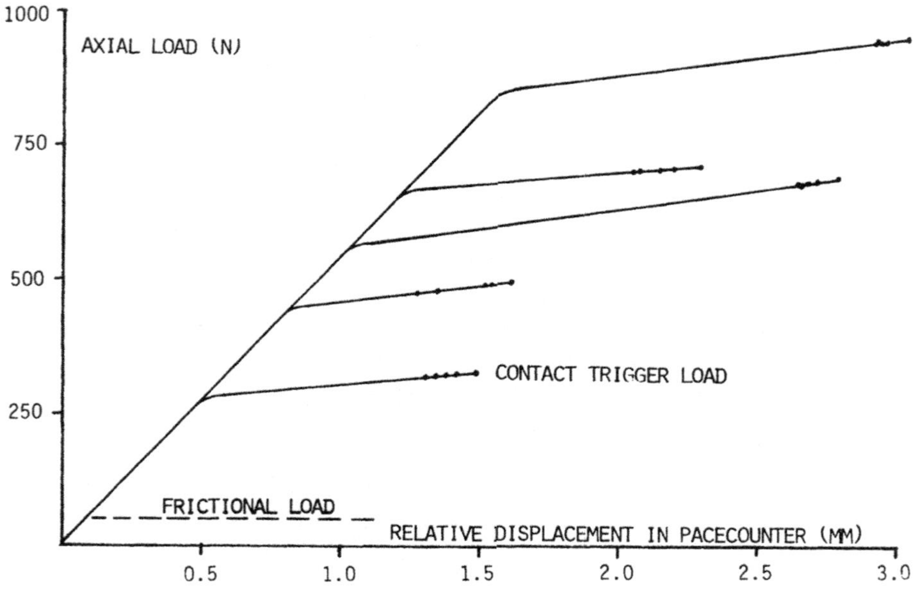

Fig 2. Repeatability of pacecounter trigger load settings

A second test consisted of determining the sensitivity of the device to combined lateral and axial loads which would occur during heel strike. In this test, a lateral load of varying magnitude was applied to the pacecounter through a cable, pulley and deadweight arrangement, in combination with the axial load. Increasing the magnitude of the lateral load, while the axial load remains fixed, has the effect of changing the orientation of the resultant load vector. The results are shown in Fig 2. The actual trigger load and the axial component of the trigger load are plotted against angle of deviation between the device and load axes. Although the resultant trigger load increases with the addition of the lateral component, the resultant and trigger loads are nearly the same, indicating that the pacecounter is insensitive to bending within its operational range.

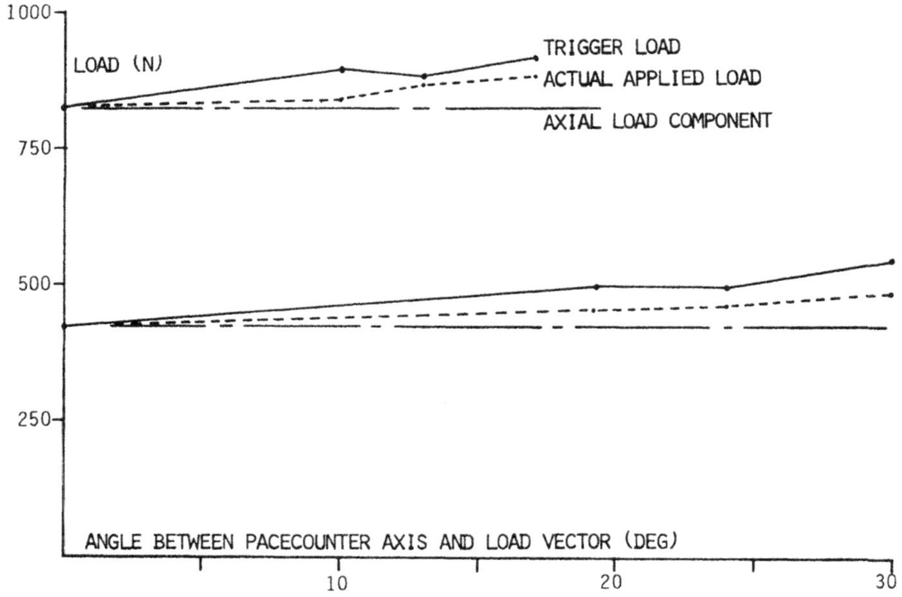

Fig 3. Effect of lateral loading on pacecounter trigger load

REFERENCES

Leiper, C., Bull, Prosth. Res., Vol. 18, No. 1, P. 253, 1981

MYOELECTRIC ANALYSIS OF NORMAL HUMAN SPINE FUNCTION

T. Carmichael*+, V. Mooney+*, A. Potvin#, S. Deivanayagam##, T. Mayer+

#Rehab. Engineering Research, Orthopedic Surgery, VA Medical Center, Dallas, Texas
+Div. of Orthopedics, U. of Texas Health Science Center at Dallas, Dallas, Texas
#Biomedical Engineering, U. of Texas at Arlington, Arlington, Texas
##Industrial Engineering Dept., U. of Texas at Arlington, Arlington, Texas

ABSTRACT

In this study, power spectral analysis of myoelectric signals was used to quantify the rate of fatigue in the erector spinae musculature of four normal human males during isometric back extension exercise. With repeated trials at load levels at 50% of maximum voluntary extension force, signals obtained from bipolar surface electrodes revealed mean and median power frequencies which decreased exponentially to final values at total fatigue ranging from 40% to 84% of initial frequencies. Fatigue time constants derived from linear regressions of the mean power frequency curves showed high repeatability and significant correlations ($r = .905$ to $.997$, $p<.001$). Median power frequency time constants showed slightly lower repeatability and correlations ($r = .853$ to $.996$, $p<.001$).

KEYWORDS

Power spectral analysis; myoelectric analysis; muscle fatigue; spine function; mean power frequency; median power frequency; back extension.

INTRODUCTION

Of all work-related injuries, approximately 12% involve the back, and tremendous time and effort are expended caring for the large population of patients with spinal pain. Estimates of human muscle fatigue, endurance and malingering during rehabilitative exercise are presently made only on a subjective basis. An objective, quantitative and reproducible method of assessment would permit clinicians to prescribe and evaluate exercise programs and determine the role of various adjuncts to muscle rehabilitation. Power spectral frequency analysis techniques of measuring mean and median power frequencies (MPF and MDPF, respectively) of myoelectric signals during fatiguing contractions have been developed for the evaluation of muscle fatigue (Lindstrom and Magnusson, 1977; Stulen and DeLuca, 1981). These myoelectric analysis techniques have been used to quantify fatigue in upper and lower extremities, but relatively few spinal fatigue studies involving myoelectric analysis have been conducted (Ortengren and colleagues, 1975).

OBJECTIVES

The objective of this study was to quantify the rate of fatigue in normal human erector spinae muscles during back extension by using myoelectric frequency power &

spectral analysis methods.

METHODS

Four well-motivated normal, healthy male subjects (ages 30 to 40) participated. The subjects stood upright with knees kept straight on a back strength measurement device which provided anterior pelvic support, and extension force was measured with a 500 lb. load cell connected by a chain to a flat padded brace placed across the scapulae. Three maximum voluntary contraction (MVC) measurements were obtained from each subject with five minutes rest between each trial. After 15 minutes rest the subjects exerted 50% of their largest measured MVC force until they could no longer maintain it. Another trial at 50% MVC level was performed after another 15 minutes of rest. Myoelectric signals were obtained from the left and right erector spinae muscles at the L3 level with two pairs of recessed Ag-AgCl bipolar surface electrodes (4 mm diameter and 11 mm separation), which were differentially amplified and recorded on an 8-channel FM tape recorder with a bandwidth of 2500 Hz and a signal-to-noise level of 40 dB. Bipolar wire electrodes were not used since they are invasive and sometimes painful, have very limited takeup volume within the muscle, and have not demonstrated very reproducible results.

Computer power spectral analysis of the myoelectric signals was performed using a DEC LSI-11/23 minicomputer by first digitizing with a 12-bit analog-to-digital converter at a sample rate of 2048 Hz with a four-pole low-pass filter set at 600 Hz and a four-pole high-pass filter at 2 Hz, and storing the digital data on dual-density floppy diskettes. Data was processed in 250 millisecond segments by multiplying by a tapered window function and transforming by a 512 point complex fast Fourier transform (FFT) to produce a power spectrum periodogram. Eight consecutive periodograms were averaged and used to obtain MPF, MDPF and total signal power for the frequency range from 12 to 512 Hertz. Fatigue time constants (Lindstrom and Magnusson, 1977) were claculated for MPF and MDPF curves as the inverse of the slope of a linear regression fit of the natural logarithm of the mean or median frequency versus time for the period of the fatiguing contraction. True rms voltages of the myoelectric signals were also computed.

RESULTS

MPF, MDPF and rms voltage data from the peak MVC trial of each subject are shown in Table 1. The force levels correspond with those obtained in a study of 220 normal male subjects that showed a mean force of 1005 newtons (S.D.=44) (Ayoub, Bethea and Deivanayagam, 1978). Right side erector spinae MPF data for two consecutive 50% MVC trials to fatigue are shown in Table 2, with the corresponding MDPF data shown in Table 3. Data from the left side erector spinae muscles demonstrated similar MPF and MDPF initial values and time constants. The MPF time constants always yielded a higher correlation coefficient (i.e., a better straight line fit) than the MDPF time constrants. Rather significant decreases in MPF and MDPF over time were found for all subjects and all 50% MVC trials (MPF decreased by 16.0% to 57.9% and MDPF decreased by 14.5% to 67.8%), indicating that the erector spinae muscles

TABLE 1 Peak MVC Trial Data for Four Normal Male Subjects

Subject Number	MVC Force (newtons)	Mean Freq. (Hz)		Median Freq. (Hz)		EMG RMS Voltage (uV)	
		Left	Right	Left	Right	Left	Right
1	1290	69.1	92.5	48.9	73.6	204	157
2	1152	93.0	93.8	81.3	78.2	127	113
3	1001	96.1	100.6	74.8	78.4	40	40
4	845	103.6	105.4	85.5	86.2	79	109

TABLE 2 Mean Power Frequency Data for Right Side Erector Spinae Muscles

Subject Number	50% MVC Trial #	Time Constant (seconds)	Correlation Coefficient (r)	Initial Frequency (Hz)	Final Frequency (Hz)	% Decrease	Endurance Time (seconds)
1	1	-238.6	.984	96.7	67.4	30.3	86
	2	-238.8	.974	88.5	67.7	23.5	75
2	1	-87.7	.997	123.4	42.3	65.7	95
	2	-90.3	.994	100.6	44.3	57.9	75
3	1	-175.6	.991	110.9	68.7	38.0	84
	2	-146.6	.983	109.5	62.6	42.8	84
4	1	-373.9	.990	111.2	68.4	38.6	206
	2	-191.5	.984	95.4	51.5	46.0	118

TABLE 3 Median Power Frequency Data for Right Side Erector Spinae Muscles

Subject Number	50% MVC Trial #	Time Constant (seconds)	Correlation Coefficient (r)	Initial Frequency (Hz)	Final Frequency (Hz)	% Decrease	Endurance Time (seconds)
1	1	-252.8	.964	72.7	51.7	28.8	86
	2	-251.2	.916	67.4	52.3	22.5	75
2	1	-82.8	.996	100.1	32.2	67.8	95
	2	-83.8	.975	78.4	32.4	58.6	75
3	1	-172.8	.973	85.3	52.5	38.5	84
	2	-134.8	.976	85.8	46.7	45.6	84
4	1	-347.9	.979	95.6	56.7	40.8	206
	2	-175.4	.981	80.5	41.1	49.0	118

fatigue considerably at this load level. Endurance time ranged from 75 to 95 seconds in three subjects, and to 206 seconds with the fourth subject. It is possible that a true MVC was not obtained foı subject #4. Straight line approximations for the MPF of the right erector spinae muscles for both trials of each subject are shown in Fig. 1. Note that the initial MPF is always lower for each subject in trial 1 than it is in trial 2. This indicates that even after a 15 minute rest, they were not fully recovered. Fatigue time constants for the second trial for each subject were within 0.1% to 16.5% of the time constants for the first trial, except for subject #4 who showed a 49% increase in fatigue rate during the second trial.

To better investigate the effects of malingering during MVC trials, two of the subjects maintained an isometric back extension trial at 20% of their MCV levels. Fatigue time constants calculated from the first minute of each trial were within 10% of the time constants determined from the two total trial times of 5 and 10 minutes, respectively. Time constant values were approximately six times larger than the 50% MVC values, indicating that the fatigue rate had decreased substantially. This suggests that submotivation or malingering could be determined for norms established for fatigue time constants, endurance times and per cent decreases in MPF and MDPF over time for given percentages of MVC force levels.

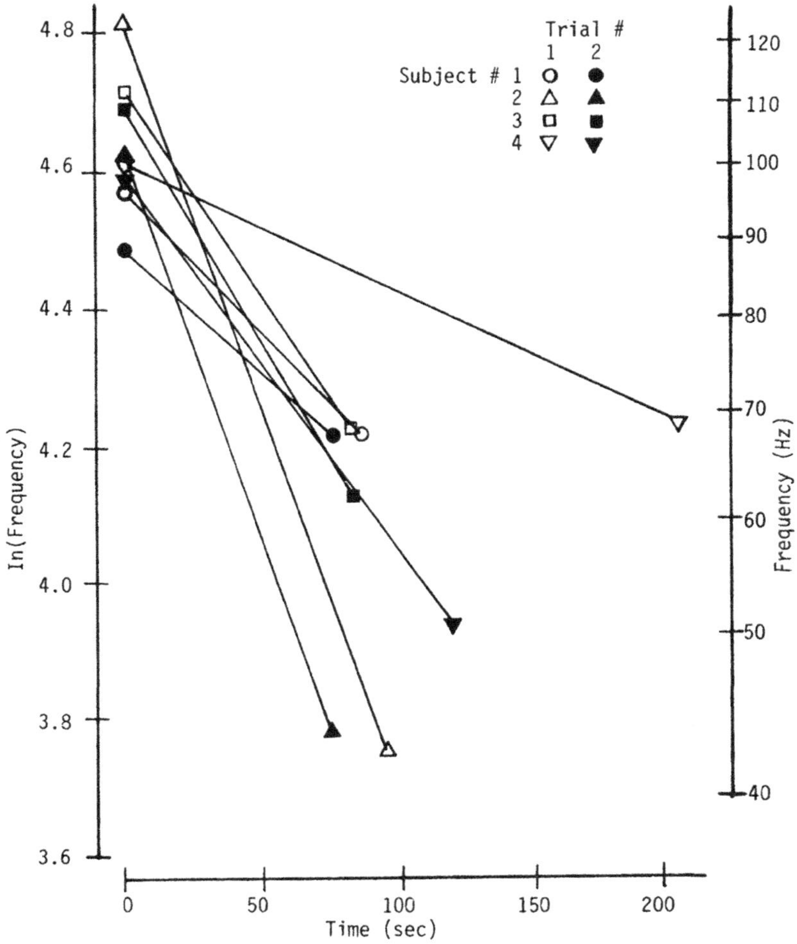

Fig. 1. Right side erector spinae MPF changes for 50% MVC contractions.

REFERENCES

Ayoub, M. M., Bethea, M. J., and Deivanayagam. S. (1978). Determination and modelling of lifting capacity. Department of H.E.W. Research Report, 262 pp.
Lindstrom, L. H., and Magnusson, R. I. (1977). Interpretations of myoelectric power spectra: a model and its applications. Proc. of the IEEE, 65, 653-662.
Ortengren, R., Andersson, G., Broman, H., Magnusson, R., and Petersen, I. (1975). Vocational electromyography: studies of localized muscle fatigue at the assembly line. Ergonomics, 18, 157-174.
Stulen, F. B., and DeLuca, C. J. (1981). Frequency parameters of the myoelectric signal as a measure of muscle conduction velocity. IEEE Trans. Biomed. Eng., BME-28(7), 515-523.

MICROPROCESSORS IN THE HOSPITAL - COMMUNICATION FOR THE ACUTELY DISABLED

A. Chesson* and R. Schubert**

*Dept. of Neurology, Louisiana State University Medical Center and Veterans Admin. Medical Center, Shreveport, LA 71130

**Dept. of Biomedical Engineering, Louisiana Tech. Univ., Ruston, LA 71270

ABSTRACT

A microcomputer interfaced system was utilized previously to aid the chronically handicapped by improving communication and environmental control. This system was adapted to acutely disabled hospitalized patients. A series of modifications of the system are detailed and observations from bedside trials are discussed. Conclusions regarding additional directions for development and changes are described.

KEYWORDS

Communication aid; microprocessor; independent living; environmental control; acutely disabled.

INTRODUCTION

The goal of this project was twofold. The first was to adapt a modified version of the General Purpose Interface (GPI) for independent living (Cronk, 1980) for patients in an acute hospital setting. The second was to identify those aspects of the system needing additional study and modification. The GPI utilized a microcomputer as an intelligent interface between the chronically handicapped user and the environment, providing better communication and environmental control. Our project adapted that system to provide improved communication for hospitalized patients with severely impaired communication abilities.

METHODS

The GPI system consists of a 32 K microprocessor based computer, cassette tape recorder, dot matrix printer, special control box and software. The control box plugs into the computer without requiring modifications, receives control signals from both the user and computer and, via the computer's signal, controls the user's environment including electrical appliances, lights, telephones, etc. The software allows the user to direct the computer via keyboard or remote switch. Because the computer's interpretation of the user's input is performed with software, many selection methods may be utilized. These selection techniques have been discussed by Vanderheiden (1975).

Three patients from the V.A. Medical Center in Shreveport, La. were tested utilizing this device. All had neurologic disabilities which included major difficulty with communication. The device was discussed and each patient agreed to participate in this trial. Patient instruction was performed at the bedside and repeated as needed. Information about usage and problems with utilization were obtained from prolonged direct observation as well as from the patient, his medical personnel and his family.

RESULTS

The first patient was a 57-year-old man with post infectious polyneuritis (Landry-Guillain-Barré Syndrome) who had total respiratory, cranial nerve and motor paralysis with the exception of minimally retained movements of his right foot. Communication was naturally severely limited. Our device was modified to translate his minimal foot movement into computer activation. The computer displays the menu selections on the CRT screen and then begins to highlight the selections one at a time. If a switch activation occurs while a particular selection is being highlighted, then that item (letter, phrase, control action, etc.) is printed at the bottom of the screen. An attached printer can produce hard copy. If no activation of the switch occurs during several scans, the computer assumes a "wait" mode of operation. The computer resumes scanning when the switch is activated. This system thus provided the ability for the patient to operate his nurse call and control his TV, as well as "talk" via letter, word, and phrase selection.

Because of his severe respiratory muscle involvement, he had to be maintained on a ventilator for 6 months. During this time he also did not have adequate strength for pointing, writing, head nodding, or other usual communication techniques. The vocabulary for phrase selection was modified on a timely basis to include phrases frequently utilized by the nurses and the respiratory therapists, and to allow him to communicate better with his family regarding family members and his business.

Difficulties encountered in using the equipment with this patient were numerous. Difficulty with equipment positioning was significant because of his need for frequent position changes. His absent eye movements meant he had to be directly facing the computer screen for visualization. His severely limited ability to manipulate an interfacing device complicated his training process as did his reduced endurance. Our equipment was added to a large amount of medical support equipment at the bedside in the Intensive Care Unit, which hampered the ability to leave the computer in an activated mode. Finally, resistance from nurses and physicians in utilizing the equipment was high because of long loading time and the difficulty positioning the switch and the computer screen.

The second patient utilizing the equipment was a 28-year-old male who was seen following his second resection of an extensively invasive temporal-occipital astrocytoma. His deficits included three extremity spastic paralysis with contractures, a moderate spastic paresis of his remaining arm, and nearly absent vocalization from a pseudobulbar palsy. Computer activation was performed by grasping a pressure sensitive switch. He was unable to use keyboard selection because of lack of fine coordination. His mild expressive and receptive aphasia made complex selection processes useless. He was able to select appropriate words from a simple selection menu. Additional problems were encountered due to visual difficulties and patient positioning problems. The patient had both decreased visual acuity and visual field defects. When attempting to use a row and column scanning technique, he tended to lose the highlighter as it passed into the non-seeing portions of his visual fields. He also had limited positions of comfort because of his contractures and spasticity. Centering the computer screen in his preserved islands of vision and positioned close enough to compensate for his reduced visual acuity was difficult. His trials were eventually abandoned because of his deteriorating mental status.

A 63-year-old physician was the third patient tested. A remote and a recent cerebrovascular accident had resulted in a left hemiparesis, a mild right arm spastic monoplegia, swallowing difficulty and an inability to verbalize due to a pseudobulbar palsy. All of these combined to produce major communication difficulty. The patient was left-handed. The spasticity of his non-dominant hand prevented legible writing at a rate compatible with reasonable communication. Utilizing our system as described above was frustratingly slow to this intelligent individual. He felt a manual alphabet board, although very slow, was faster than an on/off activation switch. He was, however, able to use his spastic non-dominant hand in such a manner that a direct selection process could be utilized from the keyboard in a one finger "hunt and peck" type format. Utilizing this format improved speed of selection considerably.

DISCUSSION

Using this system in a hospital setting where patients' ability to function changes frequently presents a bigger challange than anticipated from past experience with stable handicapped patients. In this setting the full reprogramming ability of the microprocessing system must be utilized. The fact that the computer often has to be moved away from the bedside to make room for other medical equipment is a major problem. The program must be reloaded from the cassette each time the power is disconnected. The 5 minutes loading time is frustrating for busy medical personnel. Faster means of loading the program could include a floppy disc or storing the program in an integrated circuit (EPROM). The former would allow loading in seconds, but would increase cost; the latter would allow quick access, but would limit capabilities.

Observations of patients suggested the need for variable rates of presentation of the material. To adjust for this, selection modes were incorporated allowing the patient to be able to both speed up and slow down the curser rate. Patients with sensory deprivation had difficulty telling when the switch was activated. The program was revised to display a bar in the bottom corner of the screen when the computer acknowledged the patient's switch activation.

Because the hospitalized patient's communication needs may change over a short period of time, the GPI should be provided with a means whereby the "words" and the "phrases" communication menu can easily be changed by the medical personnel. Because relatively few personnel could be expected to be proficient in computer use, this feature should be "user friendly," that is, the computer should provide clear, simple instructions on how to alter the menu with provisions to correct accidental, unwanted changes. Because the menu is stored only in the computer's memory (RAM), whenever the computer is turned off the menu changes will be lost unless stored in some secondary device such as a cassette tape or floppy disc. The cassette disadvantage is the long loading time as previously noted. Floppy disc drive with its short storage and retrieval time can provide a highly useful although expensive alternative. Battery backed up RAM could also offer some advantages.

It is felt that occasional reprogramming of the computer will be made necessary by a particular patient's unique and possibly changing needs. An exciting, but as currently unexplored method of accomplishing such reprogramming is the use of acoustic modems to transmit revised versions of the program from a host computer to the patient's computer. Softwear to allow communication between computers over the telephone line is already available and similar software employing an RS-232 communication link between a GPI and a minicomputer has been successfully tested.

CONCLUSIONS

The challenge of using a GPI in these 3 individuals with diverse deficits, etiologies and surrounding circumstances, but all with severely impaired communication ability has led us to conclude the following:

1. The microprocessor is well suited as a communication and environmental control aid and is adaptable to an acute hospital setting. The space occupied by the cluster of equipment produced some problems at the bedside in a crowded intensive care environment. Problems with positioning, in relation to a patient needing frequent position changes, and placement in relation to patients with visual limitations will need to be further addressed.

2. Program flexibility will be mandatory. Changing patient requirements and the wide variety of disabilities found in acute hospital settings necessitates frequent program revisions. The use of acoustic modems is seen as a potential method for reprogramming.

3. Various activation devices and selection modes will have to be innovatively utilized to meet the patient's changing needs and maintain program flexibility. We have found that intelligent selection of devices by persons familiar with the patient's physical limitations and abilities followed by bedside trials and coupled with careful positioning is best. A supervised structured training period should be developed to reduce patient and medical staff frustration.

4. Visual limitations appear to be a potentially significant problem. The trend in small computers is to more smaller characters per screen area. In the acute hospital setting patients with visual and positioning limitations may require large size characters. A program for visual testing which could be performed at the bedside using the computer appears to be an effective solution.

5. Phrase selection can speed up communication in patients with very limited ability to manipulate a computer interface. The program designed should allow medical and nursing staff, as well as the patient, to enter phrases that are needed at that phase in the patient's care. The process of changing phrases should be clear, simple and brief.

6. A disc drive is essential in promoting interaction and acceptance between busy medical personnel and the system. In our trial a program taking 5 minutes to load was not well utilized by busy personnel. For maximum utilization, the program should have virtually instantaneous access capability.

ACKNOWLEDGEMENT

This work was supported by a grant from the Division of Vocational Rehabilitation of the State of Louisiana.

REFERENCES

Cronk, S. and R. Shubert (1980). A general purpose interface (GPI) for independent living. *Proceedings 33rd ACEMB Meeting*, Washington, D.C., p 209.

Vanderheiden, G. and K. Gilley (1975). *Nonvocal Communications Techniques and Aids for the Severely Physically Handicapped.* Univ. Park Press, Baltimore.

TOWARDS EMPLOYMENT OF THE SEVERELY PHYSICALLY HANDICAPPED

Roy W. Schubert, William W. Moore and Stanley R. Cronk

Department of Biomedical Engineering
Louisiana Tech University
Ruston, Louisiana 71272

ABSTRACT

This paper reports efforts to place a severely physically handicapped person into a gainful employment situation. The underlying philosophy is that an inexpensive microprocessor based computer can be used as an effective interface between the worker and the workspace. The task consisted of client screening by a VR counselor, physical capabilities testing, mockup workspace, and final specification of the workspace. It was concluded for one client that mouth stick input is as fast as single finger input from an able-bodied population. More importantly the results suggest that a smaller keyboard positioned just below the CRT would require less head and neck movement. A multitasking computer operating system was designed and implemented on a 6502-based microcomputer.

KEYWORDS

microcomputer; handicapped; employment; multitasking; evaluation; physically handicapped; 6502

INTRODUCTION

The Vocational Rehabilitation efforts at Louisiana Tech include the assessment of available motions of severely physically handicapped persons through the Availiable Motions Inventory (AMI) that was developed by Wichita State University. An AMI evaluation results in a report that describes, in engineering terms, the client's range of motions and forces. These are to be used to match this individual to an industrial job. Several clients, mostly high level quadriplegics, have performed poorly because of their high level of injury. Yet these people, bright and alert, desire self-sufficiency. Consequently an evaluation procedure was devised that indicates the potential for employment using data processing typing skills, where input is limited to head pointer or mouthstick. This paper summarizes the test and procedures that evolved from this effort and the modifications needed in one test case. The need for several special tools to manipulate items in the workspace is discussed.

METHODS

The initial screening was performed by a Louisiana Vocational Rehabilitation counselor and a request was made to the Department of Biomedical Engineering to assist in workspace modification. The client was a post-polio quadriplegic, a

Certified Public Accountant, and had contact with several small businesses who were willing to let her keep their books. Because of the potential equipment costs, it was important for Louisiana Vocational Rehabilitation to know if the client could interact with a computer and what constraints might apply. Consequently a systems analysis was performed on the home-work scenario. A series of keyboard-based timed tests were devised that made use of various sized keyboards which were placed at various angles with respect to the CRT. The utility of finger guards was also explored. These results were compared with similar data from a normal population who were restricted to single finger input and were equally as familiar with the keyboard layout as the client. The client was then placed in a mock work station to find the total work effort needed while an engineer, ocupational therapist, physical therapist and a counselor observed the several major motions associated with the hypothetical job.

RESULTS: SYSTEM ANALYSIS

The systems analysis assumed that the client would be in her home and unattended for a large percentage of time. It was further assumed that an aid or representative of the small business would visit periodically to exchange verbal and written data. Additionally data and inquiries could arrive or depart by phone. The home-work needs identified were:

1) Non-fatiguing accurate keyboard interaction.
2) Manipulation of floppy disks.
3) Movement of data sheets into and out of the visual field.
4) The entry of financial data and typing of resultant forms, quarterly tax reports and form letters.
5) The ability to dial the phone.
6) The ability to turn on/off lights, TV, radio, fan etc.
7) The ability to easily alternate between the financial and environmental software.
8) Signal a "help" alarm and lock/unlock the house doors.

RESULTS: HARDWARE/SOFTWARE

The above eight items were translated into the following hard/software needs:

1) Assessment of keyboard manipulation skills.
2) Special mouthstick-operated floppy disk manipulator to move floppy disks and opens/closes the disk drive door.
3) A mock work station to evaluate equipment placement.
D) A paper holding carousel.
4) Commercial availability of financial and word processing software packages that require only single keystrokes for interaction.
5) A General Purpose Interface (GPI) to handle environmental control, manage home security and interface to a menu driven phone (Cronk and Schubert, 1980; 1982).
6) A multitasking operating system to quickly and easily switch between items 4 and 5 above.

The actions taken by the Biomedical engineering department were:

1) Administer the AMI and Visual Screening test to identify usable movements and visual impairment respectively.
2) Design and administer KEYTEST and TYPETEST to evaluate speed and accuracy of data entry for single letters and sentences respectively.
3) Start design of special mouth sticks.
4) Find source for material handling devices.
5) Procure business software appropriate to client's needs.
6) Incorporate the GPI into the system to handle environmental control, phone and security.
7) Design/modify an operating system to allow background/foreground multitasking so

that the client could switch between the business software and the GPI system easily.

The department was unable to find any software/hardware to permit multitasking on a small easily obtainable 6502-based machine. Pseudomultitasking by multiple read/writes to/from disk were judged too slow. Thus a totally RAM contained routine was considered. The microcomputers operating system was modified to allow limited multitasking, by using the 6502's non-maskable interrupt to activate a machine language routine that preserves the status of the resident program (zero page and screen RAM) after checking the status of the IEEE-488 port. The Basic and machine language programs, variables, arrays and strings were not moved. Instead alternate pointers were forwarded to the operating system. The one problem is that all software must be screened to assure that all files involved in disk or printer data transfer are either closed or have unique file numbers. This concept thus defined a programming philosophy for GPI's modification. A single pushbutton switch added to the keyboard allowed the system to "flip-flop" between the business and GPI software quickly, while at the same time preserving the state of each. The routine uses less than 2 Kbytes.

RESULTS: TESTS

The AMI evaluation revealed that the client while using a mouth stick could operate push switches easily but slide and rotary switches with difficulty. Proximity to switches was an important factor.

KEYTEST and TYPETEST data from the client and a similar able-bodied population are shown in Table 1 and Fig. '

TABLE I. TEST RESULTS: TIME BETWEEN KEYS (sec)

TEST	CLIENT	KEYBOARD	GUARD	MEAN	STD. DEV.
KEY	HC	LARGE	NO	1.82	0.42
KEY	HC	SMALL	NO	1.39	0.34
TYPE	HC	SMALL	NO	1.38	1.01
KEY	SAMPLE	SMALL	NO	1.48	0.39

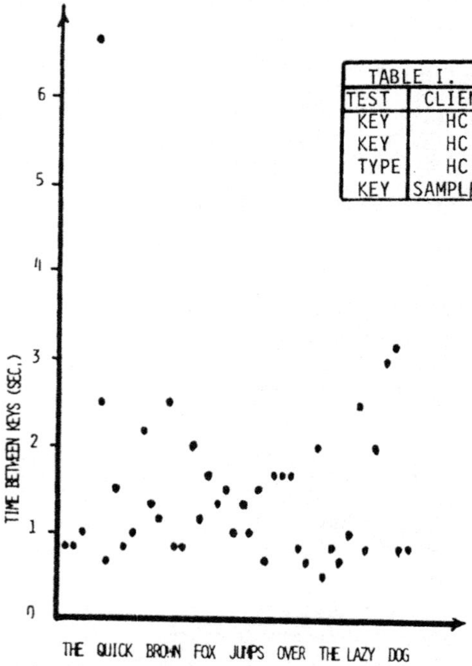

Fig. 1. TYPETEST results.

The GPI was modified to run on a disk-based system and trimmed to allow only direct keyboard selection.

DISCUSSION

The AMI results demonstrated that the client wwas not suited for industrial employment. The KEYTEST and TYPETEST, however, suggest that her speed of data entry and accuracy make her a good candidate for a data processing type of employment. While the data do not show a strong indication that the small keyboard is better than the large keyboard, it was observed by all present during the test that the small keyboard positioned just under the CRT produced much less head and neck movement and therefore less fatigue. The disk drive was positioned so that the insertion/removal of disks required equal movements backward and forward from the mean keyboard position. The client could open and close the drive doors without modification, but this required a large lateral force and a strong bite on the mouthstick. Small tabs mounted on the door reduced the motions to fore/aft because the AMI test identified this as her best movement. Upon questioning, the client preferred the small keyboard and did not want a key guard. She was not very happy with the disk drive tab, even though all present judged it less laborintensive. All evaluaters concerned suggested using the tabs to reduce the possibility of dental problems. The preference of the small keyboard has been observed in other clients, even those who still posess hand movement, because of reduced movement and the close proximity of a place to rest/brace the fingers or other parts of the hand (Schubert, 1982).

The key feature in the final system is the multitasking operating system that permits quick selection between two sets of software. This appears to be the first report of multitasking on a small machine, but its potential has been realized by others (Vanderheiden, 1981). The multitasking is not universally adaptable because of the need to screen for open files that are visible to the intelligent disk drive. In this situation, where one of the programs is under internal developement, the sacrifice is well worth the improvement in performance. We suspect that this multitasking system will make available a large amount of commercially available software to severely handicapped users while at the same time allowing partial independence and possibly gainful employment.

ACKNOWLEDGEMENTS

This effort was supported by a grant from the Division of Vocational Rehabilitation, State of Louisiana.

REFERENCES

Cronk, S., and R. W. Schubert (1980). A General Purpose Interface (GPI) for Independent Living. Annual Conference on Engineering in Medicine and Biology, 33, 209.

Cronk, S., and Roy W. Schubert (1982). An adaptive Computer-Aided Independent Living System for Severely Handicapped. First Southern Bioengineering Conference, June 1982.

Schubert, Roy W. (1982), Unpublished Observation.

Vanderheiden G. (1981). Practical Applications of Microprocessors to Aid the Handicapped. Computers, 14, 54-61.

Session 3
BIOLOGICAL SIGNAL PROCESSING

REAL-TIME PROCESSING OF ELECTROPHYSIOLOGICAL EVENTS AT THE NEUROMUSCULAR JUNCTION

S. Laxminarayan, L. Michelson, P. Goldstein,
Department of Scientific Data Processing,
T. Argentieri, J.J. McArdle,
Department of Pharmacology,
University of Medicine and Dentistry,
100 Bergen Street, Newark, NJ-07103.

ABSTRACT

A parameter of considerable interest in synaptic physiology is the average number of quanta of neurotransmitter (Acetylcholine, ACH) released from the pre-synaptic nerve terminal. The results of such quantal analysis have important implications for numerous neurological disorders in man such as the muscular dystrophies and myasthenias. In practice, the quantitative assessment of the transmitter release involves recording the electrical events from the post-synaptic or end-plate region of single muscle fibres with an intracellular electrode. These events include both the spontaneously occurring miniature end-plate potentials (mepps) and the evoked end-plate potentials (epps) due to nerve stimulation. Accurate estimation of the quantal information requires a knowledge of the statistical properties of these electrical events. The visual characterization and analysis of such data from paper charts are often much too time consuming and inefficient. This paper is concerned with digital computer methods for the real-time data acquisition and processing of the mepp and epp data generated in the laboratory. The relevant software is developed using a HP 1000 series digital computer which is part of a distributed network system currently in operation at the University of Medicine and Dentistry in Newark, New Jersey (UMDNJ). A host of real-time application software is developed which include post-processing of computed statistical estimates by applying necessary techniques to correct for effects such as membrane capacitance and non-linear summation, drift etc.

KEYWORDS

Neurotransmitter, quantal analysis, miniature end-plate potential, end-plate potential, distributed network system, ON-LINE processing, non-linear summation.

INTRODUCTION

The analysis of electrical events in the developing nerve-muscle system provides significant insights into the principles underlying the functioning of chemical synapses and their role in the development of the mature neuromuscular system. The results of these studies have important implications for numerous neurological disorders in man such as muscular dystrophies and myasthenias. One of the major areas of research which has gained tremendous importance in recent years concerns the phenomenon of chemical transmission across the neuromuscular junction (Del Castillo and Katz, 1954).

Transmission at this synapse is believed to occur by the mediation of a chemical (Acetylcholine, ACH) which is released from the nerve terminal in multimolecular packets called quanta. A parameter of significant interest in synaptic physiology is the average number of quanta of transmitter released from the pre-synaptic nerve terminal (Del Castillo and Katz, 1954). In practice, the quantitative assessment of transmitter release involves recording electrical events from the post-synaptic or end-plate region of single muscle fibres with an intracellular electrode. Initially, miniature end-plate potentials (mepps) need to be recorded. These are due to the spontaneous release of single quanta of ACH and therefore their mean amplitude reflects quantal size (q). Following the mepp recordings, the nerve is stimulated and the resulting end-plate potentials (epps) are measured. The typical features of the waveforms generated in these studies are characterized by low amplitudes, fast rise times and pulses occuring at random intervals. For example, the mepps have an average amplitude of approximately 0.5 mV with a duration of less than 5 msec. and a rise time of about 1 msec. These mepps occur spontaneously at frequencies ranging from 0.5-4/sec.

In the evaluation of the quantal information, accurate estimation of the statistical properties of the peaks corresponding to mepps and epps is an extremely crucial phase of experimental research. The visual detection and characterization of the data from paper charts or film records are much too time consuming and inefficient. Therefore digital computer techniques provide a potential tool in extracting such information. In this paper we describe appropriate methodology for the real-time ON-LINE processing of the epp and mepp data, using a Hewllet Packard HP 1000 digital computer. The system is part of a distributed network currently in operation at the University of Medicine and Dentistry (UMDNJ) in Newark, New Jersey (Michelson, Laxminarayan, Goldstein and others, 1982).

ELECTROPHYSIOLOGICAL METHODS

The electrophysiological methods for the measurement and recording of data are detailed elsewhere (McArdle, 1975). The extensor digitorum longus (EDL) and soleus muscle from Wistor rats were isolated and removed along with a 1.5-2 cm length of attached nerve. The muscles were then pinned to a Sylgarel 184 (Dow Corning) lined plexiglass chamber and continually superfused at a rate of 8-10 ml/min. with a physiological Ringers solution containing (mM): NaCl, 135; KCl, 5; MgCl, 1; $CaCl_2$, 2; $NaHCo_3$, 15; Na_2HPo_4, 1 and glucose, 11. This solution was bubbled with 95% o_2 - 5% Co_2 to maintain a pH of 7.3-7.4. Nerve-cell synapses (neuromuscular junction) were located in surface cells using a glass microelectrode filled with 3.0 m Kcl and having a tip resistance of 8-10 megaohms. These electrodes were positioned with JENA^R micromanipulators and connected by chloride coated silver wires to a DC coupled pre-amplifier (W.P. Instruments, M701). The output of the pre-amplifier was delivered to a Tektronix 565 oscilloscope for monitoring as well as to a Hewllet Packard HP 1000 series mini computer for analysis. The microelectrodes were considered to be focally positioned at the neuromuscular junction when the mepps and epps had an average rise time of one millisecond or less.

MATERIALS AND HARDWARE

The ON-LINE monitoring and data processing of the experimentally generated electrophysiological data are carried out in a distributed network environment (Michelson, Laxminarayan, Goldstein and others, 1982). The hardware configuration consists of a Hewllet Packard HP 1000 series mini computer as the central host system with a remote node located at the site of the experiment in the pharmacology laboratory. The system at the remote node is based on a second HP 1000 series computer and is equipped with 16 channels of 12-bit analog to digital converters, digital output registers, a line printer and a Tektronix digital to analog scope. For the purpose of testing the software performance, a Grass polygraph amplifier is utilized. The remote system is core-based, running under the real-time RTE IV M environment where as the system environment of the central node is based on the RTE IV B operating system. Initial data acquisition and the ON-LINE processing of acquired data are performed at the local node and the results are subsequently transferred as appropriate data files onto the host system for later retrieval and further processing if necessary.

SOFTWARE METHODOLOGY

Overall Strategy

The central aspect of the processing algorithm, written in Fortran, is the ON-LINE detection of the peaks corresponding to the miniature end-plate and end-plate potentials recorded at the synaptic region of the muscle fibre. The analog signals which are derived continuously from the glass microelectrodes are digitized at sample rates ranging from 5000-10000 samples/second, after applying suitable signal amplification. The data are passed through analog filters prior to digitization, in order to avoid aliassing effects (Bendat and Piersol 1972). Typical analog data sections comprising mepps and epps are illustrated in Fig. 1. The

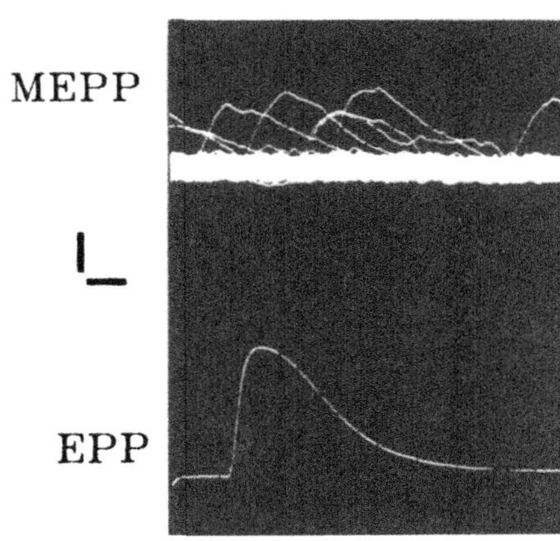

Fig. 1. Representative intracellular recordings from muscle fibre which are subjected to ON-LINE computer analysis. Top: Miniature end-plate potentials recorded at the neuromuscular junction. Bottom: End-plate potential evoked in response to nerve stimulation. Calibration:
Horizontal: mepp = 2 msec, epp = 2 msec
Vertical : mepp = 0.5 mV, epp = 10 mV

digitized samples are transferred as a continous sequence of data segments to a temporary dual-partition buffer. The size of the buffer is pre-determined depending upon the sample rate and signal duration used in the analysis. When one of the sub-partitions of the data buffer is full, the peak detection algorithm is invoked while simultaneously acquiring and transferring the next successive data segment into the other sub-partition. Both the data acquisition and the processing algorithms are controlled by a supervisory program by means of a technique called class I/O. In the RTE environment, this scheme of class I/O permits ultra high speed and secure communication between multiple groups of programs.

Identification of Mepps

The basic criterion used for the detection of the peaks corresponding to mepps is based on the computation of the maximum value between two threshold crossings. An automatic threshold level is initially established by estimating the standard deviation of the incoming data segment. From a computational point of view, the incoming signal is effectively seen as a "noisy" base-line with its mean value representing an estimate of the Resting Membrane Potential (RMP). Since the mepp amplitudes are much less than the maximum resolution of the analog to digital converter, considerable signal amplification is applied in order to identify the occurrence of mepps.

The maximum peak amplitudes of the original mepps in the data are usually in the order of 0.5 mV and are observed to be well above the noise levels of the RMP. Therefore the detection of mepps is achieved by setting a threshold level to a real multiple of the standard deviation of the incoming data. This threshold level is automatically computed for each digitized segment of data. In the present scheme, 4 standard deviations have been found to be adequate for identifying the mepps. Finally, the detected mepps are subjected to further "acceptance" procedures based on the frequency and mepp amplitudes, in order to avoid detection of false mepps.

Identification of Epps

In contrast to mepps, the end-plate potentials (epps) are generated as evoked responses to nerve stimulation. The stimulus required to evoke the muscle is controlled by the computer program. The software is designed such that the stimulation rates can be varied over any desired range. Depending upon the rate of stimulation of the nerve, the length of the incoming data segment is fixed such that an epp is picked out as the local maximum value of that current segment. Corrections are applied to the initial portion of the data segment to eliminate the presence of the stimulus artifact prior to computing the local maximum. The resting membrane potential is estimated by calculating the mean value of a part of the data segment preceding the epp.

Processing Software

Once the epps and mepps are detected and their amplitudes and RMP computed, a number of application programs are incorporated which permit the estimation of the required quantal and other parameters. These include the quantal size, the mean quantal content of the epps, the probability of transmitter release, estimate of the N-statistic etc. The amplitudes of each of the epps and mepps are corrected for the effects of membrane capacitance and non-linear summation using correction factors derived from the equivalent circuit for the post-synaptic membrane (Martin, 1976). The software is designed such that estimates of peak amplitudes, interval histograms etc. can be either generated locally at the remote node or stored as disk files in the central node.

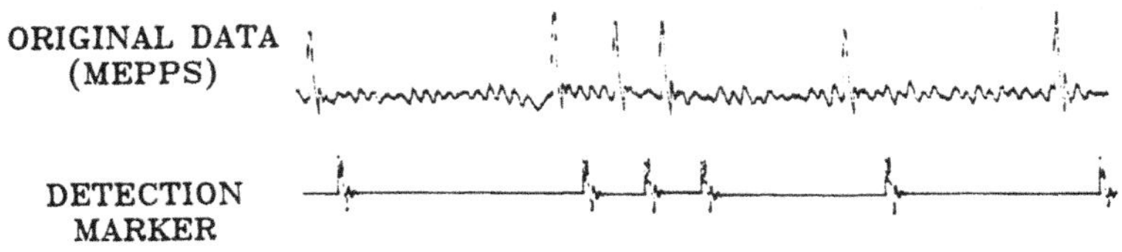

Fig. 2. Example illustrating the detection of mepps.
Top: A segment of original data containing mepps.
Bottom: Computer generated "markers" to illustrate the detection of mepps.

RESULTS

The system performance was evaluated by analyzing different types of known waveforms such as triangular and saw-tooth functions. As an example of the present output capabilities in the detection of mepps, the program sends an output signal which generates a "marker" on a polygraph, whenever a mepp is encountered. An example illustrating the detection of mepps is illustrated in Fig. 2. The results of analyzing records containing sequences of 50-100 mepps have yielded an accuracy of the order of 98-99%. In contrast to mepps which occur spontaneously, epps are obtained by using the computer to trigger the nerve stimulation. The resulting waveforms are extremely well defined and controlled and the detection criteria used in the present program have been found to be more than adequate, in the detection of epps.

ACKNOWLEDGEMENTS

The authors are deeply indebted to Ms. Marijke Zwakman for the meticulous typing of the manuscript. The work reported in this paper was supported in part by a grant from the National Institute of Health (NS 11055-09).

REFERENCES

Bendat, J. and A. Piersol (1972). Random data: measurement and analysis procedures. John Wiley and Sons, London, New York.

Del Castillo, J. and B. Katz (1954). Quantal components of the end-plate potential. J. Physiol., 124, 560-573.

Michelson, L., S. Laxminarayan, P. Goldstein, J.J. McArdle, A.B. Ritter, W.N. Duran, J. Blackwood and H. Alexander (1982). In M. Schwartz (ed.). Application of computers in medicine. IEEE Publication. In press.

Martin, A.R. (1976). The effect of membrane capacitance on non-linear summation of synaptic potentials. J. Memb. Biol., 59, 179-187.

McArdle, J.J. (1975). Complex end-plate potentials at the regenerating neuromuscular junction of the rat. Exp. Neurol., 49, 629-638.

A Method for the Measurement of Acoustical Transmission in Deafness

P. Gomez and V. Rodellar

C/O Electrical Eng. Dept.
University of Maryland
College Park, MD 20742

On leave from:
Facultad de Informatica
Departamento de Electronica
Universidad Politecnica
Madrid - SPAIN

ABSTRACT

In this paper a transmission line model for the middle ear is presented with a possible extension also for the inner ear. This model can be used for the characterization of transmission problems within the ear. A method for the necessary measurements is explored. The results may be used in the construction of more accurate hearing aids.

KEYWORDS

Transmission line model, transmission-reflection coefficients, scattering matrix, linear prediction, PARCOR algorithm, deafness characterization.

TRANSMISSION MODEL FOR THE EAR

The middle and inner ear structures have been modeled by several authors as transmission line-like acoustical circuits (Zwislocki, 1963; Flanagan, 1965). This suggests the possibility of the existence of mismatch along the line which would lead to the presence of reflected energy coming back to the outer ear from within the inner sections. Although several researchers (Boer, 1981) have disregarded the presence of such reflected waves in the ear mechanics, some others (Kemp, 1980) have been able to measure in the outer canal a certain kind of reflected wave after a previous excitation. The presence of reflected waves in the outer canal will allow to extract some information about the transmission parameters in the middle ear, and within some constraints it will be also possible to do the same with the inner ear. In figure 1 is shown the transmission line model (b) associated with each one of the main sections of the ear (a). $u_k^+(t)$ and $u_k^-(t)$ are the forward and backward volume velocity wavelengths measured in the outer canal, $Z_o(x)$, $Z_M(x)$ the transmission line impedances vs. distance and Z_{ct} the end impedance at the oval window. Suppose we divide the outer and middle sections of the ear into a set of N subsections of fixed length d along the x axis. For the kth subsection the distance at its beginning and at its end will be k.d

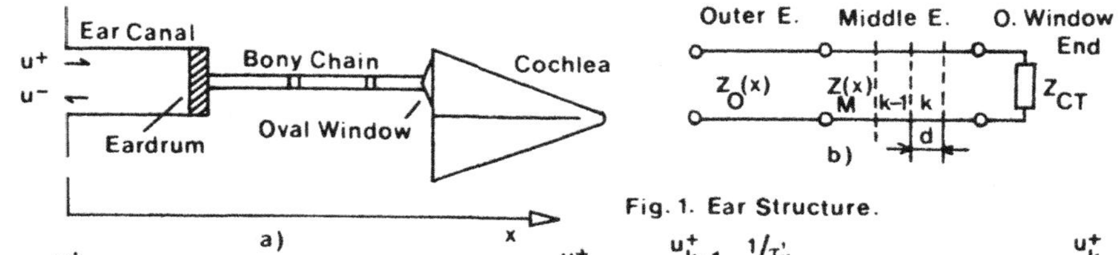

Fig. 1. Ear Structure.

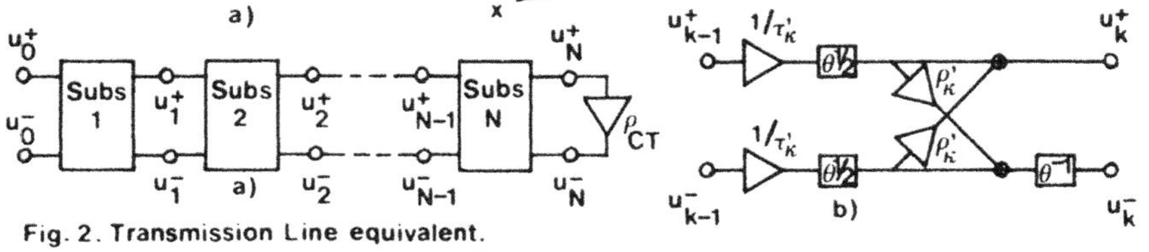

Fig. 2. Transmission Line equivalent.

and (k+1).d with k scaling from 1 to N. If N is big enough then we can suppose that the impedance of a given subsection is constant with distance $Z_k(x)=Z_k$. We can define in this way N+1 reflection and transmission coefficients between two successive subsections (k, k+1) as

$$\rho_{k,k+1} = (Z_{k+1} - Z_k)/(Z_{k+n} + Z_k) \quad ; \quad \tau_{k,k+n} = 2Z_k/(Z_{k+n} + Z_k) \tag{1}$$

The model for the propagating wave between the two sections can be set then as the cascade of N subsections (fig. 2a) each one of them having an equivalent flow graph given in fig. 2b and with the scattering matrix description:

$$\begin{bmatrix} u_k^+ \\ u_k^- \end{bmatrix} = \frac{\vartheta_k^{\frac{1}{2}}}{\tau'_k} \begin{bmatrix} 1 & \rho'_k \\ \rho'_k \vartheta_k^{-1} & \vartheta_k^{-1} \end{bmatrix} \begin{bmatrix} u_k^{+-1} \\ u_{k-1}^- \end{bmatrix} \tag{2}$$

Here u_k^+ and u_k^- are the amplitudes of the volumetric velocity at the end of the kth subsection. u_{k-1}^+ and u_{k-1}^- are the same variables at the end of the (k-1)th subsection; ρ'_k and τ'_k are the reverse transmission and reflection coefficients, defined as

$$\rho'_k = -\rho_k = -\rho_{k-1,k} \quad ; \quad \tau'_k = 1+\rho_k \tag{3}$$

and ϑ represents the delay and attenuation introduced by one subsection. In the case that losses are considered ϑ can be expressed as

$$\vartheta_k = e^{\alpha k d} \cdot e^{i\beta k d} \tag{4}$$

with α_k accounting for the losses in the kth subsection and β_k being the propagation constant in the same subsection. In our case, with no losses $\alpha_k=0$. Following Wakita (1973) the effect of only one subsection can be described by the graph and matrix given in fig. 3a. In fig. 3b the inverse graph and matrix are given. These matrices are well known in speech analysis-synthesis (Makhoul, 1975) and geophysical prospection (Robinson, 1980). If we know the excitation (u_o^+) and the response (u_o^-) in the meatus, through successive applications of the deconvolving matrix (PARCOR algorithm), we can estimate the set of ρ_k by the following expression

$$\rho_k = -\left[\sum_{i=1}^{N+n} u^+_{k-1}(n)u^-_{k-1}(n)\right] \Big/ \left[\sum_{i=1}^{N+n} (u^+_{k-1}(n))^2 \sum_{i=1}^{N+n} (u^-_{k-1}(n)^2\right]^{\frac{1}{2}} \tag{5}$$

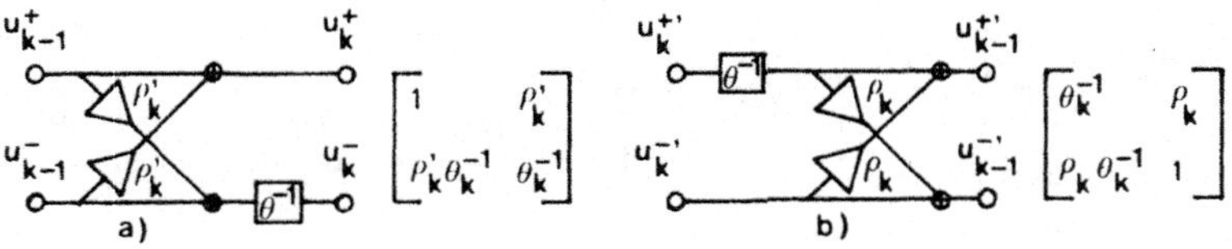

Fig. 3. Reduced matrices.

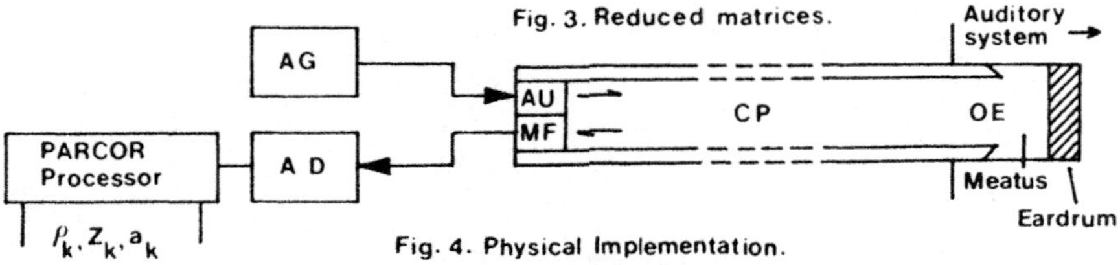

Fig. 4. Physical Implementation.

with i=1...n the sample index for the digitalization of the signals and n the total number of samples in one record. When we know the set of reflection coefficients ρ_k the outer and middle ear transfer functions vs. frequency can be calculated in a straight forward manner.

POSSIBLE EXTENSION OF THIS METHOD TO COCHLEAR MECHANICS

In a first approach to modeling cochlear mechanics it has been possible to determine a cascade model for the cochlea quite similar to the one given for the middle ear (fig. 1a). This shows the possibility for the PARCOR algorithm to be applied also for the characterization of transmission patterns inside the cochlea. For this model (Rodellar, 1982) the transmission-reflection coefficients between each one of the incremental subsections will be given by the following expressions

$$\rho_k=(\gamma_k Y_{p,k+1}+\gamma_{k+1}Y_{p,k})/(\gamma_k Y_{p,k+1}-\gamma_{k+1}Y_{p,k});\qquad(6)$$

$$\tau_k=2\gamma_{k+1}Y_{p,k}/(\gamma_k Y_{p,k+1}-\gamma_{k+1}Y_{p,k});\quad \gamma_k=\sqrt{H_{t,k}+H_{v,k}}\qquad(7)$$

$$H_{t,k}=(j\omega\sigma+R_{ft,k})Y_{p,k}/S_{t,k}\ ;\ H_{v,k}=-(j\omega\sigma+R_{fv,k})Y_{p,k}/S_{v,k}\qquad(8)$$

Here γ_k is the propagation constant for the kth subsection, $Y_{p,k}$ the admittance of the partition membranes, Rft,k and Rfv,k the viscosity constants in the scalas, St,k and Sv,k the area in the scalas and w and σ the frequency and perilymph density. All these parameters are quantized vs. distance from Zwislocki's model (1963). If we can efficiently estimate the transmission-reflection coefficients for each subsection, we will be able to get some information about the variation of these parameters vs. distance.

DESCRIPTION OF THE MEASUREMENT METHOD

The suggested method of measurement may be implemented as shown in fig. 4. An acoustic generator (AG) produces an impulse-like pressure waveform which is fed through a coupling pipe (CP) into the outer ear (OE). The reflections coming from within the ear are picked up by a microphone (MF) which has also registered the incident wavelet. These signals are digitized and processed by the PARCOR processor which gives us the set of reflection coefficients as the output. Alternatively the output can be the set of acoustic impedances of each one of the subsections or the transmission pattern vs. frequency associated with them. This pattern will be in a first approach the most desirable output, for it can provide a graphic report about the transmission losses in damaged bony chains. A correction for the acoustical characteristics of the microphones and couplers in use must be provided. The PARCOR processor design and its implementation is currently under study.

DISCUSSION

Through this paper the possibility of taking measurements about the behaviour and frequency selectivity in the middle and inner ear has been explored. The proposed method is based on a model of the ear as a transmission line composed of N cascaded elementary subsections; for each subsection a scattering matrix is given. In this way we should be able to (1)estimate the transmission characteristics of the inner ear by non-invasive means; (2)design spectrum-correcting hearing-aids with a high degree of accuracy (Gomez, 1982): (3)explain some of the not yet well known frequency selectivity mechanisms within the cochlea; (4) measure the influence that each of the cochlear parameters has in the cochlear mechanics. Some of the problems which may appear through this treatment can be caused by the non-linear behaviour observed in some circumstances by several researchers in both middle and inner ears.

The authors wish to thank Professor R.W. Newcomb for introducing us to this research topic and for all his contributions and patience. We are sure without them this work would not have been possible.

REFERENCES

Boer, E. de. (1981). Digital Solution of Cochlear Mechanics Problems. Journal of the Acoustical Society of America, 69, 1369-1373.

Flanagan, J.L. (1965). Speech Analysis Synthesis and Perception. Academic Press, New York.

Gomez, P., Rodellar, V. and Newcomb, R.W. (1982). A PARCOR Characterization of the Ear for Hearing Aids. (Prepared for publication).

Kemp, D.T. (1980). Towards a Model for the Origin of Cochlear Echoes. Hearing Research, 2, 533-548.

Makhoul, L. (1975). Linear Prediction: A Tutorial Review. Proc. IEEE, 63, 561-580.

Robinson, E.A. and Treitel, S. (1980). Maximum Entropy and the relationship of the Partial Autocorrelation to the Refection Coefficients of a Layered System. IEEE Trans. on Acoust., Speech and Signal Processing, 28, 224-235.

Rodellar, V. and Gomez, P. (1982). A Recursive Incremental Model for Non-invasive Cochlear Measurements. The Annual Meeting of the Acoustical Society of America. Evanston, Il. April, 27.

Wakita, H. (1973). Direct Estimation of the Vocal Tract Shape by Inverse Filtering of Acoustic Speech Waveforms. IEEE Trans. on Audio Electroacoust., 21, 417-427.

Zwislocki, J. (1963). Analysis of some Auditory Characteristics. In R.D. Luce, R.R. Bush, E. Galanter (Ed.), Handbook of Mathematical Physiology, Vol. 3, John Wiley & Sons, New York, pp. 1-97.

ON-LINE DIGITAL FILTERING APPLICATION USING WINDOWING TECHNIQUE

R. Laxminarayan*, L. Rajaram**, S. Laxminarayan***,
L. Michelson*** and J. McArdle***
*EEG Department, Institute for Epilepsy Research,
"Meer en Bosch", Heemstede, Holland
**N.M.K.R.V. College, Jayanagar, Bangalore, India
***Departments of Scientific Data Processing and Pharmacology,
University of Medicine and Dentistry, Newark, NJ-07103, USA.

ABSTRACT

There are a number of areas in biomedical research which call for high performance digital filters having extremely sharp cut-off features and zero phase-shift characteristics. Of the several methods that are available for designing a digital filter, the windowing technique provides a very effective method in achieving such characteristics. The central aspect of the windowing method is that the finite weighting sequence of the desired filter (with which the input data is convolved to obtain the filtered output) is modified by means of a smoothing function in order to obtain a flat frequency response, zero phase shift and a high filter discrimination. A particular smoothing function that is suited for applications in biomedical research, is the Dolph-Chebyshev function based on a modified form of Chebyshev polynomials. In this method, the design criteria are such that for a pre-selected discrimination, the cut-off rate of the achieved filter side bands for a given number of weights, is a function of the sample rate used to acquire data. The ON-LINE implementation of these filters for applications in biomedical research, using a mini computer are discussed in this paper.

KEYWORDS

ON-LINE digital filtering; Dolph-Chebyshev function; windowing method; biomedical application; convolution.

INTRODUCTION

In biomedical research, there are a number of areas which often call for high performance digital filters having extremely sharp cut-off features and zero phase-shift characteristics. For example, in studies related to the estimation of the dynamic lung compliance in neonates, digital filtering using such sharp filters is shown to be very effective in removing the cardiac effects from measured transpulmonary pressure and lung volume data (Laxminarayan and others, 1978). The design of a digital filter is approached on the basis of the filter specification either in time domain or in frequency domain. The time domain approach to synthesize the desired filtered output involves forming the convolution of the input signal with the discrete impulse response function of the desired filter. In frequency domain, the convolution is replaced by multiplication operation and the filtered output is recovered by inverse transforming the product.

Of the several methods that are used in the designing of a digital filter, the windowing technique provides a very effective method for achieving the desired sharp cut-off and zero phase-shift characteristics. The central aspect of the windowing method is that the finite weighting sequence of the desired filter is modified by means of a smoothing function in order to obtain a flat frequency response, zero phase-shift and a high filter discrimination. A particular smoothing function that is well suited for certain applications in biomedical research is the Dolph-Chebyshev function based on a modified form of Chebyshev polynomials. In this method, the design criteria are such that for a pre-selected discrimination, the cut-off rate of the achieved filter side bands for a given number of weights is a function of the sample rate used to acquire data (S. Laxminarayan, 1969, 1978). We discuss in this paper the effectiveness of the use of the method in biomedical research and the ON-LINE implementation of the method on a mini computer.

NUMERICAL PROCEDURE OF FILTER DESIGN

The numerical procedures for digital filter design based on the Dolph-Chebyshev function as the smoothing function are given by Laxminarayan (1969, 1978). In essence the computational steps consist of

1. transforming the desired filter frequency response function into its impulse response function (IRF) in time domain,
2. multiplication of the IRF with the inverse Fourier transform of the Dolph-Chebyshev function and truncation of the product to synthesize the required weighting function, and
3. convolution of the input waveform with the weighting function to obtain the filtered output.

DOLPH-CHEBYSHEV FUNCTION

The Dolph-Chebyshev function which is a modified form of the Chebyshev polynomials, is given by

$$S(K) = \frac{\mathrm{Cos}(P\ \mathrm{Cos}^{-1}\ (z_0\ \mathrm{Cos}(PI*K/P)))}{\mathrm{Cosh}(P\ \mathrm{Cosh}^{-1}\ (z_0))} \tag{1}$$

$K = 0, 1, 2, \ldots.., P/2, PI = 22/7$

where P represents the number of terms in the weighting function and the denominator represents the ripple factor, R in the filter. The frequency resolution of the desired filter frequency response function is given by $f_r = S/P$ where S is the sample rate used in the acquisition of data.

The Dolph-Chebyshev function has a main lobe whose width is as small as possible for a given side-lobe level and a given number of terms. The width of the main lobe, B, is given by

$$B = \frac{2}{PI}\ \mathrm{Cos}^{-1}(1/z_0) \tag{2}$$

Therefore, for a given cut-off rate, B follows and hence z_0 is calculated as

$$z_0 = 1/\mathrm{Cos}(PI*B/2) \tag{3}$$

It can be further derived that

$$P \geq \frac{1.46}{B} \cdot \log (2R) \qquad (4)$$

FILTER DESIGN PARAMETERS

It is seen from equations 2-4 that the advantage of using the Dolph-Chebyshev function is that for a pre-selected filter discrimination, we can optimize the cut-off rate of the achieved filter side bands and the duration of the weighting function. Some typical estimates computed from the above equations, for a 40 dB discrimination and a set of 128 weights are given in Table 1, for different sample rates.

TABLE 1

Sample Rate (Samples/Sec)	Achieved Cut-off Rates (dB/Hz)
10	152.40
20	76.20
50	30.48
100	15.24
200	7.62

Table 2 illustrates for a pre-chosen sample rate of 20 samples/seconds and 40 dB discrimination, some typical estimates of the cut -off rates and the leakage bandwidths for various durations of the weighting function.

TABLE 2

P	B.S	C
256	.262	152.40
128	.525	76.20
64	1.05	38.10
32	2.10	19.10

P=number of weights, B.S=leakage bandwidth (Hz), C=achieved filter cut-off rates = 40/B.S (dB/Hz).

ON-LINE DIGITAL FILTER ALGORITHM

The central aspect of the digital filter algorithm is the convolution operation of the input waveform with the weighting function. Initially, the frequency response characteristics of the desired ideal filter are chosen and the Dolph-Chebyshev function is appropriately computed to achieve these desired characteristics. The required weighting function is subsequently generated. Once this is done, the actual filtering procedure of the input data is carried out ON-LINE in real time by convolving the incoming data with the weighting function. The software is developed on a PDP 1103 digital computer, with the filtered data continuously transferred to a 40 megabyte disk storage.

On a PDP 1103 digital computer, typically the instruction execution time of a multiplication operation (Extended Arithmetic Instructions, EAI) is of the order of 32 microseconds, while the ADD time is about 4 microseconds. Therefore for a weighting function consisting of 128 weights, the entire convolution operation is achieved in about 5 milliseconds, enabling the filtering operation to be efficiently carried out ON-LINE, on incoming data which is sampled at rates of up to 200 samples/second. The need for higher sampling rates would of course imply selection of appropriate design parameters in the Dolph-Chebyshev function.

APPLICATIONS

The effectiveness of the use of the digital filter design method discussed here, is illustrated by considering transpulmonary pressure and lung volume data of neonates. These respiratory data are used to determine an estimate of the dynamic compliance of lungs (Laxminarayan and others, 1978). Since the measured data are corrupted by the presence of cardiac effects, digital filtering is found to be extremely effective in eliminating these effects, thus facilitating accurate determination of lung compliance. ON-LINE filtering of these data provide considerable advantage in clinical research.

Other applications include the characterization of respiratory patterns observed in the EKG recordings of infants who have had a "near-miss" for the Sudden Infant Death Syndrome and the analysis of electrophysiological events recorded at the neuromuscular junction.

ACKNOWLEDGEMENTS

The authors are deeply indebted to Ms. Marijke Zwakman for the meticulous typing of the manuscript.

REFERENCES

Laxminarayan S. (1969). Data analysis using digital filters. DAC Technical Note 4, ISVR, Southampton University, England.

Laxminarayan S., A.J.G. Spoelstra, P. Sipkema and N. Westerhof (1978). Transpulmonary pressure and lung volume of the cat and newborn: removal of cardiac effects with a digital filter. Med. of Biol. Eng. (GB), 16, 397-407.

TIME-DOMAIN ANALYSIS OF RESPIRATORY ELECTROMYOGRAM

Nirmal K. Mishra and S. Rengarajan

Electrical and Computer Engineering
California State University, Northridge, CA 91330

ABSTRACT

The electromyogram (EMG) is associated with the neuromuscular activation of a contracting muscle. This EMG, as a filtered response of several superposed neural pulse trains, with varying pulse rate, recruitment, and recording procedure, appears as an interference pattern. Various quantification techniques emphasize the frequency-domain analysis and the measurement of statistical parameters and their correlation with the pulse-frequency and the muscle force. Respiratory EMG, as recorded in the diaphragm, appears as an almost periodic signal and usually appears during the inspiratory phase only. Furthermore, it appears to represent the superposition of pulse-frequency modulated action potential trains with possible recruitment. This paper presents the filtering technique, providing a running average of the rectified EMG, as a necessary time-domain analysis of this signal. Such a filtered signal preserves the information concerning the pulse-frequency of the neural signal. It can, therefore, be useful in studying the dynamic relation between the muscle's mechanical response and its neural input.

KEYWORDS

Filtered EMG; running average; pulse-frequency modulation; respiratory EMG.

INTRODUCTION

During activation, the electrical activity of the contracting muscle, is usually picked up by either needle or surface electrodes. In either case, this electromyogram (EMG) represents a complex interference pattern. There are several reasons for this complexity. Some of these reasons are now recognized as the orientation of the recording electrodes especially in relation to the depolarizing fibers in a motor unit, variations in the amplitude of each action potential, randomness in the inter-spike interval, temporal recruitment of activated nerve fibers, muscle force and the rate of contraction in a muscle. Recently Richfield and others (1981) have reviewed various methods of quantifying the EMG especially when recorded by indwelling electrodes. Their review includes statistical parameters by modeling EMG as a stochastic process. Such parameters usually include mean value of the rectified signal, the rms value, and the power spectrum. The detailed mathematical justification of some of these procedures is provided by De Luca (1979). A significant component of such parameters is the mean firing rate of the contracting muscle. The supporting electrophysiological data is obtained by the contracting muscles during constant force and isometric conditions.

The assumption of mean firing rate is not valid under dynamic conditions with time varying mechanical response of the muscle. Such responses occur in the voluntary activity of the muscle. During spontaneous breathing, the inspiratory muscles, especially the diaphragm, exhibit an almost periodic contraction of the muscle yielding a time-varying muscle force. Bigland and Lippold (1954) established a relationship between the muscle force and the average value of the rectified EMG. While this does not establish the dynamic relation between the muscle force and the changing frequency of the action potential train, the rise of pulse frequency from 10 pps to almost 50 pps in the phrenic nerve during inspiration is well accepted. Furthermore, the phenomena of temporal recruitment of the fibers and the mechanics of the respiratory muscles is well known. With the existing physiological data, it is possible to analyze the EMG of respiratory muscles with two assumptions. These are: the existence of a pulse-frequency modulated neural signal and the dynamic relationship between the mechanical output of the contracting muscle and the incoming pulse-frequency modulated signal.

TIME-DOMAIN MODEL

Under steady-state condition, the neural discharge may be characterized by a train of Direac delta impulses yielding another train of motor unit action potentials as recorded by the electrodes. The location of these electrodes and the filtering effect of the muscle tissues determine the shape of each of the action potentials. Due to steady-state discharge, the inter-pulse interval may be assigned an average value, T, corresponding to its reciprocal, the average value of the pulse rate, R. If this inter-pulse interval is not constant, it may represent a random number T_K.

With $t_k = T_1 + T_2 + T_3 + \cdots + T_K$ and $t_0 = 0$ the Dirac delta impulse train may be written as

$$p_i(t) = \sum_{k=0}^{N} \delta(t-t_k) \tag{1}$$

where the suffix i represents the ith nerve fiber in a nerve trunk. Similarly, the recorded signal from a single motor unit becomes

$$e_i(t) = \sum_{k=0}^{N} h_i(t-t_k) \tag{2}$$

In Eq. (2), $h_i(t)$ represents the impulse response of the muscle tissues and the recording system for the ith motor unit. Since the recorded signal is a superposition of n number of such action potential trains

$$e(t) = \sum_{i=1}^{n} \{e_i(t)\} \tag{3}$$

The resulting EMG, e(t), is shown in Fig. 1.

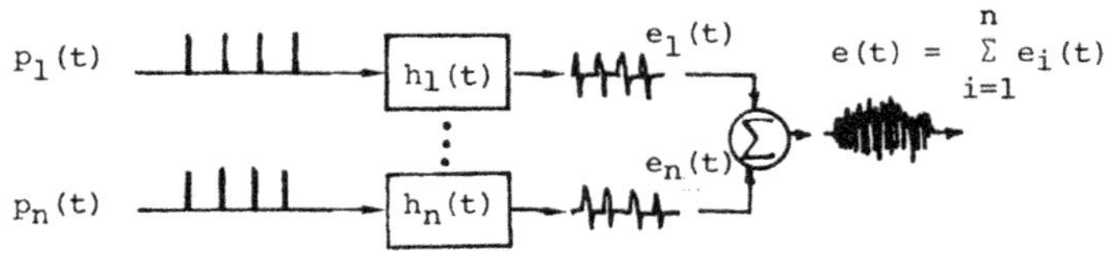

Fig. 1. EMG as a superposition of impulse response trains.

The existence of pulse-frequency modulated signal implies that the assumption of a single average value of the inter-pulse interval, T, or the mean firing rate, R, are not constant quantities. The actual variations of T or R depend on the nature of the modulating signal as well as the addition of biological or instrument noise. For steady-state discharge, the variations in T or R may be assumed to be very small. With this assumption, the mean rectified value of a single motor unit action potential train can be expressed as

$$E\left\{|e_i(t)|\right\} = \lim_{T\to\infty} \frac{1}{2T}\int_{-T}^{T} \sum_{k=0}^{N} |h_i(t-t_k)|\,dt \tag{4}$$

If the time averaging is done over a train of series of impulse responses, the summation term in Eq. (4) may be replaced by a convolution integral between the pulse train and h(t). Furthermore, if the pulse-frequency modulation does not justify a constant mean value of the pulse rate, the mean value of the pulse rate may be represented by r(t) during any small time interval. With this piece-wise averaging, Eq. (4) can be modified as

$$E\left\{|e_i(t)|\right\} = \int_{t-T}^{t} r(t')\,|h_i(t-t')|\,dt' \tag{5}$$

The above equation represents the running average of the rectified signal detected from a single train of the impulse responses. De Luca (1979) had obtained a similar expression for the mean rectified value where the mean rectified value is proportional to r(t). Making similar approximations, the time-varying mean rectified value becomes

$$E\left\{|e_i(t)|\right\} = r(t)\int_{t-T}^{t} |h_i(t-t')|\,dt' \tag{6}$$

The composite EMG is a superposition of several action potential trains as shown in Eq. (3). If it is desired to obtain the running average of the rectified signals due to such a superposition, then

$$E\left\{|e(t)|\right\} = E\left\{\left|\sum_{i=1}^{n} e_i(t)\right|\right\} \tag{7}$$

This equation shows that the superposition of the action potential trains occurs before the rectification. Clearly, the result would not be the same as the addition of the means of the rectified signals. The estimation of this difference remains to be investigated. Tentatively, on the basis of the findings by De Luca (1979), it is assumed that this difference, due to signal cancellation before rectification, are not significant. The running average of the rectified signal, therefore, is approximated as

$$E\left\{|e(t)|\right\} \simeq E\left\{\sum_{i=1}^{n} |e_i(t)|\right\}$$

$$\simeq r(t)\int_{t-T}^{t} \sum_{i=1}^{n} |h_i(t-t')|\,dt' \tag{8}$$

Equation (8) shows that the running average includes the mean firing rate, r(t), during a specified time interval T. Furthermore, since the rectified signal appears as a scaling factor, such a moving average retains this important information. The recovery of this information requires that the EMG signal be first rectified and then smoothed by using an appropriate running average filter. It is possible to use a third-order RC filter to achieve the result. The time interval for the running average depends on the time constant of the RC filter. Such a filter has been described by Mishra (1972a).

PROCEDURE AND RESULTS

As presented earlier (Mishra, 1972b), EMG was recorded by indwelling electrodes in the diaphragm of an anesthetized dog. The rectified EMG was filtered and compared with other mechanical variables such as the flow and in some cases the transdiaphragmatic pressure. The filter selected was a third-order running average filter (Philbrick/Nexus - 1968) with time-interval adjusted for 50 msec. The filter yields an idealized impulse response which is flat during the average time interval. Figure 2(a) shows the typical unrectified EMG of the diaphragm and the flow. Figure 2(b) shows the record of filtered EMG and the unfiltered and unrectified EMG. The presence of cardic spike also appears both in the filtered and in the unfiltered EMG.

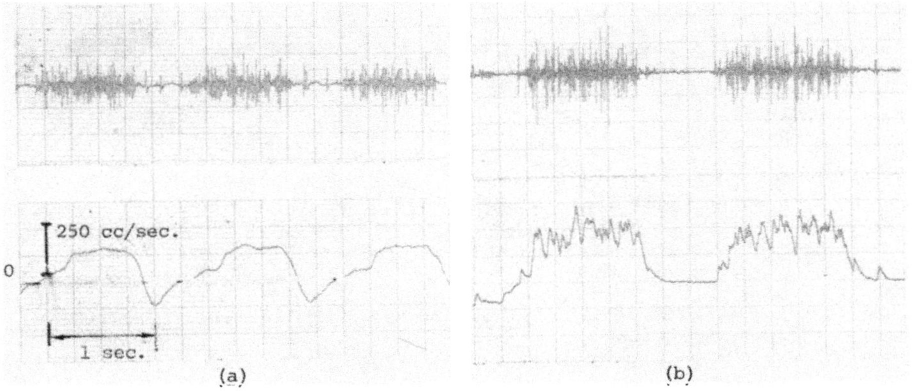

Fig. 2. (a) Typical EMG signal (top trace) and flow (bottom trace), (b) rectified and filtered EMG (bottom trace).

Figures 3(a) and (b) show the results of the filtered EMG generated by stimulating the phrenic nerve by artificial stimulation using a pulse-frequency modulator.

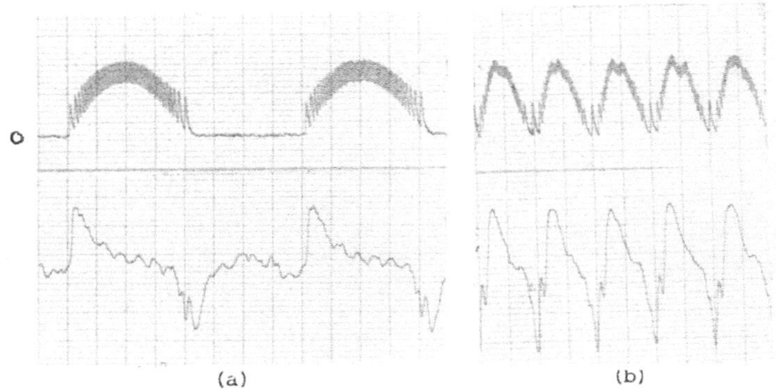

Fig. 3. (a) Filter response (top trace) with 0.5 HZ half-wave modulation and flow (bottom trace), (b) Full wave modulation (1.0 HZ) and flow.

DISCUSSION

The respiratory mechanics as shown by the flow signal depends on the dynamic response of the inspiratory muscle. The dynamic response, in turn, depends on the pulse-frequency of the incoming neural signal. This signal is known to exist during inspiration, at least under normal quiet breathing. Furthermore, the gradual rise in the pulse frequency is also well recognized. If the muscular response is related to this pulse frequency, it is appropriate that the quantification should recover this information. The relationship between the mechanical events in respiration and the filtered EMG provide the necessary tool to explore such a relationship. Since the muscle acts like a low-pass filter, and ignores the high frequency content in the EMG, it is appropriate to select a running average by selecting a filter with appropriate integration time. This choice of integration time is an important one. It cannot be too large to lose the necessary dynamics of the EMG, neither can it be too small to preserve the unnecessary high-frequency contents in the EMG. While this model does not take into account the temporal recruitment of the motor units, it does assume a linear superposition of active units.

It is felt that the running average provides a significant tool with which the EMG-muscle dynamics can be explored.

REFERENCES

Bigland, B. and O.C.T. Lippold (1954). J. Physiol. 123, 214-224.

De Luca, C.J. (1979). IEEE Trans. Biomed. Engg., BME-26, 313-325.

Mishra, N.K. (1972a), An Investigation of the Neuromuscular Command in the Control of Respiration, Doctoral Thesis, New Jersey Inst. of Technology, Newark, N.J.

Mishra, N.K., A.U. Meyer, and R.V. Lourenco (1972b), Proc. V HICSS, Uni. of Hawaii, 250-252.

Philbrick/Nexus Research (1968), Application Manual for Operational Amplifiers, Boston, Nimrod Press.

Richfield, E.K., B.A. Cohen, and J.W. Albers (1981). IEEE Trans. Biomed. Engg., BME-28, 506-514.

Session 4
SPINE BIOMECHANICS

RESPONSE OF THE INTERVERTEBRAL JOINT TO TORSIONAL STRESS

D.A. Bowman and J. F. Lafferty

Wenner-Gren Biomedical Engineering Laboratory
University of Kentucky
Lexington, KY 40506

ABSTRACT

The torsional creep response of the intervertebral joint was obtained for lumbar segments of Rhesus monkey spines submitted to two levels of axial compression. The viscoelastic response of the joint are well represented by a Kelvin unit model of the same form developed for the creep response of the joint in shear.

KEYWORDS

Intervertebral joint; torsion; viscoelastic response.

INTRODUCTION

Knowledge of the response of the intervertebral joints to external loading is important if the forces and bending moments developed in the spine are to be related quantitatively to possible modes of injury or abnormal motions of the joint. The use of biodynamic models has gained widespread acceptance as a means of evaluating human tolerance limits and the probability of injury in environments of severe mechanical stress. Numerous investigations have studied the response of the spinal column and its constituent elements under axial tension and compression, in torsion, and in bending; however, only a limited number of investigators have considered the response of the spine to shear stress. The empirical model of the response of the disc to a step input, developed by Kelley, Lafferty and Bowman (1982), is the only attempt to provide an analytical representation of disc shear properties. The present study utilizes the same experimental and modeling techniques to investigate the response of the intervertebral joint to torsional stress.

MODEL ANALYSIS

Analysis of the response of the intervertebral to shear stress resulted in an analytical representation of the form

$$\varepsilon / \sigma = E_1 + E_2 t + E_3 \left[(1 - \exp - t_1/\tau_1) + (1 - \exp - t_2/\tau_2) \right] \qquad (1)$$

where ε = strain = deformation/disc height
σ = stress = force/disc area
t = time
E and τ are constants

The analysis of the shear response of an elastic isotropic material defines the shear modulus G as

$$\varepsilon/\sigma = \frac{1}{G} \quad (2)$$

The analysis of the same material subjected to torsional stress shows that

$$\frac{\theta J}{TH} = \frac{1}{G} \quad (3)$$

where $\theta J/TH$ is the torsional equivalent ε/σ and

θ = angular deformation in radians
J = polar moment of inertial of cross sectional area (m^4)
T = applied torque (N·m)
H = specimen height normal to the plane of the applied moment (m)

The shear modulus G is the same as defined in Eq. 2.
Although the intervertebral joint is a complicated structure that is neither purely elastic nor isotropic, some degree of correspondence between shear response and torsional response (as indicated by Equations 2 and 3) is expected inasmuch as the disc structure responds to all types of stress in the same basic mode. It was, therefore, anticipated that the same Kelvin unit model (represented by Eq. 1) developed for the shear response would characterize the torsional response. One would expect, however, that the change in force components relative to the applied stress would result in different values for the constants.

METHODS AND MATERIALS

Test specimens, consisting of two vertebrae and the intervening disc with facets intact, were taken from fresh-frozen Rhesus monkey spines. The vertebrae were clampted in holding fixtures with the superior vertebra constrained to prevent angular displacement about the spinal axis but free to move inthe longitudinal direction. The torque was applied to the inferior vertebra by a MTS Servoram controlled by the Cromemco microprocessor. The torque was applied as a step function in increments of 1.875 Newton-meters up to failure. The creep response was observed for 90 seconds at each value of torque and the specimen was allowed to relax for a minumum of 180 seconds between tests. One series of tests (8 specimens) was conducted with an axial load of 12.945N and a second series (6 specimens) employed an axial load of 27.66N. The geometrical data (disc height, disc area and volume) were obtained by sectioning each specimen in the longitudinal direction and using a Talos digitizer to record the coordinates of the disc.

RESULTS AND DISCUSSION

Typical responses of the intervertebral joint subjected to a constant torsional stress is shown in Fig. 1 where the angular displacement (in radians) is plotted vs. time. The plots show the effect of torque level as well as that of axial compression is more pronounced at the higher torque levels indicating that the articular surfaces of the facet joints are forced into closer proximity, thereby increasing resistance to increased angular deformation.

Normalization of the data according to Eq. 3 should permit the results for a given axial load to be represented by a function independent of the applied stress and specimen size, i.e.,

$$\frac{\theta J}{TH} = f(k_1, k_2 \text{---} k_i, C_1, C_2 \text{---} C_i, t)$$

where k_i = shear modulus for the ith elastic element (N/m^2)

C_i - damping coefficient for the ith viscous element ($N \cdot S/m^2$)

t - time (s)

The average values of $\theta J/TH$ displayed a standard deviation on the order of $\pm 35\%$ of the mean. We concluded that the imposition of the specific axial load to specimens obtained from subjects of widely varying size and weight contributed to the magnitude of the standard deviation. The creep modulus was, therefore, divided by the axial load expressed as a percent of body weight of the donor. This procedure reduced the standard deviation of the low level axial load data to 22%. However, standard deviation of the high level axial load data increased to 42%. The creep function is then defined as

$$K = \frac{\theta J}{THW} \quad (4)$$

where W = (axial load/body weight) x 100

Use of the model in the form of Eq. 1 provided a representation of the results to within $\pm 2\%$ of the averaged data. These results are tabulated in Table I and presented graphically in Fig. 2 for the two levels of axial preload. The analytical model (represented by the continuous curve in Fig. 2) is

$$K = \frac{1}{k_3} + \frac{t}{C_3} + \frac{1}{k_1} [1 - \exp(-k_1 t/C_1)] + \frac{1}{k_2} [1 - \exp(-k_2 t/C_2)] \quad (5)$$

for which the values of k and C are shown in Table II for the torsional tests as well as the shear tests.

Comparison of the model parameters indicate a high degree of consistency between the shear and torsional results, i.e., the time constants k/C are identical in the two cases and in both cases $k_1 = k_2$. These results are not surprising when one considers the fact that the intervertebral disc resists all stress modes through tensile loading of the fibers of the annulus fibrosis. The differences in the magnitudes of the corresponding k and C values reflect (1) the changes in the force component directed along the annulus fibers as the stress mode is changed and (2) the effect of the articular facets in the torsional loading.

The results obtained under this grant provide an analytical model suitable for use in computer simulations of spinal response to shear and torsional stresses. The analytical representation is not, however, unique in that other mathematical functions (and mechanical models) can also respresent the temporal response of the system. The model used is not considered to be the best representation of the system because increased response time requires an increased number of Kelvin units to accurately represent the response. This results in a series of k and C values to represent the physical characteristics of the material. We are of the opinion that a more simple physical model with time dependent elastic and/or viscous coefficients exists. Such a model would be more representative of the actual material and would provide discrete elastic and viscous element characterization. Our initial efforts to formulate such a model are promising and its application to the general response of the system is under investigation.

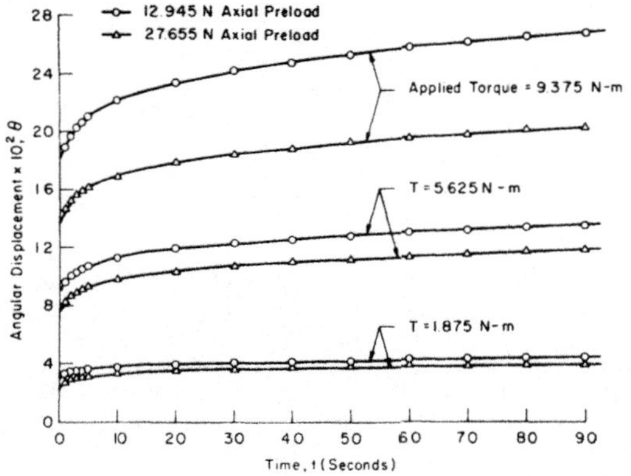

Fig. 1. Angular deformation with constant torque.

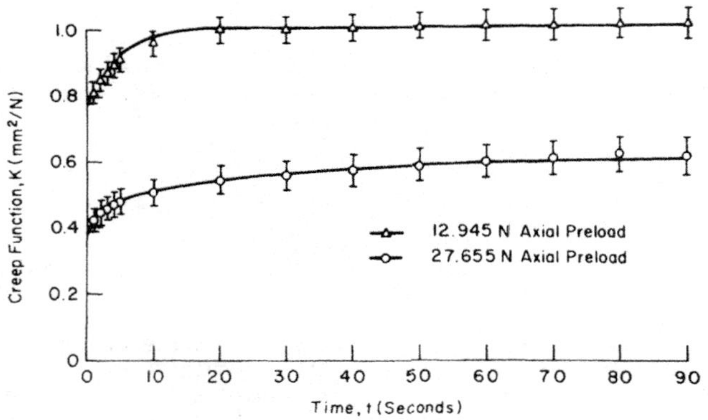

Fig. 2. Comparison of analytical and experimental results.

ACKNOWLEDGEMENT

This work was sponsored by the U.S. Air Force Office of Scientific Research under Grant AFOSR 78-3488.

REFERENCE

Kelly, B. S., J. F. Lafferty and D. A. Bowman (1982). ASME J. Biomech. Engr. Accepted for publication.

A FINITE ELEMENT ANALYSIS OF THE VISCOELASTIC INTERACTION BETWEEN THE NUCLEUS PULPOSUS AND ANNULUS FIBROSIS WITHIN THE HUMAN INTERVERTEBRAL JOINT

Capt Leslie J. Allen* and Anthony N. Palazotto**

*Graduate Student
**Professor
Aeronautics and Astronautics Department
Air Force Institute of Technology
WPAFB, Ohio 45433

ABSTRACT

An understanding of the mechanical properties and behavior of the intervertebral disc is critical to several current areas of research. Among these are the study of the effects of extreme gravitational forces on the air crews of high performance aircraft, the related problem of spinal injuries due to aircraft ejection and the study of disc degeneration. This study was undertaken in order to construct a realistic analytical model of the intervertebral joint using the finite element approach.

Experiments have shown that healthy intervertebral joints exhibit creep when subjected to axial load. A previous study of this time dependent behavior employed a simplified visco-elastic model of the disc which neglected its inhomogeneity. Though the homogeneous model successfully simulated the externally observed response of its joint, it could not adequately portray the internal response of the disc, especially the interaction of the annulus fibrosis and the nucleus pulposus.

The focus of this investigation is the behavior and interaction of the nucleus and annulus under loading, noting in particular, the role of the annulus as a restraining mechanism on the nucleus. An axisymmetric finite element model is employed which incorporates a linear visco-elastic constitutive relation for the annulus and nucleus based upon a three-parameter Kelvin solid. The visco-elastic constants are found by matching one dimensional axial experimental data with this two dimensional model. The nucleus, which is composed of a series of concentric lamellae of collagen fibers, has been modelled as a 12 ply structure of orthotropic material. The cortical bone, trabecular bone, and the bony end plate are assumed to be isotropic and linearly elastic.

Results indicate that the orthotropic material properties of the annulus have a significant impact upon its time dependent behavior and should be included when modelling the joint.

KEYWORDS

Viscoelastic, Biomechanics, Finite Element

INTRODUCTION

An understanding of the mechanical properties and behavior under loading of the human intervertebral joint is pertinent to several current areas of research. Among these are the study of the effects of extreme gravitational forces on the aircrews of high performance aircraft, the related problem of spinal injuries following aircraft ejection, the treatment of congenital spinal deformities, and the development of suitable materials for disc replacement.

Although a sizeable amount of data has been gathered from experimental testing of vertebral segments, the complexity of the structure has prevented the accurate determination of material property constants for the various regions of the joint. The most complicated problems in analysis are posed by the annulus fibrosis. Composed of a series of concentric lamellae of collagen fibers, the orthotropic nature of the annulus requires that at least seven distinct material constants be determined for a linear analysis (Belytschko, 1974). Information of this type is very limited at the present time.

Experimental investigations into the load deflection behavior of the intervertebral joint were carried out by Brown (1957), Nachemson (1960), and Rolander (1966). Galante (1967) examined the effects of inhomogeneity and anisotropy on the tensile properties of the annulus fibrosis. Markolf (1972) conducted axial compression and tension tests on thoracolumbar discs and noted significant differences in response between the lumbar and thoracic discs. Kazarian, et al. (1976) reported the creep characteristics of intervertebral joints under constant axial load. More recently, Kazarian and Kaleps (1979) presented a method for determining the Young's Modulus and viscosity coefficient for a three-parameter-solid model based on their earlier data. Burns (1980) developed a mathematical analysis technique for determining unique parameter values for the three-parameter-solid model using experimental strain data.

Finite element analysis techniques have been successfully employed by several researchers to model the joint. Spilker (1980), Belytschko (1974), and Kulak (1976) used axisymmetric finite element models to study the time independent behavior of the disc. Belytschko's, et al (1974) model assumed an incompressible nucleus in a hydrostatic state of stress with an orthotropic, inhomogeneous, annulus fibrosus. This study led to the conclusion that the material anisotropy of the disc could not be omitted from the analysis without causing large errors. It was also found that a linear analysis could not account for the differences between thoracic and lumbar disc behavior or the differences between tensile and compressive response. Kulak, et al (1976) expanded on the earlier work, using a nonlinear, elastic constitutive relation for the annulus to study the aforementioned differences.

Most recently, Hinrichsen (1980) employed an axisymmetric finite element model which included material inhomogeneity and visco-elasticity to study the creep response reported by Kazarian. The disc was idealized as a single, homogeneous, isotropic, linear, visco-elastic substance. This model successfully reproduced the externally observed response of the joint but due to its simplifications, could not adequately portray the internal response of the disc, in particular, the interaction of the annulus fibrosis and the nucleus pulposus.

ANATOMY OF THE INTERVERTEBRAL JOINT

The human spine is a flexible column composed of a series of 24 segmented bones called the vertebrae. Each vertebra consists of a ventral body and a set of posterior elements. One of these elements is a spinous projection, or process, which extends downward and backward. These processes may be felt as a series of bumps down the back. In addition, there are two transverse processes, one to either side, which provide attachment for muscles and ligaments. The vertebral body itself is composed of soft, trabecular bone encased in a thin shell of relatively hard cortical bone. On the upper and lower surfaces of the body are thin plates of cortical bone called bony end plates.

Between each pair of vertebrae is a cartilaginous intervertebral disc. The discs vary in size, shape, and thickness at different spinal levels. The lumbar discs are the thickest, while those in the thoracic region are relatively thin. The disc is composed of a soft center called the nucleus pulposus which is encircled by a tough, flexible, fibrocartaliginous ring called the annular fibrosis.

The nucleus pulposus is composed of a collagenous lattice enmeshed in a mucoprotein gel. Its water content ranges from 88 percent at birth to approximately 69 percent in middle age. Experiments carried out by Nachemson (1960) indicate that the nucleus in normal discs behaves hydrostatically.

The annulus fibrosis is formed by a series of concentric encircling lamellae. The lamellae consist of collagen fibers which have two well defined areas of orientation. The fibers run in a single direction, alternating from the previous one and aligned at approximately 30° to the horizontal.

Exterior to the disc proper are two ligaments; the anterior and posterior longitudinal ligaments. Although these ligaments transmit considerable force during complex loading, they do not appear to be significant as load carrying structures under axial loading only. Therefore, in order to focus on the disc interaction, the ligaments are usually removed during axial testing of the disc.

GENERAL APPROACH AND ASSUMPTIONS

The vertebra is modelled as an axisymmetric body which is symmetric with respect to the centerline of the intervertebral disc. Only the body itself is modelled, based upon the assumption that little if any load is transmitted by the anterior and posterior longitudinal ligaments during axial loading. Pictured in Fig 1A are two adjacent vertebrae and their intervertebral disc. The dotted lines indicate the truncated portion of the joint used in the finite element model. Fig 1B is an enlarged diagram of the model showing the various regions of the joint. The outer radius of the nucleus pulposus is taken to be 0.707 times the outer radius of the disc.

Kazarian's research (1976) indicates that the vertebral body alone does not exhibit creep under constant axial load. Therefore, it is assumed that the trabecular bone in the vertebral core and the cortical bone in the outer shell and end plates are linear, elastic materials.

Since the vertebral body does not appear to contribute to the observed creep behavior of the joint, it is assumed that the disc is the visco-elastic medium. However, at the time of this analysis, it was unknown whether the visco-elastic response should be attributed to the nucleus, the annulus, or both. As a first quess, it was assumed that the nucleus alone was the visco-elastic medium. The assumption was based upon Kazarian's (1979) work with degenerated discs. The nuclei of these discs exhibited varying degrees of herniation and/or desiccation, mainly due to ageing. Kazarian reported that unlike healthy discs, the degenerated discs did not exhibit creep response. Therefore, the annulus was originally assumed to be linearly elastic. During the course of the investigation, however, it was decided to model both the nucleus and the annulus visco-elastically.

An existing finite element program was adoped for use in this investigation. The original program was written by Hinnerichs (1980) to handle plane-stress, plane-strain problems in isotropic materials. Hinrichsen (1980) modified the program to accommodate axisymmetric structures and to account for visco-elastic response in inhomogeneous materials. In order to account for the orthotropic properties of the annulus, the program was again modified.

The visco-elastic model used in this analysis is a three-parameter solid composed of an elastic spring coupled with a Kelvin solid (Zienkiewicz, 1968). This can be symbolically represented with its spring stiffness E and a parallel arrangement of a dash-pot and spring.

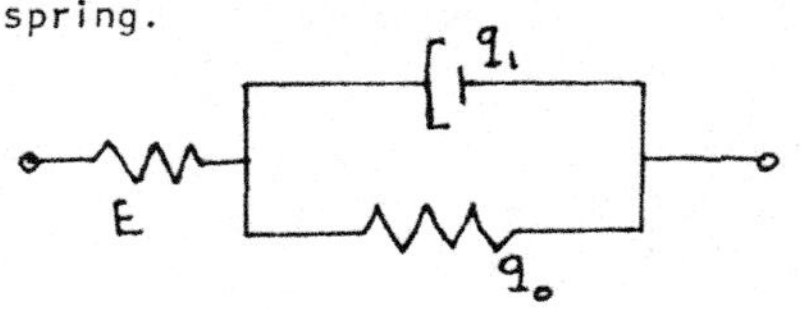

Since the instantaneous elastic deformation is dependent only on the spring, it can be separated from the creep deformation and solved by the finite element method. If the constitutive relation for the three-parameter solid is integrated over time, the resulting equation is

$$\{\Delta\varepsilon_c\} = \frac{\Delta t}{q_1}\left[\frac{2}{3}[A]\{\sigma\} - q_0\{\varepsilon_c\}\right] \qquad (1)$$

$\{\sigma\}$ is the axisymmetric stress vector, $\{\varepsilon_c\}$. The creep strain vector and $\{\Delta\varepsilon_c\}$is the change in strain due to creep over a small time interval where [A] is given by

$$[A] = \begin{bmatrix} 1 & -1/2 & -1/2 & 0 \\ & 1 & -1/2 & 0 \\ & & 1 & 0 \\ & \text{Symmetric} & & 1/4 \end{bmatrix} \qquad (2)$$

The quantity $\{\Delta\varepsilon_c\}$ may be considered an initial imposed strain $\{\varepsilon_0\}$. Since initial strains enter into the governing finite element equations in the force term $\{F^e\}$, a method of solution to the visco-elastic problem now presents itself. As time is increased by small increments, Δt, the governing equation is solved at each time step for the values of $\{u\}$, $\{\sigma\}$, and $\{\varepsilon_c\}$. The values for $\{\sigma\}$ and $\{\varepsilon_c\}$ are then substituted into Eq (1), time is incremented, and the process continues until a maximum time is reached.

CONCLUSION

The mechanical behavior of the intervertebral joint can be realistically modelled through the use of the finite element method. By matching one-dimensional experimental displacement data, material parameters are determined which provide a good overall description of the time dependent behavior of the joint.

The inclusion of the orthotropic material properties of the annulus in the model caused significant stress redistributions throughout the disc. These redistributions indicate the importance of the annulus as a restraining mechanism on the outward flow of the nucleus under axial loading. It is concluded that the orthotropic nature of the annulus is a factor which should be included when modelling the intervetebral joint.

The results of this study indicate that both the nucleus pulposus and the annulus fibrosis are visco-elastic materials. The initial assumption that the experimentally observed creep response of the joint was due to the nucleus alone proved unworkable. It was found that the creep behavior of the joint could be simulated only if both the nucleus and the annulus were modelled visco-elastically. Due to the limitations of the curve matching technique, unique values for the visco-elastic parameters in these regions cannot be determined. Determination of unique parameters would require at least two-dimensional displacement data, which at present is unavailable.

Examination of the displacement profiles and stress redistributions within the model provides insight into the internal behavior of the joint under axial loading. It was found that the cortical bone which encases the vertebral body behaves essentially like a shell subjected to internal pressure. On the other hand, the geometry of the thoracic disc is such that the annulus behaves like an axially loaded ring. Under small loads, the inner lamellae of the annulus do most of the work in restraining the outward flow of the nucleus. It appears that when the joint is subjected to axial compression, the load is transmitted mainly through the cortical shell and end plates to the nucleus, which then acts as the primary load carrying structure of the joint.

REFERENCES

Belytschko, T., Kulak, R.F., Schultz, A.B. and Galante, J.O. (1974). "Finite element stress analysis of an intervertebral disc," *Journal of Biomechanics*, Vol 7, pp 277-285.

Brown, T., Hanson, R.J. and Yorra, A.J. (1957). "Some mechanical tests on the lumbosacral spine with particular reference to the intervertebral disc," *Journal of Bone and Joint Surgery*, Vol 39A, pp 1135-1164.

Burns, M.L. (1979). "Analytical modelling of load deflection behavior of intervertebral discs subjected to axial compression," AFOSR-TR-79-0795.

Galante, J.O. (1967). "Tensile properties of the human lumbar annulus fibrosis," *Acta Orthop. Scand.*, Supplement 100.

Hinnerichs, T., (1980). "Viscoplastic and creep crack growth analysis by the finite element method," PhD Dissertation, presented to the Aeronautics and Astronautics Department: Air Force Institute of Technology.

Hinnrichsen, R.L. (1980). "A viscoelastic finite element model of the human intervertebral joint, Thesis, Wright-Patterson Air Force Base, Air Force Institute of Technology.

Kazarian, L.E. (1976). "Creep characteristic of the human spinal column," *Orthop. Clinics of North America*, Vol 6:1.

Kazarian, L.E. and Kaleps, I. (1979). "Mechanical and physical properties of the human intervertebral joint," AMRL-TR-793.

Kulak, R.F., Belytschko, T.B., and Schultz, A.B. (1976). "Nonlinear behavior of the human intervertebral disc under axial load," *Journal of Biomechanics*, Vol 9, pp 377-386.

Markolf, K.L. (1976). "Deformation of the thoracolumbar intervertebral joints in response to external loads," *Journal of Bone and Joint Surgery*, Vol 54-A, No. 3.

Nachemson, A.L., (1960). "Lumbar intradiscal pressure," *Acta Orthop. Scand.*, Supplement 43.

Rolander, S.D., (1966). "Motion of the lumbar spine with special reference to the stabilizing effect of posterior fusion," Thesis, *Acta Orthop. Scand.*, Supplement 90.

Spilker, R.L. (1980). "A simplified finite element model of the intervertebral discs," *Spine*, Vol 5, No. 2.

Zienkiewicz, O.C., Watson, M. and King, I.P. (1968). "A numerical method of visco-elastic stress analysis," *Int Journal of Mechanical Science*, Vol 10, pp 807-827.

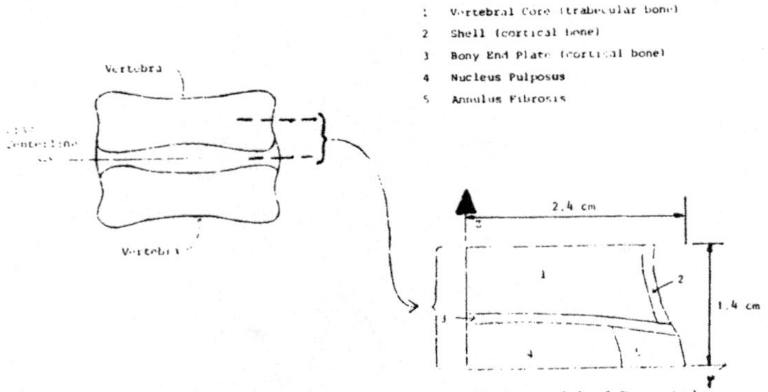

Fig 1A Side View of Two Adjacent Vertebrae and the Intervertebral Disc

Fig 1B Model of Truncated Joint.

A MODEL FOR ANALYSIS OF HEAD-NECK DYNAMICS WITH MUSCULAR RESPONSE

J. L. Williams* and T. B. Belytschko**

*Department of Mechanical Engineering,
Northeastern University, Boston, Ma. 02115
**Department of Civil Engineering,
Northwestern University, Evanston, Il. 60201

ABSTRACT

A three-dimensional analytical model of the cervical spine is described. Some results of a simulation of sideways impact acceleration are compared with experimental tests on living human subjects. A special feature of the model is the ability to include the effects of muscular contraction.

KEYWORDS

Human impact acceleration; side impact; spine; neck muscles; stretch reflex response; head-neck model.

INTRODUCTION

Discrete-parameter models of the head and neck system have been proposed in the past (Reber and Goldsmith, 1979; Huston, Huston and Harlow, 1978), but have not been validated for situations other than frontal impact. Described here is a three-dimensional model of the head and neck which includes extensive details of the neck musculature and other soft tissue structures and which has been tested in both frontal and sideways impact acceleration. The basic tool for analysis is a previously developed large deflection finite element code, capable of handling both material and geometric nonlinearities (Belytschko, Schwer and Klein, 1977).

MODEL DESCRIPTION

The vertebrae from T1 to C1 are treated as rigid bodies connected by deformable elements representing the intervertebral discs, articular facet joints, ligaments and muscles. Beam elements are used for the discs, and nonlinear spring elements for the ligaments. A six-node isoparametric pentahedral element was developed to model the articular facets, which were found to play a crucial role in the cervical spine. Special attention is paid to the directional properties of the facets. Geometric data for the cervical spine and stiffness data for motion segments from C1 to T1 were estimated from the work of Liu and colleagues (1981). A simplified representation of the body is used for the region below T1. The thoracolumbar spine is modeled by three beam elements representing the regions T1-T10, T10-L3 and L3-S1.

Emphasis is placed on accurately representing the neck musculature. Twenty-two major muscle groups are included, mostly extensors and lateral flexors. Each major muscle group may insert onto as many as seven or more different bones, thereby requiring several elements for adequate representation. To model the correct line of action during large head and neck motion, a muscle element was developed, which can have up to ten intermediate sliding nodes. The element has only axial stiffness and is fixed at both ends to rigid bodies in the model. Between both ends the element is constrained to slide around intermediate nodes representing underlying tissue in the neck.

A three-parameter viscoelastic model, as proposed by Apter and Graessley (1970), is used for the stress-strain law of the muscle element. The mechanical response of the muscle model is given by

$$\sigma + (\eta/E_2)\dot{\sigma} = E_1\epsilon + (E_1 + E_2)(\eta/E_2)\dot{\epsilon}, \tag{1}$$

where σ is the stress and ϵ is the strain. The strain is defined as

$$\epsilon = (\ell - \ell_o)/\ell_o, \tag{2}$$

where ℓ is the existing length and ℓ_o is the instantaneous unstretched length of the muscle. The instantaneous unstretched length, ℓ_o, and the viscoelastic parameters, E_1 E_2 and η, vary with the state of contraction of muscle. The state of contraction is assumed to depend on the concentration of some molecules within the muscle cells. This concentration varies with time as

$$\dot{n} = k_2\epsilon - k_3 n + S(t), \tag{3}$$

where k_2 and k_3 are rate constants and S(t) is a step function which is used to stimulate the muscle to contract at the time of the stretch reflex response. For example, in side impact of a relaxed individual, the neck muscles begin to contract, by means of the stretch reflex response, at about 40 ms after impact.

SIDE IMPACT SIMULATION

To test the validity of the model, frontal and sideways impact accelerations were simulated and the results compared with tests on living human subjects (Ewing and Thomas, 1972; Ewing and colleagues, 1978). In the experimental tests, volunteers were strapped into a sled and subjected to peak sled accelerations of up to 7.5 G in the side impact tests and 10 G in the frontal impact tests. Detailed data on head-neck motion were obtained by means of accelerometers and high-speed cameras. Excellent agreement is obtained between the model results and the experiments, both for frontal and for sideways impact acceleration. Only the side impact simulation is discussed here.

The simulation described here corresponds to the experimental test at 7 G, in which the sled was accelerated linearly to 7 G at 40 ms, followed by a linear decay to zero at 140 ms. In the first simulation, the muscles of the neck remain passive throughout the 200 ms run; in the second, the lateral flexors of the head and neck on one side are stimulated to contract at 40 ms. Ten different muscle groups are allowed to contract at 40 ms. While all of these are lateral flexors, some also function as extensors of the head and neck. The deformed configurations of the spine are shown in Fig. 1(a) and 1(b) for simulation with the passive muscle model and the stretch reflex model, respectively.

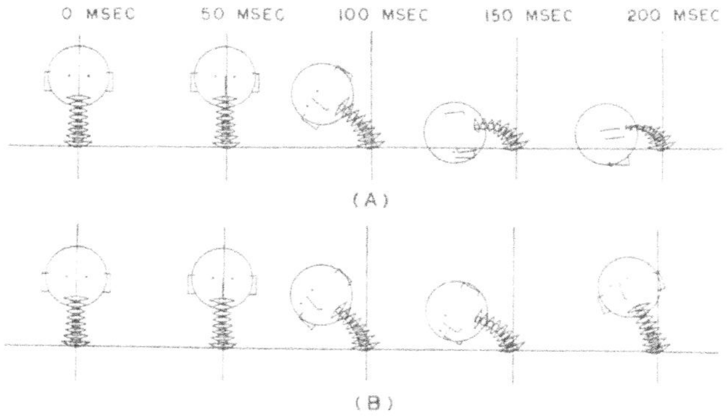

Fig. 1. Response of head-neck model, combined with lower spine model, to sideways impact acceleration
(a) with passive muscles
(b) with stretch reflex response

Because the muscles are stimulated at 40 ms and another 100 ms is required before the muscles reach full contraction, the effects of the stretch reflex response are not noticed until about 120 ms. The effect of muscular contraction is further shown in Fig. 2 where the predicted vertical displacement of the head is compared with the measured displacement. Muscular contraction is shown to have a considerable effect on the amount of head-neck displacement even at high impact accelerations.

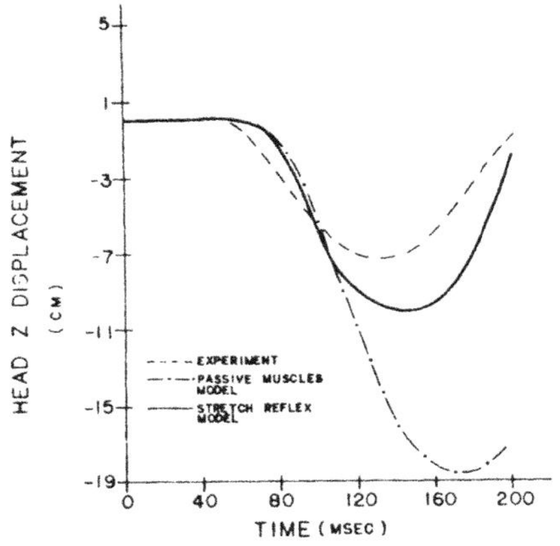

Fig. 2. Vertical displacement of the head during sideways impact acceleration.

ACKNOWLEDGEMENTS

This work was supported by the Air Force Aerospace Medical Research Laboratory (Contract Number F33615-78-C-0523) and by the National Institutes of Health (5T32GM07350-04).

REFERENCES

Apter, J. T., and W. W. Graessley (1970). A physical model for muscular behavior. Biophys. J., 10, 539-555.

Belytschko, T., L. Schwer, and M. J. Klein (1977). Large displacement, transient analysis of space frames. Int. J. for Num. Methods in Eng., 11, 65-84.

Ewing, C. L., and D. J. Thomas (1972). Human head and neck response to impact acceleration. Naval Aerospace Medical Research Laboratory, Pensacola, Florida. (NAMRL Monograph 21).

Ewing, C. L., D. J. Thomas, L. Lustik, G. C. Willems, W. H. Muzzy III, E. B. Becker, and M. E. Jessop (1978). Dynamic response of human and primate head and neck to $+G_y$ impact acceleration. Naval Aerospace Medical Research Laboratory, Pensacola, Florida (DOT HS-803-058).

Huston, R. L., J. C. Huston, and M. W. Harlow (1978). Comprehensive three-dimensional head-neck model for impact and high-acceleration studies. Aviat. Space Environ. Med., 49, 205-210.

Liu, Y. K., K. W. Krieger, G. Njus, K. Ueno, and K. Wakeno (1981). Investigation of cervical spine dynamics. Aerospace Medical Research Laboratory, WPAFB, Ohio (AMRL-TR-138).

Reber, J. A. and W. Goldsmith (1979). Analysis of large head-neck motion. J. Biomechanics, 12, 211-222.

VISCOELASTIC BEHAVIOUR OF THE HUMAN SPINAL DISC AND THE VERTEBRA

Harcharan Singh Ranu

Biomedical Engineering, Louisiana Tech

University, Ruston, Louisiana 71272

ABSTRACT

Pressures in the nucleus pulposus and under the annulus fibrosus have been measured, using the specially developed miniature pressure transducers, when the vertebral segment was subjected to physiological loading. Strain data and stress-relaxation were also measured. Results show that pressures are linear with loading. Pressure variations within the nucleus are due to disc degeneration.

KEYWORDS

Nucleus pressure: annulus pressure; disc degeneration; vertebral strain; stress-relaxation; annulus strength; miniature transducers.

INTRODUCTION

The problem of low back pain is as old as mankind. Almost eight out of ten people in the universe suffer from this agony. In the United States alone somewhere around 75 million Americans are victims of back pain problems and 6 million or so are added to this list each year.

It is known that with advancing age (40/up) the nucleus pulposus loses its gel like characteristics, as a result of this the annulus fibrosus is no longer able to redistribute the stresses uniformly. Because of this, the disc bursts through the posterior end of the surrounding annulus, thus putting pressure on spinal nerves. In order to understand the load bearing mechanics of the spinal disc, it became essential to understand the interrelation between the different regions of the nucleus and the annulus, when subjected to simulated in vivo loading. Therefore, this investigation was carried out.

DEVELOPMENT OF AN INTRADISCAL PRESSURE TRANSDUCER (IDPT) AND AN INTERVERTEBRAL PRESSURE TRANSDUCER (IVPT)

A needle tipped with an ultra miniature pressure transducer, diameter 2.03mm, was developed in order to measure the intradiscal pressure. The calibration of the needle was carried out using a dead-weight tester. It was capable of measuring intradiscal pressure up to 3450kPa (500 psi).

In order to map the pressure distribution under the annulus fibrosus of a spinal disc and its adjacent end plates an intervertebral pressure transducer was developed and fabricated. It was a diaphragm-type pressure transducer with a slender box section. The intervertebral pressure transducer was calibrated as described by Ranu, Denton and King (1979) and its response was linear up to a pressure of 4830kPa (700 psi). The IVPT's and holder (a partial ring) has unique features to control the depth of penetration under the annulus to attach multiple units to the spinal segment.

PREPARATION OF SPECIMENS

Sections of spine were obtained from the unembalmed human bodies within a short period of death. Their suitability for experimental use was determined by taking radiographs. After this the lumbar sections of the spine were stripped of all muscles and making sure that none of the ligaments, bone and discs were damaged.

Preparation of the segments began by moulding the top and bottom ends of the segment in an alloy (Ostalloy m.p. 47.3^{o}c).

The disc was carefully dissected and the IVPT's were inserted to a controlled depth of penetration. The whole assembly was kept moist with normal saline solution and was covered by a plastic sheet before testing.

In order to implant the instrumented needles into the disc a specially developed probe was used. This guided the initial insertion of the needles and also helped to control the depth of penetration of the needles into the nucleus pulposus.

Strain gauges were mounted on the anterior, lateral right and lateral left sites of the vertebra. They were prepared as described by Ranu and King (1980).

TEST PROCEDURES

A quasi-static compressive load was applied on the specimen using a Riehle testing machine via a specially designed loading head which could control the eccentricity of the applied load. In order to establish the line of action and the eccentricity of the applied load, a three-dimensional loading platform equipped with a six-axis load cell was incorporated into the experimental test set up. The eccentric loading of the segment was carried out by moving the platform in controlled increments along the two orthogonal axes. The rate of loading used was approximately 0.13mm/S. The entire segment was loaded to a pre-determined physiological load. During each test, the following data were recorded.

(1) Intradiscal pressures as a function of time.
(2) Intervertebral pressure as a function of time for the anterior and lateral edges of the vertebra and at a depth of 3.175mm.
(3) Strain as a function of time was also obtained for the anterior, lateral right and lateral left sites of the vertebra.
(4) Applied compressive load as a function of time.
(5) Moments in two orthogonal axes as a function of time.
(6) Stress-relaxation of the entire segment as a function of time.

After the test, the discs were sectioned and their degeneration was accessed. The scheme used by Galante (1967) was adopted for classification purposes: Group I (no degeneration) to Group IV (severely degenerated). During this procedure it was further observed that this classification scheme could be improved. Therefore, if the disc degeneration did not exactly match with Galante's classification. Then a further + ve or - ve sign was assigned to the graded discs.

Annulus specimens from the anterior, posterior and both the lateral sides of the discs were also tested in tension to failure. The method and techniques used for this study were as described by Ranu (1975, 1979 and 1981).

RESULTS AND CONCLUSIONS

Typical results of the investigation are presented to illustrate the behaviour of the spinal segment from a 53 year old male cadaver having a disc degeneration of +2. Figures 1 to 6 show the variation of load, moments X and Y, intradiscal pressures, strains, intervertebral pressures and stress-relaxation with time. It is interesting to note that Figs. 1 to 5 all are showing the same response to loading as well as to stress-relaxation. However, no stress-relaxation in annulus pressures is observed (Fig. 6).

Figure 4 shows the pressures in different regions of the nucleus. This pressure variation is 0.2MPa due to a compressive load of 1708N. Figure 7 illustrates that the relationship between these pressures is linear.

Average pressure developed in the nucleus and under the annulus are 0.8MPa and 0.82MPa and vertebra strain 900 μ for a load of 1708N, when moments of 35NM in x-direction and 8NM in y-direction are applied.

Stress-relaxation results show (Figs. 1 to 5) that in 76 seconds, load relaxed by 420N, nucleus pressure by 0.13MPa, strain by 190 μ and moment by 8NM.

Strength of the annulus from the posterior side has been found to be 40% weaker when compared with the anterior and lateral sides of the disc.

Conclusions drawn from this study are: (a) Pressures developed in different regions of nucleus and under the annulus are linear with applied compressive load. (b) Pressure variations within the nucleus exist and this could be due to the disc degeneration. (c) Nachemson (1960) postulated that pressure in the nucleus is three times the annulus pressure. But these findings do not agree with his relationship. (d) It is established that the annulus from the posterior region of the disc is the weakest point in the disc, and therefore, may contribute to disc prolapse from the posterior.

ACKNOWLEDGEMENTS

The author is grateful to Drs. A. I. King and H. K. Ranu for their generous support

REFERENCES

Galante, J. O. (1967) Tensile properties of the human lumbar annulus fibrous. Acta Orthop. Scand., Suppl. 100.
Nachemson, A. (1960) Lumbar intradiscal pressure. Acta Orthop. Scand., Suppl. 43.
Ranu, H. S. (1975) Effects of ionizing radiation on the mechanical properties of skin. Doctoral Thesis, CNAA, London.
Ranu, H. S. (1979) The mechanical and structural response of skin to irradiation J. Biomech. 12, 601-608.
Ranu, H. S. (1981) Effects of fractionated doses of X-irradiation on the mechanical properties of skin: a long-term study. In R. Marks and P. A. Payne (Ed.), Bioengineering and the Skin, MTP Press, Boston, pp. 15-21.
Ranu, H. S., R. A. Denton, and A. I. King (1979) Pressure distribution under an intervertebral disc-an experimental study. J. Biomech., 12, 807-812.
Ranu, H. S., and A. I. King (1980) Correlation of intradiscal pressure with vertebral endplate pressure. Engineering Aspects of the Spine Conference, I. Mech. E. Publication, 1980-2, 37-42.

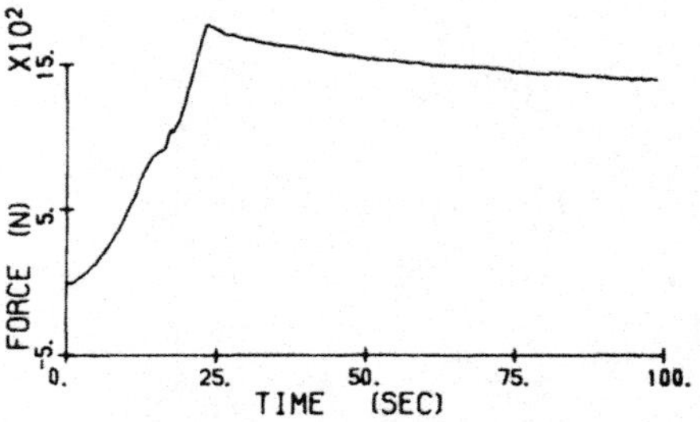

Fig. 1 Variation of load against time for a lumbar spinal segment.

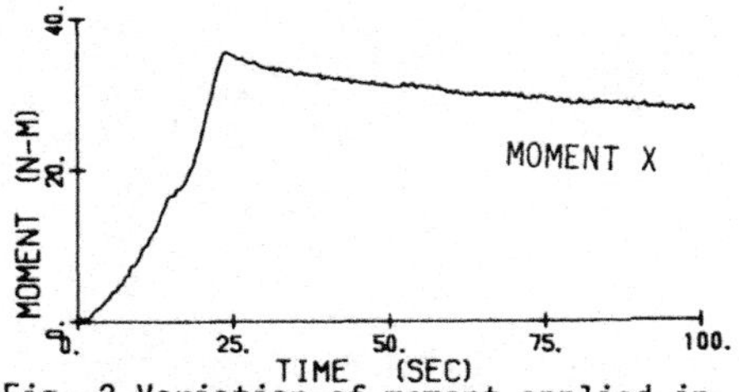

Fig. 2 Variation of moment applied in the anterio-posterior direction against time.

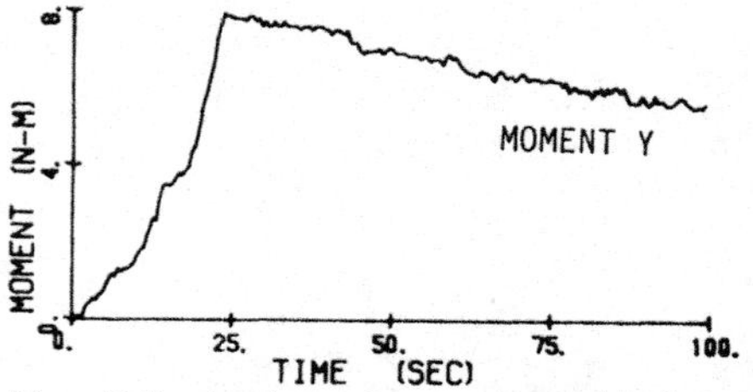

Fig. 3 Variation of moment applied in the lateral direction aqainst time.

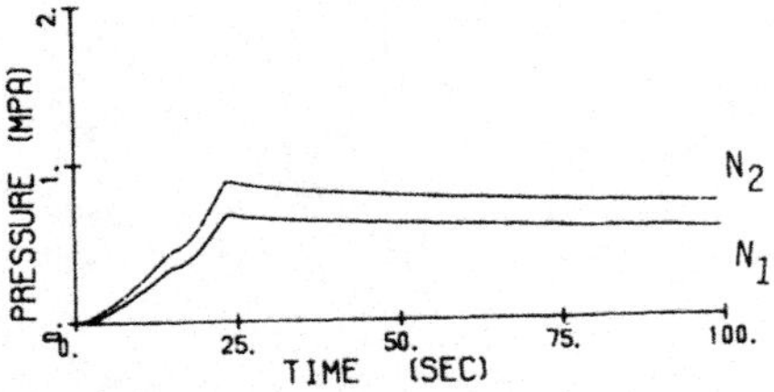

Fig. 4 Variation of pressures in different regions of nucleus against time.

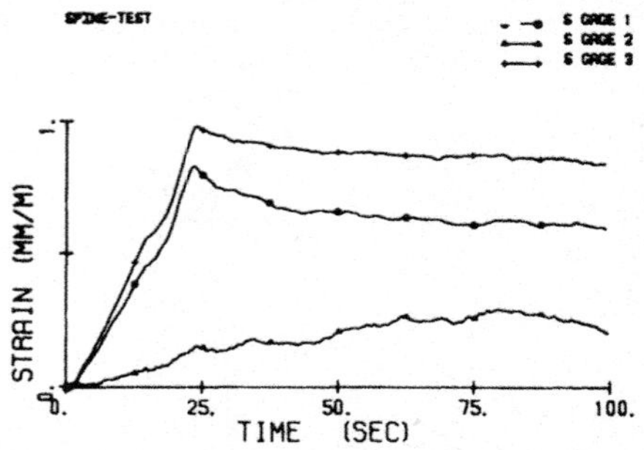

Fig. 5 Variation of strain against time.

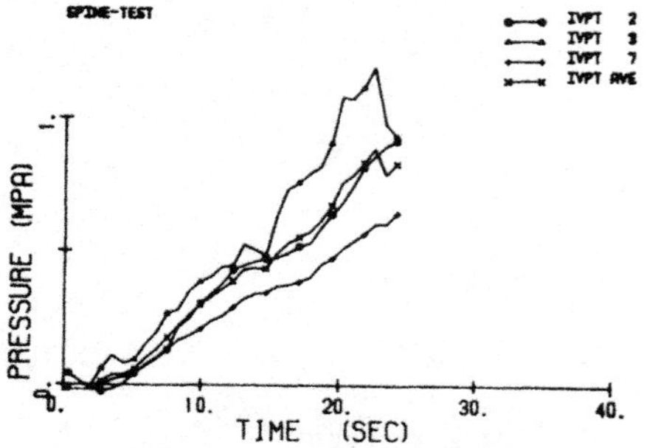

Fig. 6 Variation of annulus pressure against time.

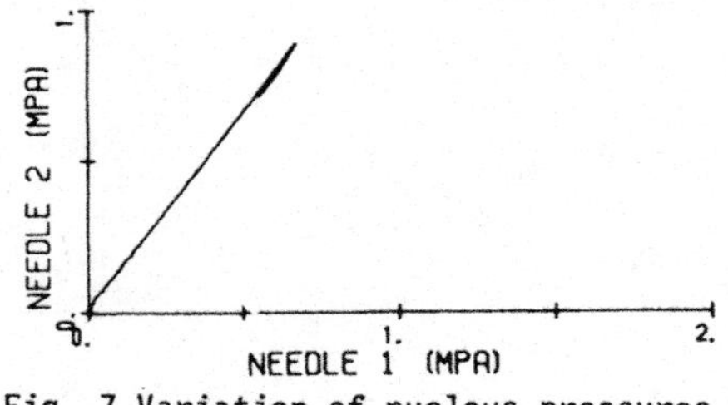

Fig. 7 Variation of nucleus pressures N_2 against N_1.

DISTRIBUTION OF IN VIVO LATERAL BENDING STIFFNESSES IN IDIOPATHIC SCOLIOTIC SPINES

by

Richard Romick, G. Jayaraman
Dept. of Mechanical Engineering
Michigan Technological University
Houghton, Michigan 49931

and

Rae Jacobs, Orthopaedic Surgery, Kansas Univ. Med. Center
Kansas City, KS 66108

Scope:

A biomechanical model developed to analyze progressive idiopathic scoliotic spines led to an optimization scheme for determining lateral bending stiffness, based upon Wolff's law. Stiffness distributions for four adolescent patients, whose progressive idiopathic scoliosis was corrected surgically with Harrington distraction instrumentation, exhibited the largest stiffnesses near the curve apex and decreasing values on either side. Mean stiffness parameters for the scoliotic spines are comparable to those reported by Miller and Skogland for adolescent cadaver spinal motion segments. However, stiffnesses near the curve apex are more on the order of those reported by Schultz et al. and Panjabi et al. for adult cadaver motion segments.

Biomechanical Spinal Model:

Scoliotic spines were modeled as large deflection linkage systems in the frontal plane, predicting response to tractions applied by Harrington distraction instrumentation. The linkage system is composed of both rigid link elements, representing vertebral bodies, and interconnecting torsional springs, representing the intervertebral substance. The torsional springs characterize the motion segment lateral bending stiffnesses. Linkage boundary conditions consisting of simple support on one end and a roller support on the opposite end allowed only displacements along the axis of the distraction rod.

Equilibrium equations for the corrected spinal configuration are derived based on minimum total potential energy (with respect to the corrected linkage angles). The linkage boundary conditions were included implicitly in the formulation of the following equilibrium equations:

$$Q_i = \frac{\partial \Pi}{\partial \psi_i} = 0 = K_i(\psi_i - \theta_i - \psi_{i-1} + \theta_{i-1}) - K_{i+1}(\psi_{i+1} - \theta_{i+1} - \psi_i + \theta_i)$$

$$+ \frac{\partial \psi_n}{\partial \psi_i} [K_n(\psi_n - \theta_n - \psi_{n-1} + \theta_{n-1}) + K_{n+1}(\psi_n - \theta_n) + 1/2\ L_n F_n \sin\psi_n]$$

$$+ 1/2\ L_i\ (\cos\psi_i \sum_{j=i}^{n-1} P_j + \sin\psi_i \sum_{j=i}^{n} F_j) \ :\ i = 1 \text{ to } n-1 \qquad (1)$$

where: $$\frac{\partial\psi_n}{\partial\psi_i} = -\frac{L_i \cos\psi_i}{L_n \cos\psi_n} \qquad (2)$$

If the lateral bending stiffnesses are known for the scoliotic spine, then corrected spinal configurations can be determined from the application of corrective tractions. However, the stiffnesses are surrounded with uncertainty and are also the parameters most affecting the progressive nature, as well as the treatment of, the scoliotic deformity. The equilibrium equations cannot be used directly to determine the bending stiffnesses, because an under-determined system of equations results, with "n-1" equilibrium equations for determining "n+1" stiffnesses. As a result, an optimization scheme was developed governed by Wolff's law. Wolff's law indicates that the stiffnesses should be distributed along the spine to provide a minimum gradient. The gradient is expressed as:

$$G = \sum_{i=1}^{n} (K_i - K_{i+1})^2 \qquad (3)$$

The stiffness should also satisfy the equilibrium equations while minimizing the gradient, which provides that a linear combination should also be a minimum, i.e.:

$$H = G + \sum_{i=1}^{n-1} \lambda_i Q_i \quad \text{and}$$

$$\frac{\partial H}{\partial K_j} = 0 = 2(-k_{j-1} + 2K_j - K_{j+1}) + \sum_{i=1}^{n-1} \lambda_i \frac{\partial Q_i}{\partial K_j} \ :\ j = 1 \text{ to } n+1 \qquad (4)$$

The above "n+1" equations and "n-1" equilibrium equations provide a linear system of equations to solve for "n+1" bending stiffness parameters and "n-1" Lagrangian multipliers (λ_i).

Scoliotic Spinal Data:

Clinical data, in the form of pre- and post-operative spinal configurations was obtained for five patients with progressive idiopathic scoliosis undergoing surgical correction with Harrington distraction instrumentation. Spatial location of the vertebral body centers for thoracic and lumbar spine identified the scoliotic configuration, as determined by a three-dimensional radiograph technique. Least squares curves were fitted to the vertebral center data in the frontal plane to construct linkage representations and describe the spinal curvatures. Corrective tractions were measured at the time of surgery using an instrumented distraction rod.

In vivo lateral bending stiffness distributions were calculated for four of the five scoliotic spines. Unacceptable stiffness distribution for the fifth case (E) was attributed to difficulties in constructing spinal linkage representations consistent with the model. Lateral bending stiffness distributions for specimens A, B, C, and D are shown in figures 2, 4, 6 and 8 respectively, and the curvature distributions are shown in figures 3, 5, 7, and 9 respectively.

The stiffness distributions found show that the largest stiffnesses appear near the curve apex, and decrease to either side. Additionally, the spinal curvature distributions appear to be related to the corresponding lateral bending stiffness distributions. For example, consider the stiffness and curvature distributions for specimen B (Fig. 4 and 5). The region between the pre- and post-operative curvatures are in accordance with the stiffness distribution, except near the ends of the intermediate distraction rod (T_6 to L_2). The same trend follows in the remaining three specimens (A, C and D).

Having the in vivo lateral bending stiffness distributions allows for quantitative clinical evaluation of scoliosis patients. Additionally, with appropriate clinical procedures, it should be possible to evaluate the in vivo bending stiffnesses of scoliotic spines to ascertain the progressive nature of the scoliotic deformity.

References:

1. Ghista, D. N. et al, "Biomechanical Analysis of Scoliosis Correctional Devices: Specifications for the Corrective Forces in Terms of the Achievable Corrections", 1981 Biomechanics Symposium. ASME, Vol. 43, p. 241.

2. Miller, J. A. A., and Skogland, L. B., "A Comparison of the Mechanical Behavior of Adolescent and Adult Spinal Motion Segments", paper presented at Engineering Aspects of the Spine, May 1980, The Institution of Mechanical Engineers, Westminster, London.

3. Panjabi, M. M. et al, "Three-Dimensional Flexibility and Stiffness Properties of the Human Thoracic Spine", J. of Biomech., 1976, Vol. 9, p. 185.

4. Schultz, A. B. et al, "Mechanical Properties of Human Lumbar Spine Motion Segments, Part I", J. of Biomech. Eng., Feb. 1979, p. 46.

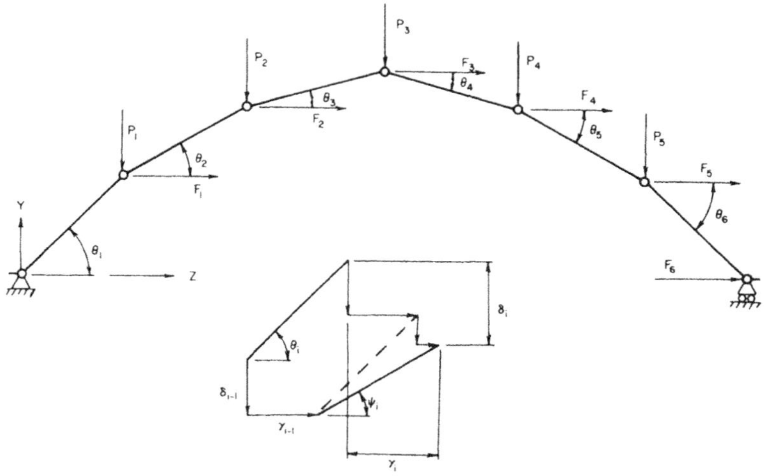

Fig. 1. Scoliotic Spinal Linkage Representation

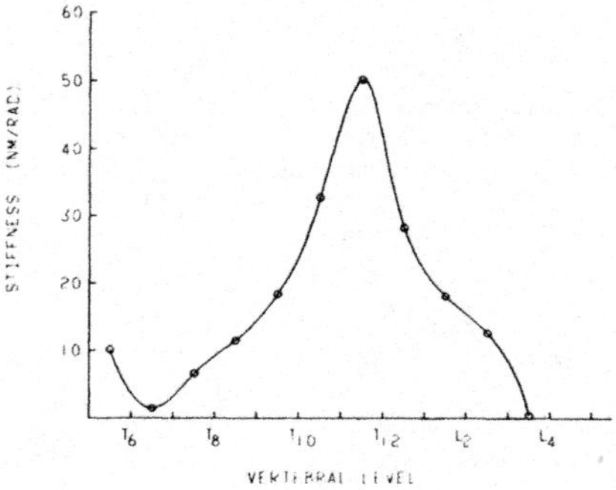

Fig. 2. Stiffness Distribution: Specimen A

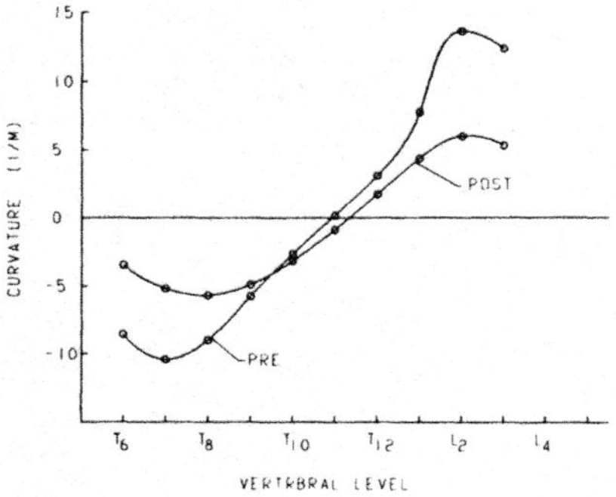

Fig. 3. Curvature Distribution: Specimen A

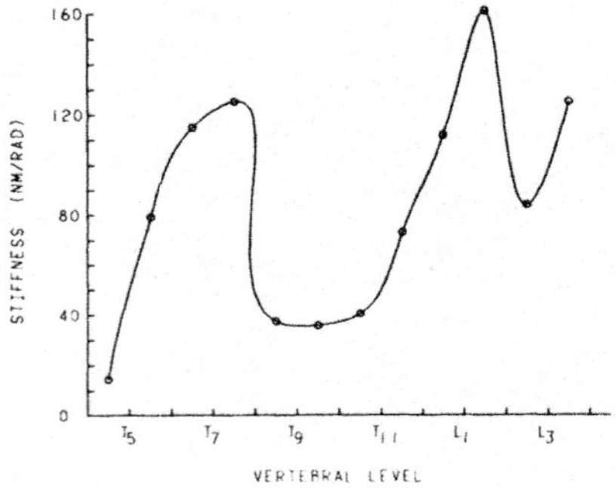

Fig. 4. Stiffness Distribution: Specimen B

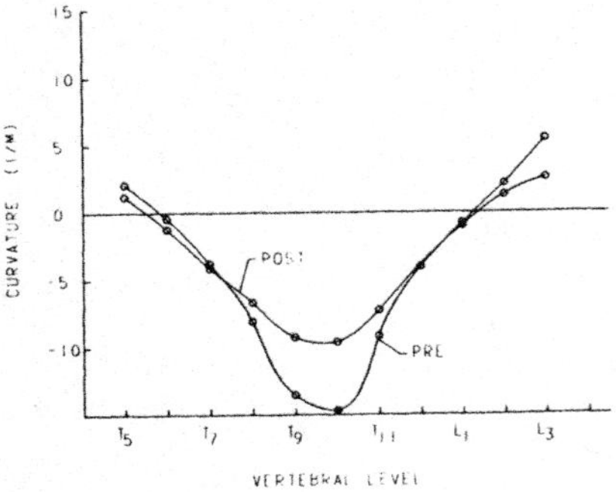

Fig. 5. Curvature Distribution: Specimen B

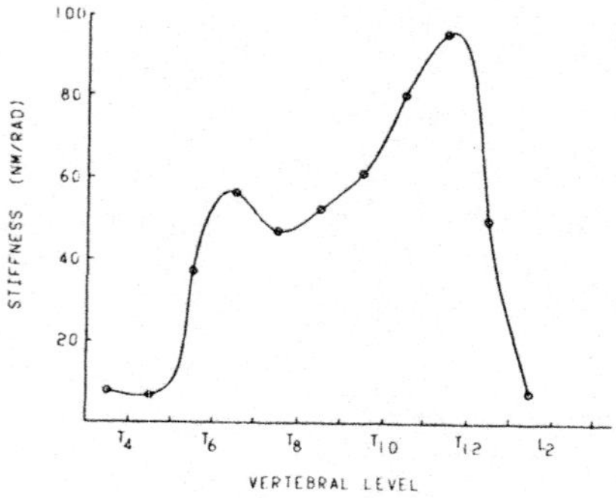

Fig. 6. Stiffness Distribution: Specimen C

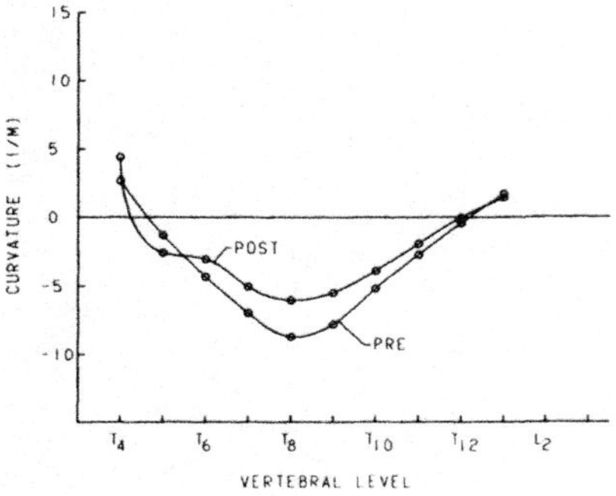

Fig. 7. Curvature Distribution: Specimen C

AN X-RAY TRANSLUCENT CERVICAL ORTHOSIS

M. Strauss* and J. O'Reagan**

*University of Texas at Arlington, Biomedical Engineering Program, Arlington, Texas
**Rehabilitation Engineering Center, Charlottesville, Virginia

ABSTRACT

The XRT (X-Ray Translucent) is a new cervical orthosis that rigidly immobilizes the injured cervical spine. The design allows for the quick application of the XRT in the emergency room and for its sole removal (several weeks or months later) prior to final medical discharge. This is accomplished by fabricating the entire cervical region and most of the thoracic region of the XRT from X-ray translucent plastic, providing for an unobstructed view for radiologic diagnostics. The XRT is constructed in two halves so that prior to surgery, the anterior or posterior half may be swung out of the intended surgical area, while the patient's head is still being supported by the opposing half.

To the patient, the XRT is a light weight orthosis allowing for easy care of activities of daily living, while being less confining and conspicuous than the conventional Halo brace. Most clothing, unmodified, can be worn over the XRT.

KEYWORDS

Cervical orthosis; cervical brace; spine immobilization; neck injury; neck brace.

INTRODUCTION

Each year approximately 10,000 people incur a spinal cord injury (Kalspeek, 1980). Some of these injuries will be fatal, some will produce permanent deficits, and others will ultimately heal completely. The survivors of these injuries will require the immobilization of their injury site for the duration of the healing period, normally three to five months.

The potential severity of the injury increases as the level of the injury site is located more rostrally. The most rostral level of injury will occur in the cervical spine, that part of the spine located in the neck region. It is also this region which is most susceptible to injury due to the relatively strong thoracic spine below, and the large moment arm above (the head).

Several different orthoses are currently available to immobilize the cervical spine and head (Hart and coworkers, 1978; Johnson and coworkers, 1978). Of these, the most efficient orthosis in curtailing movement is the Halo orthosis (Johnson, and coworkers, 1977). It has been shown that the Halo is at least three times more effective in immobilizing the cervical spine than any other cervical orthosis.

Although the Halo is best in immobilizing the cervical spine, it possesses several disadvantages:

1- Part of the Halo orthosis is fabricated from metal. As seen in Fig. 1, most of the metal is located at the level of the cervical spine. This produces difficulties in obtaining roentgenograms of the cervical spine while the Halo is being worn. CT scans are completely prohibited. Thus for either case, the Halo might be removed, putting the patient in a compromising situation.

2- Some cervical injuries require surgical intervention for the procurement of maximum sensory and motor return. The Halo is an obstacle in the operating field which the surgeon must work around. Usually the Halo orthosis is removed prior to surgery or not fitted to the patient until after surgery. This also leaves the patient in a vulnerable situation for further injury.

3- The Halo orthosis is bulky enough so that only modified clothing can be worn over it.

4- The obtrusive appearance of the Halo brace is partly due to its bulkiness as well as its fabrication materials.

The goal of this project was to produce a cervical orthosis which eliminated the disadvantages of the Halo while retaining its high degree of immobilization.

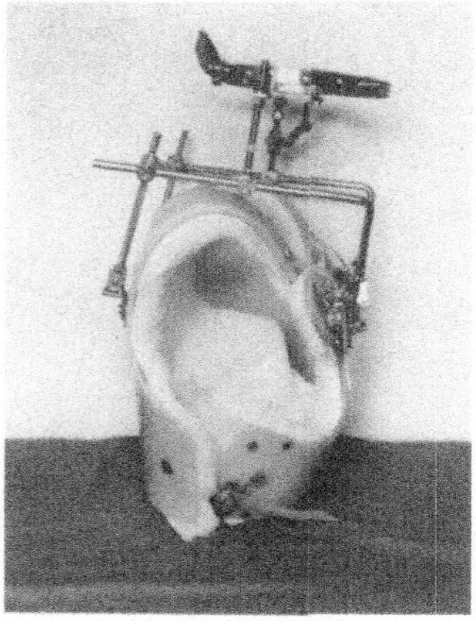

Fig. 1. Halo vest.

RESULTS

The XRT is predominatly fabricated from radiotranslucent plastic, with the metal components located well above or below the cervical spine (Fig. 2). This allows for a complete radiodiagnostic exam which may include myelograms, roentgenograms or CT scans, without obstructing the field of view.

If surgery is warranted, it may be performed on the patient with the XRT in place. The anterior or posterior half of the XRT will fold out of the operating field allowing for an unobstructed surgical approach. In addition, the opposing half continues immobilizing the cervical spine during surgery (Fig. 3).

Fig. 2. XRT cervical orthosis (anterior).

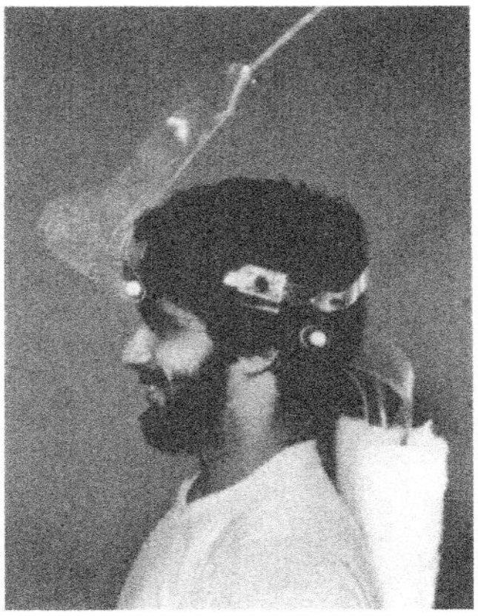

Fig. 3. XRT cervical orthosis positioned for surgery.

The XRT was designed to follow the contours of the body while allowing enough room for daily hygiene. This has two major advantages. Unlike the Halo, the XRT can be worn under unaltered clothing. The second advantage is the improved appearance of a patient wearing the XRT rather than the Halo. The Halo distorts body lines, but the XRT fits closer to the body and utilizes smooth curves and transparent materials.

DISCUSSION

It is seen that the use of the XRT has several advantages over the Halo orthosis:

1- It allows for a complete radiologic examination;
2- It permits an unobstructed surgical approach to the anterior or posterior cervical spine;
3- It allows for the use of unaltered clothing;
4- It weighs approximatly half that of the Halo;
5- It improves the patient's appearance and self image;
6- It allows for incremental anterior and posterior positioning of the head as well as flexion-extension adjustments.

Qualitatively, the XRT is rigid and is quite effective in immobilizing the cervical spine and head. Preliminary patient testing is in progress at the University of Virginia Hospital with favorable results being obtained.

ACKNOWLEDGMENTS

The authors would like to express their gratitude to the Department of Health/RSA for providing the two year Rehabilitation Engineering Traineeship program, grant #44-T-81435, from which the XRT resulted. Thanks are due to Mr. Richard Rosenberger for his consultations in his field of orthotics and prosthetics, Dr. Steven Reger for his professional advice and Dr. Richard Whitehill for his medical guidance and testing of the XRT. Special thanks go to Dr. Colin McLaurin who provided advice and guidance.

REFERENCES

Hart, D. L., R. M. Johnson, E. F. Simmons, J. Owens (1978). Review of cervical orthoses. Phys. Ther., 58, 857-860.

Johnson, R. M., D. L. Hart, E. F. Simmons, G. R. Ramsby, W. Haven, and W. O. Southwick (1977). Cervical orthoses. J. Bone and Joint Surg., 59-A, 332-339.

Johnson, R. M., D. L. Hart, J. R. Owen, E. Lerner, W. Chapin, and R. Zeleznick (1978). The Yale cervical orthosis. Phys. Ther., 58, 865-871.

Kalsbeek, W. D., R. L. McLaurin, R. L. Harris, S. H. Benjamin III, and M. J. Douglas (1980). The national head and spinal cord injury survey: Major findings. J. Neurosurgery Supp., 53, S19- S31.

Moratz, V. A. (1979). Adapting shirts to fit over a Halo vest. Amer. J. Occ. Ther., 33, 524-525.

Session 5
CARDIOLOGY

ASYNCHRONOUS ONSET OF MECHANICAL SYSTOLE STUDIED BY COMPUTER GENERATED THREE-DIMENSIONAL FUNCTIONAL IMAGES OF THE LEFT VENTRICLE

Gautam Ray and Donald H. Schmidt

University of Wisconsin - Mount Sinai Medical Center
Milwaukee, Wisconsin

ABSTRACT

Regional asynchrony in the onset of mechanical systole is defined as the physiological phenomenon that all regions of the left ventricle do not begin to contract at the same time. Three-dimensional left ventricular images of each set of biplane cine frames were generated to study this regional asynchrony. A discrete Fourier analysis on segmental volume versus time data was performed to obtain the phase relationship between the onset of global and segmental systole. Three different types of wall segments analyzed by this method were normal wall motion in subjects with normal coronary arteries (Group A), normal wall motion in subjects with coronary artery disease (Group B), and abnormal wall motion in subjects with coronary artery disease (Group C). Wall segments in all groups were found to have synchronous as well as early and late onset of mechanical systole. Group A had a significantly ($p<.01$) smaller number of segments (7%) having late onset of systole compared to Groups B (25%) and C (29%). Only Group C had a significantly ($p<.01$) smaller number of segments (40%) having synchronous onset compared to Groups A (68%) and B (55%). Twenty to 30% of segments in all three groups were found to have an early onset of systole, with no significant difference among the groups. We conclude that (i) all segments of left ventricular wall do not contract in synchrony; (ii) segmental onset of systole is more synchronous in walls with normal motion than in walls affected by coronary artery disease; and (iii) this method provides an assessment of wall motion independent of spatial variants, and a quantitative evaluation of the physiological phenomenon on regional asynchrony in the onset of mechanical systole.

KEYWORDS

Left ventricle; segmental wall motion; asynchrony; onset of systole; three-dimensional image; Fourier analysis.

INTRODUCTION

Wall motion (WM) changes resulting from contractile abnormalities of the left ventricular (LV) myocardium are considered to be important manifestations of coronary artery disease (Gorlin et al, 1967). Generally, in a clinical setup, an evaluation of WM is done on a subjective basis with the possibility of error dependent on observer experience. Quantitative and computer-assisted methods for the evaluation of regional LVWM have been developed (Chitman et al. 1975, Shapiro et al. 1981). Global and regional ejection fraction, percentage shortening of area, Chord or radii and velocity of fiber shortening (Gelberg et al, 1979) are typical outputs of these quantitative analyses. Often descriptive terms such as akinesis, hypokinesis, dyskinesis, asynergy and asyneresis are used to quantify contractile abnormalities (Herman 1967). We want to note here that all the above qualitative and quantitative evaluations of regional LVWM are obtained from spatial analyses of wall motion.

The Regional volume change, however, is periodic in nature and is, therefore, ideally suited for temporal Fourier analysis to transform the description of the function (volume versus time) in terms of its frequency components instead of spatial coordinates. Analysis of regional changes of the radioactive counts from the LV chamber showed a variation in temporal pattern of the onset of systolic contraction (Verba et al, 1976). Since contrast ventriculography provides the best definition of the LV chamber outline; we have performed a discrete Fourier series representation of the regional volume versus time data generated from biplane cine frames to study asynchronous onset of regional mechanical systole (RMS) We define asynchrony in the onset of RMS as the physiological phenomenon that all regions of the left ventricle do not begin to contract at the same time.

METHOD OF ANALYSIS

Three-dimensional LV images were generated by digitizing each set of RAO and LAO biplane cineangiographic frames during the entire cardiac cycle. The method of reconstruction is summarized in Figure 1. The reconstructed LV image at each frame consists of 256 (16 along chord x 16 around circumference) segmental volumes. Time history of each segmental volume v_i (t) and thus the global volume V $(t)=\Sigma v_i$ (t); where i = 1 to 256 is stored. Systolic time interval is defined by $T/2 = t_2 - t_1$ where at t_1, $V(t_1)$ is the maximum global volume representing global end-diastole and at t_2, $V(t_2)$ is the minimum volume representing global end-systole. Since the cardiac cycle is not symmetrical (i.e., the systolic phase of the wave form is not

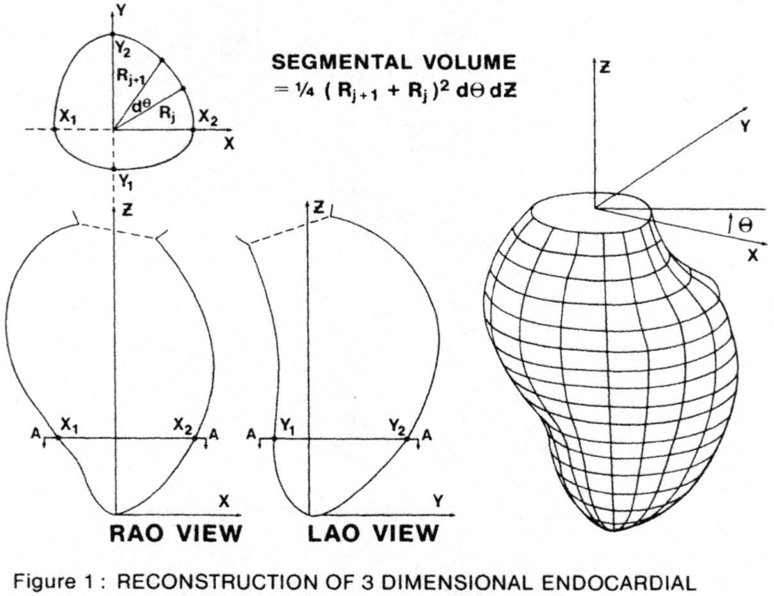

Figure 1 : RECONSTRUCTION OF 3 DIMENSIONAL ENDOCARDIAL SHAPE FROM BIPLANE CINEANGIOGRAPHY.

the mirror image of the diastolic phase), in order to study the systolic phase only, we generated a function f_i (t) such that $f_i(t) = v_i(t)$, $t_1 \le t \le t_2$ and is the mirror image of v_i (t) within $t_2 \le t \le t_3$, where $t_3 - t_2 = T/2$. We performed a discrete Fourier analysis on f_i (t) such that:

$$f_i(t) = A_0^i/2 + \sum_{k=1}^{\infty} A_k^i \ \text{Cos}\ (2\pi tk/T - \Phi_k^i) \quad \text{where,}$$

$$A_k^i = \frac{2}{T}\int_0^T f_i(t)\ \text{Cos}\ \frac{2\pi tk}{T}\, dt$$

$$b_k^i = \frac{2}{T}\int_0^T f_i(t)\ \text{Sin}\ \frac{2\pi tk}{T}\, dt \qquad (1)$$

$$\Phi_k^i = \tan^{-1}\ (b_k^i\ /A_k^i)$$

We explored only two frequencies k= 1 being the heart rate, and k = 2 a frequency double the heart rate. In general, the amplitude of higher frequency term k = 2, were insignificant and did not need further consideration.

We determined the Φ_k^g for the global behavior by analysis of V (t) by equation 1. Therefore, segmental asynchrony in the onset of RMS is defined as $\Theta_1^i = \Phi_1^g - \Phi_1^i$. A positive Θ_1 indicates early onset and a negative value indicates late onset of RMS relative to global behavior. Typically, T/2 = 60f/h = 300 msec (equivalent to ½ cycle or 180°),where h is heart rate (e.g., 80) and f is the fraction of cardiac cycle spent in systole (e.g., 0.4). Therefore, a typical 10 degree asynchrony in the onset of systole means that the segment under consideration started to contract 16.7 msec after the global onset of mechanical systole. This amounts to a duration of one frame at the rate of 60 frames per sec. Hence, we considered the onset of RMS as asynchronous for a $\Theta = \pm 5$ degrees.

RESULTS AND DISCUSSION

We analyzed 10 normal patients (normal WM and normal coronary arteries) and 10 patients with coronary artery disease (CAD). Segmental volumes were divided into three groups. Group A is the 10 normal patients that made up 2560 segmental volumes. In the 10 CAD patients, 1024 segmental volumes with normal WM made up Group B and 1536 segmental volumes with abnormal WM (by subjective evaluation) made up Group C. The percentage of segmental volumes in different groups that exhibited ansynchrony in the onset of RMS are given in Table 1. The detailed histogram is presented in Figure 2.

TABLE 1

Group	Early Onset $\Theta > +5^o$	Synchrony $\Theta = \pm 5^o$	Late Onset $\Theta < -5^o$
A	25 (4)	68 (12)	7 (3)
B	22 (4)	55 (10)	25 (4)
C	31 (5)	40 (6)	29 (5)

The numbers in parentheses represent standard deviation.

The results shown in Table 1 reveal that wall segments in all groups were found to have synchronous as well as early and late onset of RMS. Group A had significantly ($p < .01$) smaller number of segments (7%) having late onset of systole compared to Groups B (25%) and C (29%). Only Group C had a significantly ($p < .01$) smaller number of segments (40%) having synchronous onset compared to Groups A (68%) and B (55%). Twenty to 30% of segments in all three groups were found to have early onset of systole, with no significant difference among the groups.

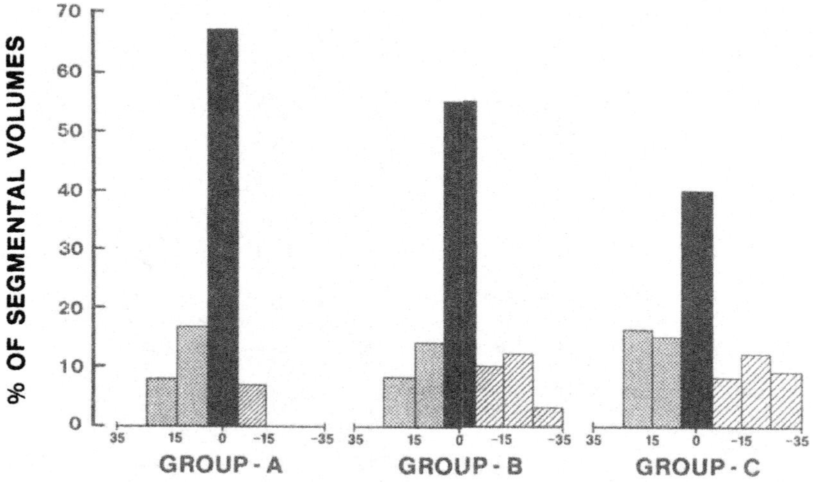

Figure 2: HISTOGRAM OF PERCENTAGE OF SEGMENTAL VOLUMES WITH SYNCHRONOUS ■, EARLY ▨ AND LATE ▨ ONSET OF REGIONAL MECHANICAL SYSTOLE. THE X AXIS OF THE HISTOGRAM IS PHASE ANGLE IN DEGREES. + IS EARLY AND − IS LATE ONSET.

From the results of this study, we conclude that (i) all segments of left ventricular wall do not contract in synchrony; (ii) segmental onset of systole is more synchronous in walls with normal motion than in walls affected by coronary artery disease; and (iii) this method provides an assessment of wall motion independent of spatial variants, and a quantitative evaluation of the physiological phenomenon on regional asynchrony in the onset of mechanical systole.

It may be mentioned here that this method of analysis required enormous effort and time for digitizing a large number of cineangiographic silhouettes making it prohibitive for routine clinical evaluation. The recent advent of digital imaging equipment with suitable interface with a large computer may make such analysis feasible for routine use.

REFERENCES

1. Chitman, B.R., et al: Circulation 52: 420, 1975.
2. Gelberg, H.J., et al: Circulation, Vol. 59, No. 5, pp. 991-1000, May 1979.
3. Gorlin, R., et al: American Journal of Medicine, 42: 512, 1967.
4. Herman, M.V., et al: New England Journal of Medicine, p 222, Aug. 1967.
5. Shapiro, E., et al: British Heart Journal: 45, pp. 264-270, 1981.
6. Verba, J.W., et al: Journal of Nuclear Medicine (Abstract) 20:p. 625, 1976.

FLOW DISTURBANCE DOWNSTREAM OF MODERATE STENOSIS IN THE CANINE COMMON CAROTID ARTERY

K.J. Hutchison and E. Karpinski

Dept. of Physiology, Univ. of Alberta, Edmonton, Alberta, Canada, T6G 2H7

ABSTRACT

A 20 MHz pulsed doppler ultrasonic velocimeter was used to obtain velocity profiles in two orthogonal axes upstream and downstream of stenoses causing 70-85% area reductions of dog common carotid arteries. The velocity jet extended from 3 to 20 diameters downstream. Spectral processing of the doppler audio signals was used to investigate the location and magnitude of disturbed flow. Jet dissipation tended to occur further downstream as the ratio of tube to stenosis N_{Re} decreased.

KEYWORDS

Doppler velocimetry; Arterial flow disturbances, Velocity profiles; Spectral analysis.

INTRODUCTION

The detection and description of flow disturbances in the arterial circulation is important to both the diagnosis and etiology of arterial disease. Detailed analysis of post-stenotic flow disturbances using hot-film and laser-doppler anemometry have been carried out in acrylic models (Cassanova, 1978; Khalifa, 1981; Yongchareon, 1979). While the hot-film anemometer has been used *in vivo* (Giddens, 1976; Nerem, 1972) it is not very suitable for detailed mapping of flow disturbances in animal experiments. The pulsed doppler ultrasonic velocimeter provides a method which may be easily applied to the *in vivo* situation, even though there are problems with the interpretation of its output.

METHODS

In the present experiments, a 20 MHz pulsed doppler ultrasonic velocimeter (Hartley, 1978) was used to obtain velocity profiles in two orthogonal axes at positions varying from 1 diameter (i.d. unstenosed artery at stenosis site) upstream to 22 diameters downstream of stenoses applied to canine carotid arteries. A micromanipulator was used to move the transducer across the artery in the horizontal plane, ultrasonic coupling being obtained by bathing the artery in saline. Range gating was used to obtain the velocity profile in the vertical plane. In addition, both audio quadrature outputs were recorded for spectral processing with the sample volume placed in at least three radial positions at each location used to obtain velocity profiles. The signals were combined in a frequency shifting circuit which subtracted frequencies due to reverse velocities from the shift frequency and added frequencies due to forward velocities to the shift frequency. The resulting signal was processed using a digital spectrum analyser (Angioscan Unigon Ind.). Further processing to carry out averaging and graphics display was made possible by interfacing the spectrum analyser with a minicomputer (PDP 11-03).

The stenosis was obtained by applying three strands of 4-0 surgical silk around the artery at a tension which caused a minor reduction in systolic blood flow as measured by an electromagnetic flowmeter. The resulting geometry was obtained after the experiment by making a silicone rubber cast of the arterial lumen with the stenosis in place and the artery stretched and pressurised to the *in vivo* values. The stenoses obtained were smooth and were quite gradual constrictions (i.e. no sudden changes in diameter or discontinuities). The cast from one experiment, in which the area reduction caused by the stenosis was 74%, is shown in Fig. 1. Area reductions ranged from 70-85%. Heart rate, arterial blood pressure and carotid artery blood flow were monitored throughout the experiment. Thus, a description of the hemodynamics downstream of a stenosis was possible.

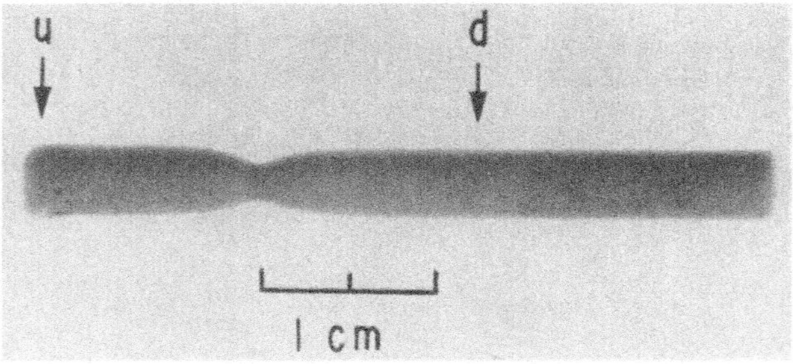

Fig. 1. Silicone rubber cast of stenosed artery. Upstream diameter (u) - 3.5 mm, downstream diameter (d) - 3.3 mm, stenosis diameter - 1.8 mm.

RESULTS

Velocity profiles on either side of the apparent hemodynamic center (midstream) are shown in Fig. 2 for 1 diameter upstream and at 1, 4 and 12 diameters downstream of the 74% stenosis depicted in Fig. 1. In this experiment, tube N_{Re} was 1100 and stenosis N_{Re} was 2100. Midstream velocity at 1 diameter downstream was double the upstream value. There was diastolic velocity reversal 2 mm to the left of midstream at this location. At 2 and 3 diameters downstream (not shown in Fig. 2), systolic flow reversal was evident in the velocity profiles, thus suggesting asymmetric flow separation. At the site of maximum flow disturbance 4 diameters downstream, midstream velocity was still increased but was close to the upstream value at 12 diameters downstream.

Corresponding spectral waveforms obtained by processing the midstream doppler audio signals are shown in Fig. 3. In each case, 15 pulses were averaged and the mode and 8 db down-points for each of the averaged spectra are plotted against time over 1 pulse interval. At 1 diameter downstream, frequencies are increased due to the velocity jet, but there is little change in spectral width, whereas at 4 diameters downstream there is marked spectral broadening in the post-peak velocity phase. Frequencies and spectral width almost return to upstream levels at 12 diameters downstream, but the waveform remains somewhat distorted in the post-peak velocity phase.

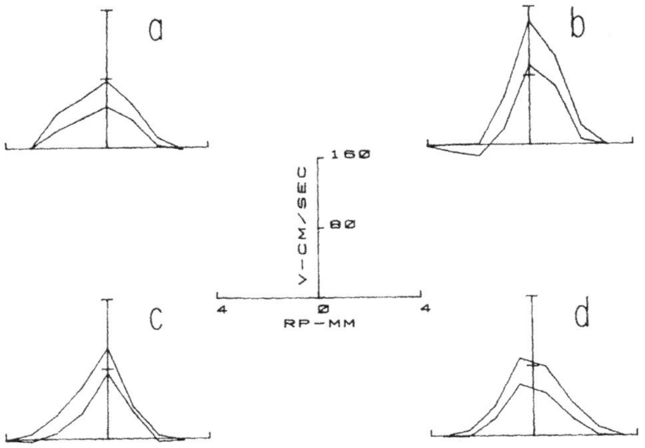

Fig. 2. Systolic and diastolic velocity profiles (a) 1 diameter upstream, (b) 1, (c) 4, and (d) 12 diameters downstream of a 74% stenosis.

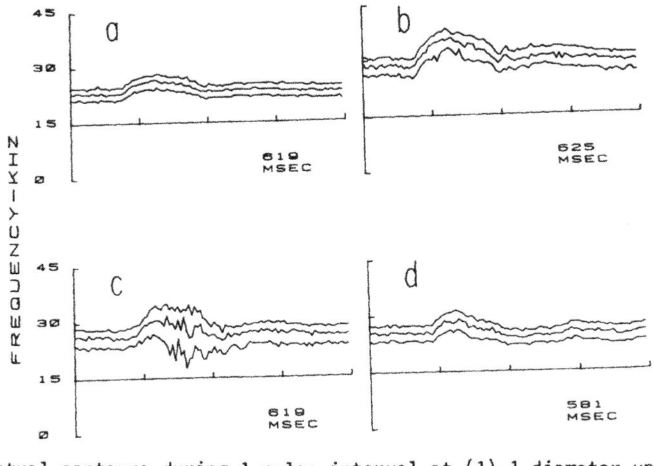

Fig. 3. Spectral contours during 1 pulse interval at (1) 1 diameter upstream, (b) 1, (c) 4, and (d) 12 diameters downstream of a 74% stenosis.

Fig. 4 summarises the results of the spectral processing for this experiment at all positions examined. Peak mode frequency and spectral width are plotted in units of velocity. Maximum velocity occurred at 1 to 2 diameters downstream, whereas maximal spectral broadening occurred at 4 diameters downstream.

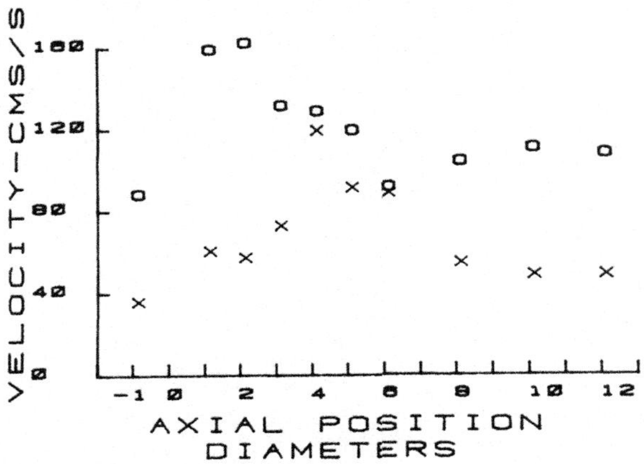

Fig. 4. Variation in peak velocity (o) and spectral width (x) with axial position.

DISCUSSION

In seven similar experiments tube N_{Re} varied from 500-1100 and stenosis N_{Re} from 990-2100, and area reduction of stenoses from 70-85%. Jet dissipation tended to occur further downstream as the ratio of stenosis to tube N_{Re} decreased. The amplitude of spectral broadening was related to both tube and stenosis N_{Re}. These results compare well with results obtained from acrylic models (Cassanova, 1978; Khalifa, 1981; Yongchareon, 1979).

ACKNOWLEDGEMENT

Supported by grants from the Alberta Heart Foundation and the Alberta Heritage Foundation for Medical Research.

REFERENCES

Cassanova, T.A., and Giddens, D.P. (1978). Disorder distal to modeled stenoses in steady and pulsatile flow. *J. Biomech.*, *11*, 441-453.

Giddens, D.P., Mabon, R.F., and Cassanova, R.A. (1976). Measurement of disordered flows distal to subtotal vascular stenoses in the thoracic aortas of dogs. *Circ. Res.*, *39*. 112-119.

Hartley, C.J., Hanley, H.G., Lewis, R.M., and Cole, J.S. (1978). Synchronized pulsed doppler blood flow and ultrasonic dimension measurement in conscious dogs. *Ultrasound Med. Biol.*, *4*, 99-110.

Khalifa, A.M.A., and Giddens, D.P. (1981). Characterisation and evolution of post-stenotic flow disturbances. *J. Biomech.*, *14*, 279-296.

Nerem, R., and Seed, W.A. (1972). An *in vivo* study of aortic flow disturbances. *Cardiovasc. Res.*, *6*, 1-14.

Yongchareon, W., and Young, D.F. (1979). Initiation of turbulence in models of arterial stenoses. *J. Biomech.*, *12*, 185-196.

ANALYSIS OF ENDOTHELIAL LESIONS CREATED USING A SUBMERGED LAMINAR JET

Mohan D. Deshpande and Marianne P. Vaishnav
Department of Mechanical Engineering
The Catholic University of America
Washington, D. C. 20064

ABSTRACT

The vascular endothelium is highly susceptible to damage by imposed shear stresses. Since the integrity of this delicate layer appears to be relevant in the transport processes associated with the atherogenic process, it is important to have an estimate of the endothelial strength. A method involving normal application of submerged saline jets was devised to estimate the strength of the vascular endothelium. In a typical experiment, a freshly excised segment of a canine thoracic aorta was cut open dorsally, mounted in a tissue rack so as to assume an approximately physiological state of stretch, and kept submerged in a bath of physiological saline at room temperature. The tissue rack was then mounted on the platform of a specially designed jet apparatus consisting of a modified microscope with one of the objectives replaced by an attachment holding a jet needle of 0.4 mm diameter. A chosen site on the segment was exposed to the jet for a period of 30 seconds. The efflux of the jet eroded the endothelium where the shear stresses exceeded the strength of the endothelium, thereby producing a characteristic lesion. This communication is based on an analysis of photographic records of six such lesions created in two canine aortic segments using laminar flows with Reynolds number values of about 1500. The average lesion diameter was found to be 0.493 mm. Using numerically calculated shear stress calibration curves obtained previously, the endothelial strength for a 30-second application of shear stress was found to be approximately 8000 dynes/cm^2. This value is consistent with the in vivo values (950 and 400 dynes/cm^2 for the erosion stress and the critical yield stress, respectively) for the canine endothelium obtained by Fry using shear stress exposures of 1 and 3 hours.

INTRODUCTION

The permeability of the blood-vascular interface to plasma lipoproteins plays an important role in the atherogenic process (1). The dominant barrier to the flux of macromolecules across the vascular wall appears to be the vascular endothelium, which is a delicate monolayer of cells. This layer is easily damaged by imposed shear stresses. Fry (2, 3) studied the strength of the vascular endothelium _in vivo_ by carefully introducing into the thoracic aorta of a dog a cylindrical plug with a channel that was specially designed to increase the blood flow velocity and the shear stress on a portion of the endothelium. From the knowledge of the longitudinal variation of the intensity of the shear stress he deduced values of "acute critical yield stress (τ_c)" and the "erosion stress (τ_e)" for the endothelium. The former corresponded to a value of the shear stress causing the first observable changes in the endothelium, and the latter corresponded to the value of the shear stress when the endothelium was completely eroded as a result of the imposed shear stress. For a shear stress application duration of 1 to 3 hours, he found the average values of τ_c and τ_e to be approximately 400 and 950 dynes/cm^2 at body temperature. He also found that the permeability of the endothelium was significantly enhanced in the presence of shear stresses approaching but below τ_c (2, 3).

In this paper we report on studies designed to estimate local variations in the endothelial strength. The need to study local variations of strength stems from the fact that the atherogenic process is a discrete process. Therefore, knowledge of the local strength of the endothelium and of the local flow fields is essential to an understanding of the development of early atherosclerosis. The method we used involved the application of saline jets normal to the endothelium of a vascular segment, and the measurement of the size of the resulting lesions. The knowledge of the lesion size along with the knowledge of the distribution of the shear stress field associated with the impinging jet permits estimation of the endothelial shear strength.

METHODS

A freshly excised segment of a canine middle descending thoracic aorta was cut open dorsally and mounted in a tissue rack so as to assume an approximately physiological state of stretch in the circumferential and longitudinal directions. Throughout the experimental period, the rack was kept submerged in a bath of physiological saline which was maintained at room temperature. The tray containing the rack and the tissue was mounted on the platform of a specially designed jet apparatus consisting of a modified microscope with turret-mounted objectives. One of the objectives was replaced by an attachment holding a jet needle of 0.4 mm inside diameter. The jet could be adjusted to deliver a desired flow rate. The bath temperature was kept constant at a room temperature value of about 23° C. The flow rate chosen for the experiments corresponded to a Reynolds number (Re) of 1500, which was defined as Re = $\bar{u}d/\nu$, where $\bar{u}$ = average velocity at the tube exit, d = tube diameter (0.4 mm), and ν = kinematic viscosity of saline. Earlier experiments using a hot-film probe indicated that the flow was laminar for values of Re up to 1500.

In a typical experiment a chosen site on the segment was carefully positioned under the jet, and the nozzle was adjusted to a height of 0.8 mm above the tissue. The tissue site was then exposed to the jet for a period of 30 seconds. The efflux of the jet eroded the endothelium where the shear stresses exceeded the strength of the endothelium, thus producing a characteristic lesion. The lesions were made visible by bathing the tissue in a solution of Evans blue dye (EBD), which is absorbed by the damaged tissue but not by the undamaged tissue. The lesions were then photographed. The diameters of the lesions in two perpendicular directions were measured and averaged. The endothelial strength was deduced from the knowledge of the distribution of the shear stress (4) on the endothelium due to jet impingement and the size of the lesion created. For further details, see (5).

RESULTS

In this study, six lesions created in two aortic segments were analyzed. The lesions generally displayed a characteristic annular configuration. The central core area was undamaged, since it was not exposed to shearing stresses larger than the endothelial shear strength. The annular region surrounding the core was the damaged region where the shearing stresses exceeded the shear strength of the endothelium. The endothelium outside the annular region was not damaged since the imposed shear stress was again below the endothelial strength. The inside and outside boundaries of the lesion corresponded to the radial distances at which the shear stress equaled the endothelial strength. For better accuracy, the outside diameters of the lesions were measured in this study. Table 1 shows the results from the analysis of the 6 lesions from the two aortic segments studied. The average endothelial strength was found to be 7962 dynes/cm^2.

TABLE 1

Segment No.	Lesion No.	Temperature	Re	Lesion Size $R_o = r_o/a_o$	Average Lesion Size for Segment	Endothelial Strength (dynes/cm^2)	Average Endothelial Strength (dynes/cm^2)
	1	23°	1506	1.280			
1	2	23°	1506	1.080	1.213	8108	
	3	23°	1506	1.280			7962
	1	23°	1493	1.240			
2	2	23°	1493	1.315	1.246	7815	
	3	23°	1493	1.183			

r_o = lesion radius (mm)
a_o = needle radius = 0.2 mm

DISCUSSION

The endothelial strength obtained by the measurement of the lesion size is comparable, in principle, to the erosion stress as defined by Fry, for the visible lesion size corresponds to the area where the endothelium is completely eroded. Numerically, the value 7962 dynes/cm^2 is quite different from the average in vivo value of 950 dynes/cm^2 obtained by Fry. The two values, however, are quite consistent. There are two reasons why the present value is higher. First, in the present study, the endothelium was exposed to shear stress field for only 30 seconds as against 1 to 3 hours in Fry's study. The fact that the erosion stress goes down as the duration of application of shear stress increases, was shown by us in another set of studies reported elsewhere (5). In the same study we have shown that an increase in temperature of the tissue considerably reduces the strength of the endothelium. This is the second reason why Fry's value is lower than ours. More detailed experiments encompassing a large number of lesions, a wide temperature range and involving shear stress application for various durations are needed to further quantify the role of variables affecting the endothelial strength. Such information should indeed be useful in further elucidating the roles of endothelial strength and permeability in atherogenesis.

ACKNOWLEDGMENTS

The support of PHS grants HL-15270 and HL-12083, and NSF grant CME8006338 is gratefully acknowledged. We thank Dr. R. N. Vaishnav for providing us access to the data on which this analysis is based.

REFERENCES

1. Fry, D. L. (1976). Hemodynamic forces in atherogenesis. In P. Scheinberg (Ed.), Cerebrovascular Diseases, Raven Press, NY, 77-95.
2. Fry, D. L. (1968). Acute vascular endothelial changes associated with increased blood velocity gradients, Circ. Res., 22, 165-197.
3. Fry, D. L. (1969). Certain histological and chemical responses of the vascular interface to acutely induced mechanical stress in the aorta of the dog, Circ. Res., 24, 93-108.
4. Deshpande, M. D. and R. N. Vaishnav (1982). Submerged laminar jet impingement on a plane. J. of Fluid Mechanics, 114, 213-236.
5. Vaishnav, R. N., D. J. Patel, H. B. Atabek, M. D. Deshpande, F. Plowman, and J. Vossoughi (1982). Determination of the local erosion stress of the canine endothelium using a jet impingement method. Submitted for publication.

TECHNIQUES FOR TWENTY-FOUR HOUR HEMODYNAMIC MONITORING IN UNANESTHETIZED RATS

T. L. Smith, T. G. Coleman, W. R. Murphy, and K. A. Stanek

Department of Physiology and Biophysics
University of Mississippi Medical Center
Jackson, Mississippi 39216

ABSTRACT

A system is described which allows an investigator to continuously monitor hemodynamics (cardiac output and arterial pressure) in unanesthetized, tethered rats for periods of 24 hours and longer. Computerized data collection is used to handle the large number of data points (1440/day) and record data in digital form for increased accuracy. This data collection system shows that highly dynamic hemodynamic variables can be accurately described by their long-term mean values and standard deviation. Further, a marked diurnal variation was observed, characterized by nocturnal increases in cardiac output, heart rate and arterial pressure.

KEY WORDS

hemodynamics; diurnal variation; rats; computerized data collection; cardiac output; continuous monitoring; conscious rats

INTRODUCTION

Hemodynamic investigations in rats are of interest since there are many genetic substrains (Okamoto spontaneously hypertensive rats; Milan hypertensive rats; Dahl hypertensive rats) and experimental models (Goldblatt renovascular hypertension; deoxycorticosteroid hypertension; subtotal nephrectomy hypertension) which are studied to gain insight into comparable human disease conditions. A basic assumption in all hemodynamic investigations is that the results of a study are indicative of the normal physiology of the model in question. This assumption may not obtain, however. The majority of hemodynamic studies in rats which reported cardiac output and arterial pressure, have used either anesthetized, surgically compromised animals, or have used measurement techniques which inherently alter the hemodynamics being studied. Anesthetics (Smith and Hutchins, 1980) and surgery (Katz and Kolin, 1938) can both affect hemodynamic measurements. Similarly, methodologies used in some types of flow determinations can also alter the hemodynamics under consideration (Flaim and co-workers, 1978).

Recent advances in biomedical technology have allowed researchers to chronically instrument rats with electromagnetic flow probes and arterial catheters and then study hemodynamics in conscious rats after they have recovered from the implantation surgery (Smith and Hutchins, 1979). The results of these studies were difficult to analyze since there was not a range of normal values with which to compare them. In addition, cardiac output and arterial pressure could not be monitored for extended periods of time (beyond a few hours) because of the requirement for tethering (telemetry is not yet feasible in animals of this size). In order to compile a range of hemodynamic values for conscious rats and assess the impact of diurnal variation upon the hemodynamics being observed, we devised a system for continuous monitoring of cardiac output and arterial pressure in conscious rats.

METHODS

Animal Preparation

Adult male Sprague-Dawley rats [302-383 grams body weight (BW)] were used in this experiment. Each animal was surgically instrumented with an electromagnetic flow probe (Carolina Medical Electronics, King, NC, model EP-100) on its ascending aorta and an arterial pressure catheter (Smith and Hutchins, 1979). All surgery was performed using pentobarbital sodium anesthesia (50 mg/kg, I.P.). The rats were allowed to recover for one week after surgery. After this recovery period the animals had regained any weight lost due to surgery and were in good health. The rats were housed in clear plastic cages (18.5 cm X 15 cm X 25 cm) after surgery and during continuous hemodynamic monitoring.

Recording Techniques

A requirement for continuous hemodynamic recording in the unanesthetized rat was that the animal not be restrained. To allow free range of movement for the rat, an electrical and hydraulic swivel (Airflyte Electronics, Bayonne, NJ, model CAY-12) was interposed between the rat and recording apparatus (Figure 1). This allowed the rat to turn in any direction within its cage. This swivel was then mounted on a carriage which could roll the length of the cage. All moving parts of the carriage/swivel assembly were mounted on ball bearings. This greatly reduced rotational and rolling resistance so that even minor movements of the rat were easily accomodated. To reduce strain on the wires of the flow probe connector, a lightweight (3.7-4 grams) plastic harness was attached to the connector and sewn to the rat's back during the surgical preparation. A flexible coiled spring protected the flow probe cable and arterial catheter within the rat's cage and transmitted rotational

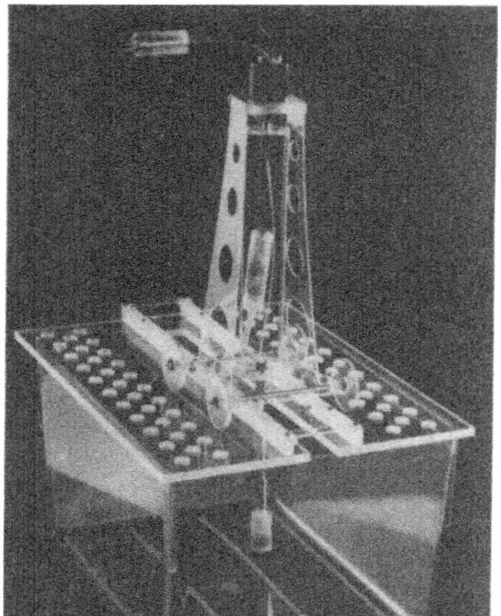

Figure 1: Electrical and hydraulic swivel/carriage assembly mounted on cage.

movements to the swivel. Rats were connected to the swivel with minimal handling and without restraint.

The electrical and hydraulic connections on the fixed end of the swivel were attached to an electromagnetic flowmeter (Carolina Medical Electronics, King, NC, model 501) and arterial pressure transducer (Century Technology, Inc., Inglewood, CA, model CP-01), respectively. Catheter patency was ensured for the duration of the recording session by the slow (0.5 ml/hr) infusion of heparinized (33 I.U./ml) Ringer's injection. The pressure transducer and flowmeter connected to a recording oscillograph (Grass Instruments, Quincy, MA, model 7D) which routed information to a digital equipment PDP 11/70 computer via two analogue to digital channels.

The computer was programmed to sample mean arterial pressure and pulsatile aortic flow for one second each minute. This second of information was then digitized at two millisecond intervals. Pulsatile aortic flow, as sampled by the computer, is illustrated in Figure 2. Note that the signal is inverted when analyzed by the computer.

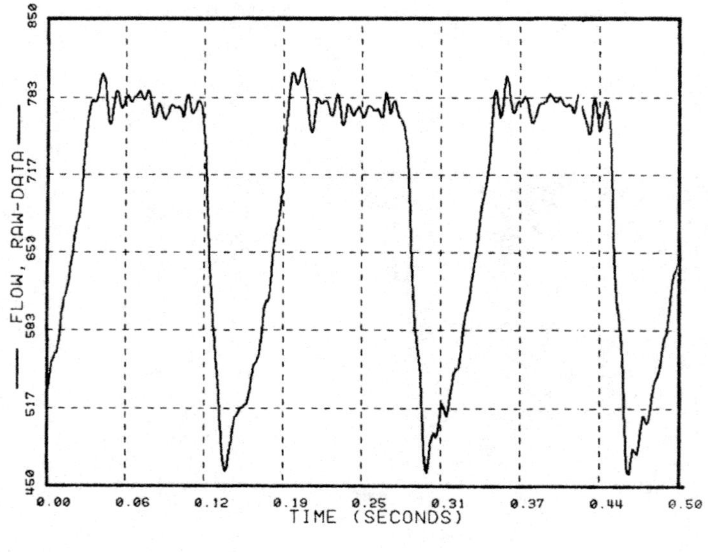

Figure 2: Ascending aortic flow signal as sampled by the PDP 11/70 computer.

Cardiac output for the sampling period was determined as follows: First, zero flow (diastolic flow) was determined and used as a baseline (approximately 783 on Fig. 2). The area under each flow waveform was then integrated to yield stroke volume, and the average stroke volume for the sampling period was determined. Cardiac output was calculated by multiplying the average stroke volume times the heart rate (derived from the average time interval between peak aortic outflow values during the sampling period). Information collected or derived from the one second sample period each minute included cardiac output ($ml \cdot min^{-1}$), cardiac index ($ml \cdot min^{-1} \cdot 100$ g BW^{-1}), mean arterial pressure (mmHg), stroke volume (microliters), stroke volume index ($\mu l \cdot 100g$ BW^{-1}), heart rate ($beats \cdot min^{-1}$), peak aortic flow ($ml \cdot min^{-1}$), and total peripheral resistance index ($mmHg \cdot ml^{-1} \cdot min \cdot 100g$ BW). Values for each of these variables were collected for 1440 contiguous minutes in this experiment.

RESULTS

The initial work with this continuous monitoring system established a set of "normal" twenty-four hour values. Using the units given above, normal values (mean ± average S.D.) for a group of 8 rats are: arterial pressure = 93 ±8; cardiac index= 29 ± 4; total peripheral resistance= 3.3 ± 0.5; heart rate= 347 ± 45. In addition, we have found that there is significant diurnal variation in mean arterial pressure, cardiac index, heart rate, and total peripheral resistance with nighttime values of cardiac index, mean arterial pressure, and heart rate being 27%, 9%, and 19% higher at night than during the day, respectively. We also found that a significant degree of the daily variance in hemodynamic data was due to minute-to-minute changes.

DISCUSSION

A system is described which allows continuous hemodynamic monitoring in unanesthetized rats. System components are commercially available and reliable as configured. The results suggest that this system has applications not only in long-term monitoring, but would also be useful in short-term (2-3 hours) protocols.

ACKNOWLEDGEMENT

This work was supported by National Institutes of Health grants HL 11678 and HL 07270.

REFERENCES

Flaim, S.F., Z.Q. Morris, and T.J. Kennedy (1978). Dextran as a radioactive microsphere suspending agent: severe hypotensive effect in rat. Am. J. Physiol., 235, H587-H591.

Katz, L.N., and A. Kolin (1938). The flow of blood in the carotid artery of the dog under various circumstances as determined with the electromagnetic flowmeter. Am. J. Physiol., 122, 788-804.

Smith, T.L., and P.M. Hutchins (1979). Central hemodynamics in the developmental stage of spontaneous hypertension in the unanesthetized rat. Hypertension, 1 (5), 508-517.

Smith, T.L., and P.M. Hutchins (1980). Anesthetic effects on hemodynamics of spontaneously hypertensive and Wistar-Kyoto rats. Am. J. Physiol., 238(7), H539-H544.

A CAROTID SINUS BARORECEPTOR AND WALL MODEL SUITABLE FOR BAROPACING*

WOLF W. VON MALTZAHN, WILLIAM R. OSBORN

BME PROGRAM AT THE UNIVERSITY OF TEXAS AT ARLINGTON

ABSTRACT

A baroreceptor model is developed from a two-layered, cylindrical wall model for arteries and a neuromime. This model is intended to be used in baropacing. It is capable of reproducing the relationship between arterial pressure and sinus nerve activity for a single baroreceptive cell. The best results are achieved when the linking between the wall model and the neuromime is taken to be the stress difference between the two layers.

KEYWORDS

Arterial wall, baropacing, carotid sinus baroreceptors, neuromime

INTRODUCTION

The carotid sinus is a distinct arterial segment at the entrance of the internal carotid artery. Many baroreceptive nerve endings are embedded in the wall there to sense arterial blood pressure. These baroreceptors do not respond to pressure directly, but rather to deformation of the wall caused by pressure. The purpose of this study is to delineate further the nature of the stimulus transfer between arterial pressure inside the carotid sinus and the firing activity of the sinus nerve.

From electronmicroscopic studies Knoche et al.(1980), Böck and Gorgas (1976), and Rees (1967)have determined that the majority of baroreceptive nerve endings are located close to the border between media and adventitia. The media in the carotid sinus consists of alternating layers of elastic membranes and muscle cells. The adventitia consists primarily of collagenous fiber bundles, connective tissue cells and a polysaccharide ground substance. Since both layers are so distinctly different, they are assumed to have different mechanical properties, as well.

* This work was supported by ORF 15-414, a research grant awarded by the Organized Research Fund at the University of Texas at Arlington.

In the past two decades numerous neuromimes have been proposed, most of which are based on the original work of Hodgkin and Huxley (1952). An excellent review of these is given in the monograph by MacGregor and Lewis (1977). Although the exact excitation mechanism of receptors is still not known, most of these models describe the relationship between a given stimulus and either the action potential or the nerve activity quite well.

THEORY AND METHODS

The arterial wall model for the carotid sinus is taken from von Maltzahn et al. (1981). It is based on the assumption that the work performed by axial forces and internal pressures is stored inside the wall material in form of elastic potential energy. When this energy is refered to the unit volume of the undeformed wall, it is called the strain energy density W. The analytical relationships between stresses and strains inside the deformed wall can be derived from W. For incompressible, axially symmetric, cylindrical arteries W depends of two strain variable only. These are either the axial and tangential Green-St.Venant strain γ_z and γ_t or the strain invariants J_1 and J_2. The strain energy density function for the anisotropic adventitia is $W = (93.12\ \gamma_t^4 + 20.82\ \gamma_t^2 \gamma_z^2 + 15.37\ \gamma_z^2)$ kPa and for the isotropic media it is $W = 3.51\ (J_1^2 + J_2^2)$ kPa. These polynomials are plotted in figure 1 as a function of radius r for two different values of internal pressure P. The axial force F is kept constant. The media lies between r_i and r_m and the adventitia between r_m and r_o. There is a pronounced discontinuity at the border between media and adventitia which appears also in the curves of the tangential and axial stresses. Since the stress gradients caused by these discontinuities are believed to be important in the exciting mechanism of baroreceptors, the difference between tangential stress in the adventitia and tangential stress in the media at the media adventitia border is praphically displayed in figure 2 as a function of internal pressure P. In the upper curve, the axial force F decreases from 0.1N to 0.01N as the internal pressure increases from 5 kPa to 25 kPa. This is based on experimental data published by Cox (1975). In the lower curve the axial force F is kept constant at 0.1N. The diagrams in figure 3 show the change of tangential strain γ_t at the media adventitia border as a function of internal pressure P. Again, in the upper curve the axial force decreases as the internal pressure increases and in the lower curve the axial force is held constant. Figures 1 through 3 describe the behavior of the carotid sinus wall model in lieu of a relative large number of equations.

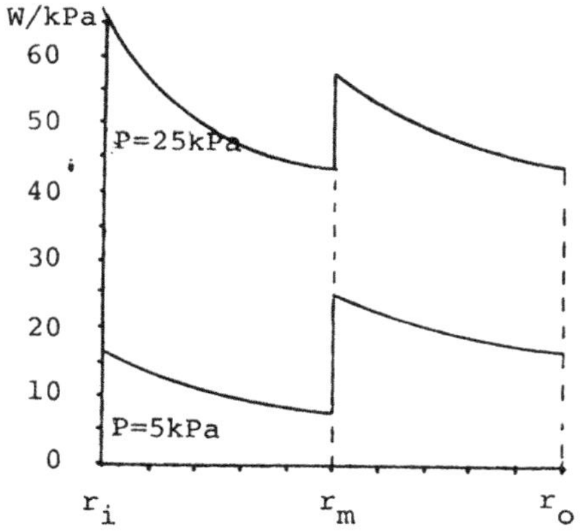

Figure 1 : Strain energy density as a function of radius and pressure.

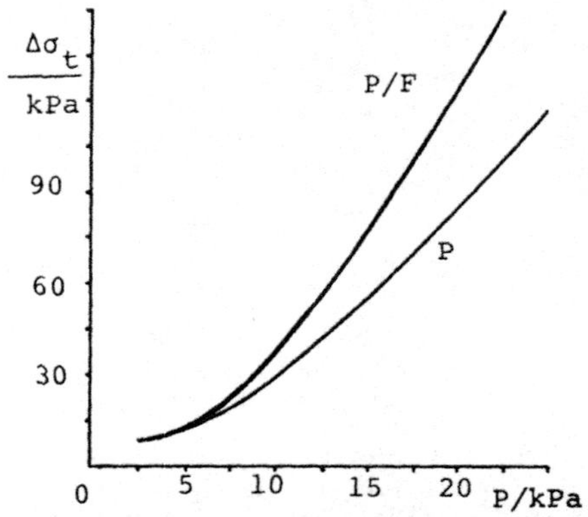

Figure 2: Relationship between internal pressure P and tangential stress difference $\Delta\sigma_t$.

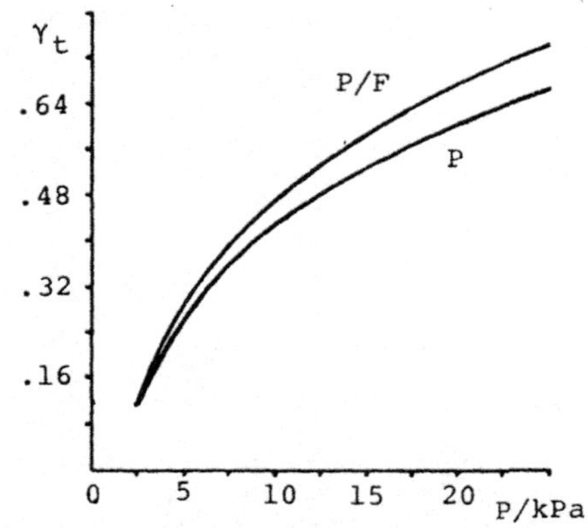

Figure 3: Relationship between internal pressure P and tangential strain γ_t.

Selection of a neuron model was guided by two ideas: (a) it should describe the relationship between nerve activity and stimulus for static and dynamic excitations and (b) it should be available as an electronic analog. The neuromime proposed by MacGragor and Oliver (1974) meets both criteria. Only very minor changes were made to produce typical responses of mechanoreceptors. The model allows to set different values for adaptation, accommodation, and threshold levels. The input variable is a non specific stimulating current SC. Although static and dynamic characteristic curves can be produced, only the static steady state activity rate as a function of SC is shown here in figure 4. There exists a distinct threshold level below which now firing occurs, an almost linear region between SC=2.8 and SC=5.2, and a saturation region in which the firing rate does not change. These are typical characteristics of baroreceptors which is well documented and excellently reviewed by Kirchheim (1976).

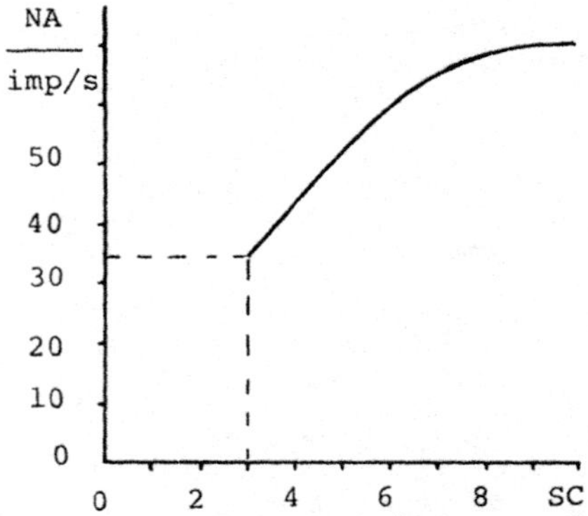

Figure 4: Nerve activity as a function of stimulating current

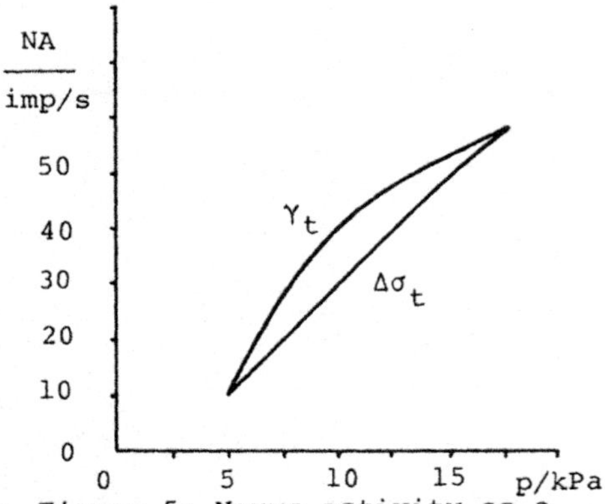

Figure 5: Nerve activity as a function of internal pressure

RESULTS

It is not yet resolved whether strain or stress is the appropriate input stimulus for baroreceptors. Therefore, both variables were used and the results are presented in figure 5. The linking between the wall model and the neuron model was established by relating 50 mmHg (6.65 kPa) to a stimulating current of SC=3.0 and 175 mmHg (23.3 kPa) to SC=6.0. This way physiological ranges for arterial pressures, energy densities of the wall and nerve activities all fall into this interval.

DISCUSSION

Koushanpour and Kelso (1972) are to our knowledge the only authors who attempt to separate mechanical properties of the wall from receptor characteristics of carotid sinus baroreceptors. Their wall model is single layered and isotropic. However, it has been clearly demonstrated (Patel and Vaishnav, 1972) that the arterial wall is anisotropic, inhomogeneous and multilayered. The model presented here incorporates the mechanical properties of the two main layers media and adventitia. This leads to the important observation that large stress gradients develop at the border between these two layers. When these stress gradients are taken as the input variables for the neuron model as shown in figure 5, the pressure nerve activity relationship is closely related to diagrams reported in the literature (Kirchheim, 1976).

Baropacing as described by Zerbst et al. (1978, 1980) has been used clinically in hypertensive patients for about one decade. Since hypertension is often accompanied by arteriosclerosis and/or atherosclerosis, it is important to consider such changes in the baropacer. The presented model is a significant step towards this goal since it incorporates measurable mechanical properties of the arterial wall.

REFERENCES

1. Böck,P. Gorgas,K. (1976), Cell Tiss. Res.,170, 95-112
2. Cox, R.H. (1975), J.Biomech.,8, 293-300
3. Hodgkin,A.L.,Huxley,A.F.(1952), J.Physiol.,117,500-544
4. Kirchheim,H. (1976), Physiol. Rev.
5. Knoche,H.,Wiesner-Menzel,L.,Addicks,K. (1980) Acta. Anat. 17, 63
6. Koushanpour,E.,Kelso,D.(1972), Circ. Res., 31, 831-845
7. MacGregor,R.J., Oliver, R.M.,(1974), Kybernetik, 16,53-64
8. MacGregor,R.J., Lewis,E.R., Neural Modelling, Plenum Press New York, London, 1977
9. Patel, D.J., Vaishnav, R.N., (1972), in Cardiovascular Fluid Dynamics, Vol. 2, ed. by Bergel, Academic Press
10. Rees,M.R.,(1967), J. Comp. Neuro., 131, 517-548
11. von Maltzahn, W.W., Wiemer, W., Besdo, (1981), J. Biomech, 14, 389-397
12. Zerbst,E., Peters, T. Koralewski, H.E.,(1980), Meeting Art. Organ, 4(2), 154

Session 6
BIOELECTRICITY

DERMAL EFFECTS OF ELECTROCUTANEOUS STIMULATION

A. Y. J. Szeto, Ph.D. and L. Mao, M. S.
Department of Biomedical Engineering
Louisiana Tech University
Ruston, Louisiana 71272

ABSTRACT

The dermatological reaction to extended electrocutaneous stimulation was investigated with respect to three different types of stimulating waveforms -- monopolar, alternating polarity, and bipolar. The experimental approach used 3 different portable stimulators worn by 3 male and 2 female subjects for 3 two-week test periods each. The 3 male subjects received continuous stimulation for 10 hours each day while the 2 female subjects received noncontinuous stimulation 12 hours a day with 30 minutes of rest after every 2 hours of stimulation. Periodic examinations of the stimulated skin site by a dermatologist, photographic records, biopsy reports, and data in subject questionnaires all indicated that these three commonly used electrocutaneous stimuli produced only mildly reactive and readily reversible skin changes. Subjects applied the most current to achieve adequate and comfortable vibratory tactile sensations when using the bipolar stimulating waveform. As a result, it produced the most skin reddening as compared with the effects of the monopolar and alternating polarity stimulation. Sensory adaptation occurred during the wearing period for all three stimulus waveforms. Periodic rests significantly reduced the amount of skin reddening and the amount of sensory adaptation.

KEYWORDS

Electrocutaneous Stimulation; Sensory Adaptation; Dermal reaction; Artificial Sensory Communications

INTRODUCTION

Electrocutaneous stimulation is the application of electrical impulses to produce comfortable vibratory sensations on the skin. Such stimulation has been successfully used for supplemental sensory feedback in limb prostheses (Prior, Case and Lyman, 1977; Szeto and Lyman, 1977), vibrotactile hearing aids (Saunders, 1973), and visual aids (Collins, 1970). Electrocutaneous stimulation has also been used to inhibit postoperative pain (Fisher, 1978) and chronic pain (Burton, 1976). Since electrocutaneous stimulation is not a natural stimulus, the skin may be damaged with heavy and prolonged use. Investigation of this possibility was the goal of the work reported here.

Three commonly used or potentially usable stimulus waveforms -- monopolar, alternating polarity, and bipolar -- were chosen based upon a survey of the literaature (Saunders, 1977; Szeto and Saunders, 1982). The experiment sought answers to the following questions:

1. What are the effects of electrical stimulation on the stimulated site?

2. If detrimental effects appear, can they be reduced or eliminated through periodic rests from stimulation?

3. Which type of stimulation produces the most comfortable and clear vibratory sensation?

EXPERIMENTAL PROCEDURES

Three different stimulators were built, one corresponding to each type of stimulating waveform. The pulse width and pulse rate of monopolar and alternating polarity electrocutaneous stimulators were adjusted to produce 150 us pulses, 30 times per second. Peak stimulus current levels were adjusted for comfort by the subjects between 0.5 mA and 25 mA. The bipolar electrocutaneous stimulator produced bursts of 10 KHz, 20 us sec wide bipolar pulses, 30 times a second with each burst containing 15 bipolar pulses. The outputs of all the stimulators were capacitively coupled to the subject's skin via silver concentric electrodes. To serve as a control, an unconnected, "dummy" electrode was worn next to the stimulating electrode. Whether the source of skin reaction was due to the electrode itself or the presence of the electrical stimulus could thus be determined.

Five subjects (3 males and 2 females) were involved in this study. During each two week test period, the subjects tested only one type stimulator. After a week or two of rest, they were asked to test another type of stimulus. The 3 male subjects used their stimulators continuously for 10 hours a day while the 2 female subjects used their stimulators noncontinuously for 12 hours a day with 30 minutes of rest following every 2 hours of stimulation. After using an elastic clo th bandage to affix the stimulating and dummy electrodes around the biceps of his nondominant arm, a subject adjusted the current level midway between his perception and pain thresholds. In anticipation of possible tactile sensitivity changes, all subjects were asked to periodically monitor the vibratory sensation level and increase or decrease the stimulating current accordingly to maintain a constant sensation level.

Subjects were also asked to daily record the following information on a questionnaire sheet:

1. the current level used at all times,
2. any periods of non-stimulation,
3. the perceived character of the tactual sensation, and
4. the condition of the stimulated skin site.

Other data used to determine skin response included daily color photographs and tissue biopsies of the stimulated site. A consulting dermapathologist examined the biopsies; his report and the opinion of a second dermapathologist have been incorporated into the results below. To ensure the safety of the subjects, their skin conditions were checked by a consulting dermatologist every other day during every test period.

RESULTS AND CONCLUSIONS

The daily data log sheets of the subjects indicated that they needed the most peak current to get clear and comfortable vibratory sensations when using bipolar stimulation. Monopolar and alternating polarity waveforms required approximately one-sixth and one-third, respectively, as much peak current (Fig. 1 & 2).

The regression lines for the <u>hourly</u> stimulation current (Fig. 1), averaged over all the subjects and the 14 days of wear, indicate that all three stimulus types had about the same rate of current increase during continuous stimulation. The slopes of current increase for bipolar, alternating polarity, and monopolar waveforms were not significantly different ($p < 0.05$) from each other but were significantly different from zero ($p < 0.01$). During noncontinuous stimulation, the slope of bipolar current increase was significantly higher ($p < 0.05$) than either the slopes for alternating polariity or monopolar current (i.e., 0.07 mA/hr versus 0.04 or 0.03 mA/hr).

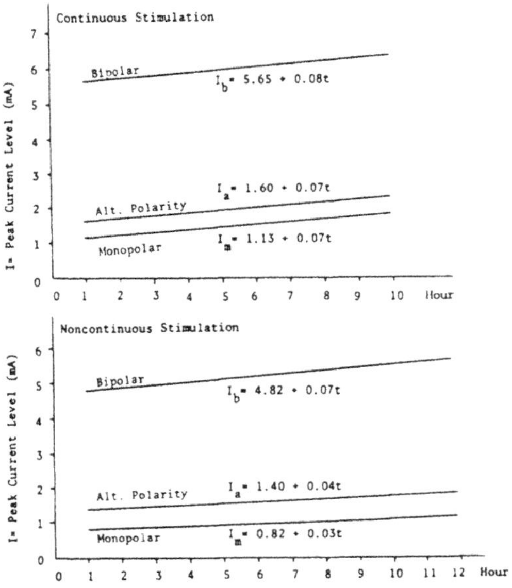

Fig. 1. Regression lines for the average daily current chosen for each type of stimulating waveform.

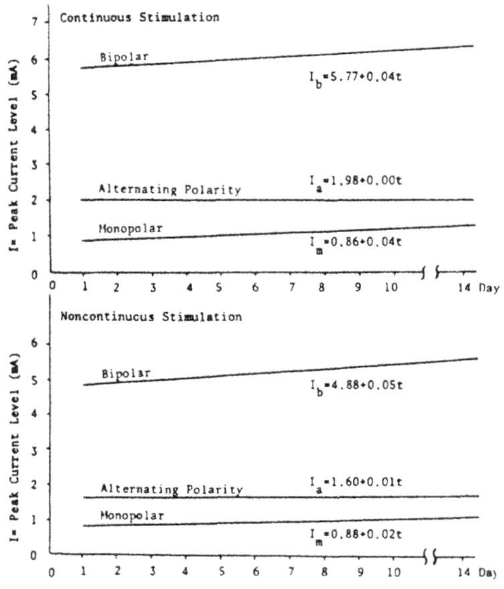

Fig. 2. Regression lines for the average hourly current chosen for each type of stimulating waveform.

Data summarized by Fig. 1 also indicate that the rate of the current increase, and hence the amount of sensory adaptation to electrocutaneous stimulation, was significantly reduced ($p < 0.05$) by periodic rests for alternating polarity and monopolar stimuli. The continuous and noncontinuous stimulation hourly current slopes for these two waveforms were significantly different ($p < 0.05$), but the corresponding slopes for bipolar stimulation were not significantly different ($p < 0.05$). Hence sensory adaptation to bipolar stimulation was altered very little by the periodic rests given.

A comparison of all the lines in Fig. 1 with the corresponding ones in Fig. 2 indicates that the hourly current level increments were greater than the daily current level increments. Assuming that the current increments reflected the subjects' decreased tactile sensitivity to the electrical stimulus, more sensory adaptation occurred in one day's wear than occurred from day to day. A partial answer for this observation may be the more complete recovery of tactile sensitivity possible with overnight rests than is possible with the more frequent but shorter rests associated with noncontinuous stimulation.

Based on the daily color photographs, the diagonses of the consulting dermatologist, and biopsy reports, the following can be said about the skin's response to electrocutaneous stimulation:

1. No skin reaction was produced by the non-stimulating, "dummy" electrode.
2. The bipolar stimulus elicited more skin reddening then did the other two stimuli.
3. The periodic rests under noncontinuous stimulation significantly reduced the amount of skin reddening.
4. The small amount of skin reddening correlated well with the amount of stimulation current applied. Bipolar stimulation produced the most skin reddening, and it also needed the most current for adequate tactile sensation.
5. The skin tolerated all three stimuli well. No significant epidermal

inflammatory changes were observed by the pathologist under the microscope.

6. The dermal responses at the stimulated site were clinically considered to be mild reactive changes in the epidermis that could be readily reversed upon cessation of stimulation.

Data recorded in the subjects' questionnaire sheets described the quality of the tactual sensation as being mild, comfortable, easily ignored, tolerable, or painless vibrations. The subjects unanimously rated bipolar stimulation as being the most comfortable but least clear of the three stimulating waveforms. Conversely, monopolar stimulation produced the clearest but least comfortable vibratory sensation.

SUMMARY

The effects of three waveforms (monopolar, bipolar, and alternating polarity) used or usable in electrocutaneous stimulation for sensory communications and/or pain control were examined. Five subjects received each type of stimulation for two weeks up to 12 hours a day. Current levels needed for adequate sensation, photographs of the stimulated skin site, questionnaires and daily logs from the subjects, periodic check-ups by a dermatologist, and biopsy reports from a dermapathologist yielded these findings:

1. All three stimulus waveforms were tolerated well. Only mild, reactive, and readily reversible skin changes were produced by the test stimuli.

2. Superficial skin reddening, though very mild, was related to the amount of peak current needed to elicit clear and comfortable vibratory sensations.

3. As expected, periodic rests reduced the amount of skin reddening and the amount of sensory adaptation to the test stimuli.

4. The amount of sensory adaptation, as reflected in the gradually increased current level used, was about the same for all three types of waveforms under continuous stimulation.

ACKNOWLEDGEMENTS

The expert assistance of Dr. Thomas D. Carey, dermatologist, and Dr. Herbert Ichinose, pathologist, are gratefully acknowledged. Project supported by grants from the Division of Vocational Rehabilitation, State of Louisiana.

REFERENCES

Burton, C. (1976). Transcutaneous electrical nerve stimulation to releve pain. Postgrad Medicine, 59, 105-128.

Collins, C.C. (1970). Tactile television: Mechanical and electrical image projection. IEEE Trans. on Man-Machine System, MMS-11(2), 85-91.

Fisher, A.A. (1978). Dermatitis associated with transcutaneous electrical nerve stimulation. CUTIS, 21, 24-47.

Prior, R.E., P.A. Case, and J. Lyman (1977). Hook opening feedback for the below elbow amputee. Pro. San Diego Biomedical Symposium, 16, 357-368.

Saunders, F.A. (1973). An electrotactile sound detector for the deaf. IEEE Trans. on Audio and Electroacoustics, 21, 285-287.

Saunders, F.A. (1977). Recommended procdures for electrocutaneous displays. In F. T. Hambrecht and J. B. Reswick (Eds.), Functional Electrical Stimulation. Marcel Dekker, New York, 303-309.

Szeto, A.Y.J. and J. Lyman (1977). Comparison of codes for sensory feedback using electrocutaneous tracking. Ann. Biomed. Eng., 5, 367-383.

Szeto, A.Y.J. and F. A. Saunders (1982). Electrocutaneous stimulation for sensory communication in rehabilitation engineering. IEEE Trans. Biomed. Eng., 29(4). (in press).

DETECTION OF SURFACE ELECTROPOTENTIALS OF HUMAN BREAST LESIONS

Don M. Morris, M.D., F.A.C.S.
Assistant Professor of Surgery
Louisiana State University Medical School at Shreveport

INTRODUCTION

In 1949, Burr and Langman pointed out the existence of a different direct current electric potential between normal tissues and malignancies of the female genital tract. The differences in the electropotentials between normal and malignant tissues were measured using special silver chloride electrodes and electrometers.[1] The National Aeronautic and Space Administration also found the surface electropotential of cells to be electronegative in different types of experimental tumors.[2] Schauble and Habal demonstrated that the surface electropotential of freshly excised malignant tumors was electronegative when compared to freshly excised normal tissues. It was also demonstrated that surface electropotential is a function of the living organism and disappears shortly after death.[3] Surface electropotentials are sometimes affected by changes in blood flow and body temperature.[4] We have developed a methodology to detect and measure the surface electropotentials of different organs and tissues in the human body. We have previously reported our initial experience with this equipment.[5] This report presents our results after making some modifications in our original machine based upon the outcome of our initial tests.

MATERIALS AND METHODS

Twenty-four women were studied on the tumor service at the Louisiana State University Medical Center in Shreveport, LA. (Table I) Informed consent was obtained from all patients prior to testing. The women that were tested had all been admitted to the hospital to undergo removal of a palpable breast lesion or a suspicious lesion discovered by mammography. Women with bilateral lesions, two lesions in the same breast, and previous breast biopsies were not excluded. Patients of all ages were included. All tests were performed by the same examiner.

The skin was first cleaned with alcohol. This was allowed to dry. The measuring electrode was placed on the skin directly over the lesion in the involved breast. Five other electrodes were used to map the breast. One electrode was placed on each nipple. One electrode was placed half-way between the nipple of the involved breast and the superior margin of the breast, the lateral margin of the breast, and the medial margin of the breast. In patients with bilateral breast masses, the electrodes were placed first on one breast and then on the other. In patients with two lesions in one breast only the electrode that was placed over the lesion being measured was changed. The electrodes were held in place with disposable adhesive collars; electrode paste was used to insure good electrical contact. Sixty seconds was allowed for the attainment of electrical equilibrium before a reading was recorded. The machine has circuits to change the ground reference point to one of three points. A reading was taken from each of these three points. Two complete sets of readings were obtained for each mass measured.

SEPs were measured with an electrometer having a high input impedance. Acceptance of the device was excellent, none of the patients complained of pain or discomfort during the measurements, and no complications associated with its use were seen. The test procedure was explained to the patients by comparing it to that used in obtaining a standard electrocardiogram.

RESULTS AND DISCUSSION

Twenty-four patients in this series had twenty-five masses tested and twenty-five breast biopsies. All patients tolerated the procedure well. Four patients were found to have fibrocystic disease at biopsy. The machine predicted two of these lesions would be cancer. Seven patients had fibrosis demonstrated by biopsy. The machine correctly identified four out of these seven lesions as being benign. Ten fibroadenomas were removed in the course of the study. The accuracy of the machine in predicting the benign nature of this group of lesions was only 50%. One patient patient's breast mass turned out to be a normal lymph node. The machine correctly catagorized this lesion. Three malignant lesions were encountered during the course of the study. All three were correctly characterized as being malignant by our device. One patient had an infiltrating duct carcinoma of the breast. The other patient had two granular cell tumors, one in each breast. These are very rare low grade malignant tumors.

Unfortunately for the purposes of the study we did not see more women with breast cancer. The machine does not appear to be useful in younger women with clinically benign lesions. Fibrosis in the breast seems to interfer with our ability to accurately characterize lesions. I think the 50% rate of correct identification of patients with fibrocystic disease is encouraging. This is the group of patients that really are difficult to manage. The machine clearly detected all patients that presented with cancer. We did not have a false negative test in the entire series. Since the only false readings obtained were false positive ones we feel the machine deserves additional testing and modification to improve the accuracy in patients with benign lesions. We are continuing our investigations in this area.

Table I

Patient	Age	#1 Reading	#2 Reading	Histology	Machine Reading	Right or wrong
BS	47	M-30.2 0-2.0 S-15.1	M-15.3 0-16.6 S-20.5	fibrocystic disease	+	W
ET	52	M+3.4 0+4.9 S-2.4	M+3.5 0+5.3 S-2.7	fibrocystic disease	-	R
LH	32	M-6.2 0+11.8 S+14.3	M-7.3 0+12.2 S+15.1	fibrocystic disease	-	R
RW	50	M-14.8 0-26.9 S-29.2	M-31.2 0-26.7 S-21.2	fibrocysitc disease	+	W
RB	35	M-12.0 0-3.8 S-3.3	M-12.5 0-4.5 S-4.0	fibrosis	+	W
PE	27	M+0.6 0-4.0 S-14.1	M+5.0 0-7.0 S-20.5	fibrosis	-	R
LT	23	M-2.0 0-26.8 S-19.0	M-8.7 0-28.7 S-23.5	fibrosis	+	W
JF	46	M-2.5 0-2.3 S+9.3	M-2.9 0-2.1 S+9.7	fibrosis	-	R
MC	58	M-5.6 0-1.9 S-1.2	M-6.5 0-2.3 S-2.8	fibrosis	+	W
IR	37	M+3.5 0-21.4 S-31.5	M+5.4 0-19.1 S-28.1	fibrosis	-	R
LW	46	M-4.8 0+8.4 S+3.6	M-3.9 0-2.4 S+2.5	fibrosis	-	R
BW	41	M+9.5 0+2.6 S+4.5	M+10.1 0+2.4 S+2.9	fibroadenoma	-	R
RA	16	M-1.5 0+3.6 S-4.0	M-0.9 0+4.3 S-4.5	fibroadenoma	-	R
MM	21	M+0.4 0-3.2 S+1.9	M-1.8 0-8.4 S+3.9	fibroadenoma	-	R

Table I cont.

Patient	Age	#1 Reading	#2 Reading	Histology	Machine Reading	Right or Wrong
CM	19	M-5.2 0-3.7 S-3.9	M-5.7 0-4.0 S-5.0	fibroadenoma	+	W
BD	22	M-1.3 0-29.2 S-46.3	M-0.9 0-17.1 S-44.2	fibroadenoma	+	W
JL	17	M+5.0 0-18.4 S-25.1	M-4.0 0-42.4 S-32.1	fibroadenoma	-	R
DJ	26	M-1.6 0-2.3 S+4.2	M-2.3 0-3.1 S+4.3	fibroadenoma	-	R
VW	21	M-18.1 0-35.8 S-53.3	M-25.9 0-40.1 S-41.6	fibroadenoma	+	W
KK	22	M-26.8 0.28.1 S-29.0	M-18.0 0-23.8 S-13.4	fibroadenoma	+	W
CW	21	M-20.4 0-24.5 S-22.0	M-19.5 0-22.5 S-17.4	fibroadenoma	+	W
BS	33	M-9.8 0-7.2 S+23.7	M-0.6 0-5.4 S+24.8	lymph node	-	R
SB	25	M-7.6 0-6.0 S-16.2	M-7.1 0-8.0 S-21.0	granular cell tumor	+	R
		M-5.5 0-13.4 S-23.2	M-2.0 0-19.1 S-26.0	grandular cell	+	R
AJ	72	M-8.6 0-1.6 S-0.4	M-9.1 0-0.6 S-2.0	Inf. ductal carcinoma	+	R

MACHINE READING + = cancer
- = not cancer

R- = machine correctly predicted status of the mass
W = machine incorrectly predicted status of the mass
M = mass
O = opposite nipple
S = same nipple

REFERENCES

Burr, H.S. and Langman, L.A.: A technique to aid in Detection of Malignancy of the Female Genital Tract, Am. Jo. Obs. Gyn. 57:274, 1949.
Committee on Aeronautical and Space Sciences Towards a Better Tommorrow U.S. Senate, 94th Congress, Second Session.
Schauble, M.K., and Habal, M.B.: Electropotentials of Surgical Specimens, Arch. Path., 90:411, 1970.
Schauble, M.K., Gullick, H.D., and Habal, M.B.: Variations in Tissue Electropotentials and their Possible Significance, Jo. Sur. Res., 12:325, 1972
Morris, D.M. and Hirschowitz B.: Electrical Monitoring of Breast Carcinoma. Jo. Bioelectricity (In Press).

CELL CONSTANT, ELECTRODE GEOMETRY AND CURRENT DENSITIES IN MEASURING TISSUE RESISTIVITY

J. Goodrich* and R. Rinaldi**

*Dept. of Chem., Louisiana St. Univ., Shreveport, Louisiana 71105
**Dept. of Anat., Louisiana St. Univ., School of Med., Shreveport, Louisiana 71130

ABSTRACT

This study was designed to determine the electrical conduction properties of bone in vivo. It was learned that erratic resistance measurements could be prevented by recognizing certain critical parameters inherent in measuring conduction using metal electrodes in tissues. It was found that voltages should be low (<0.7 volts) to reduce undesirable chemical reactions at the electrodes. Moreover, to circumvent capacitance, a frequency of above 1000 Hz was utilized. Furthermore, the geometry and size of the electrodes relative to one another is critical and requires that a cell constant be determined for each experiment. The specific resistance of bone was approximately 1000 ohm/cm^3 when the cell constant was determined for the electrode configurations used in this tissue.

KEYWORDS

Five point electrode method; cell constant; frequency effects; electrical double layer; phase angle.

INTRODUCTION

Electrical stimulation by electrodes implanted in animal and human tissue is widely used in clinical medicine. Heart pacemakers, treatment of various orthopedic conditions, brain stimulators and pain relievers are several examples. Ideal design of such electrical devices should be based on the correct electrical equivalent circuit model in the tissue stimulated. Electrical parameters of bone including resistivity are primary focuses of this work. Polarization of platinum electrodes embedded in both cortical and medullary bone was performed. A five point method was used where the outer electrodes introduce current (0.1 to 1 volt a.c. at selected frequencies from 0 to 5000 Hz) into the bones and the three inner electrodes monitor the current at different distances (1 cm and 2 cm). These spacings of 1 and 2 cm were used to observe the uniformity of resistance values obtained. The electrodes were embedded in bone cement (Surgical Simplex) with the electrodes insulated from fluids surrounding the bone with Neoprene "O" rings. The electrodes were platinum (0.5 mm diameter) and protruded from the bone cement 1.5 mm and 3 mm. Measurements were made with a high impedance voltmeter and a low current ammeter which utilized minimal amounts of energy in the measurements.

Polarization of the outer electrodes yielded resistance values that were erratic for different electrode configurations. Current vs. voltage curves yielded slopes corresponding to resistances between 170 and 1200 ohms. It was decided to construct a model cell with four electrodes to simulate the arrangement used *in vivo*. This cell was immersed in 0.01 M KCl solution of known specific conductance and electrical testing performed as above. Movement of the four electrodes with respect to one another in addition to their depth in solutions was investigated. The two inner non-polarized electrodes exhibited voltage changes (upon polarization by the outer electrodes) due to the total potential drop between their geometrical areas. When placed at 90° to the polarizing current, no voltage drop occurred as shown in Fig. 1, indicating an equipotential between them. Further movement of the electrodes in a variety of configurations led to the conclusion that resistance measurements depend on the size and orientation of all four electrodes. The current density pattern from the outer electrodes and the orientation of the inner electrodes established the resistance values observed.

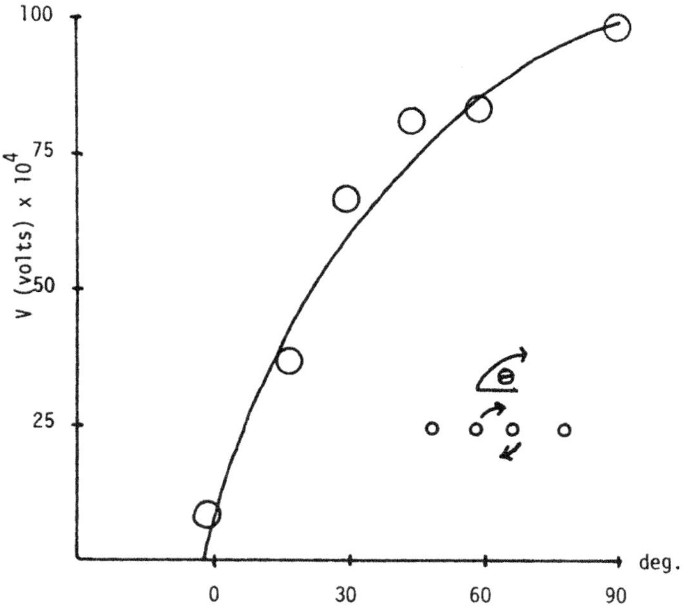

Fig. 1. Voltage dependence of electrodes on angular geometry.

The following procedure is the classical method for determining the specific conductance and cell constant of a specific electrode configuration in an electrolytic solution (Sawyer, 1974). A resistance measurement was obtained in an electrode system *in vivo* (rabbit) with a Wheatstone bridge (capacitance nulled). After removal of the electrode system intact, it was placed in a solution of known specific conductance (0.01 M KCl) and a final resistance value obtained. This procedure permitted calculation of the "cell constant" of the electrode system as well as the specific resistance (ohm/cm^3) of the tissue. The specific resistance of bone was determined to be approximately 1000 ohm/cm^3 and this value was found to be comparable for all measurements regardless of cell geometry.

Since 1000 Hz alternating current was used in obtaining specific resistance by bridge measurements, studies of frequency dependence of the system was deemed appropriate. Inner electrode voltages and current flow were measured from 0 to 5000 Hz at several polarization voltages (Fig. 2). Frequency dependence of both current and voltage below 1000 Hz as well as independence above this to 5000 Hz was observed. The dependence below 1000 Hz is attributed to capacitive impedance at the polarizing electrodes.

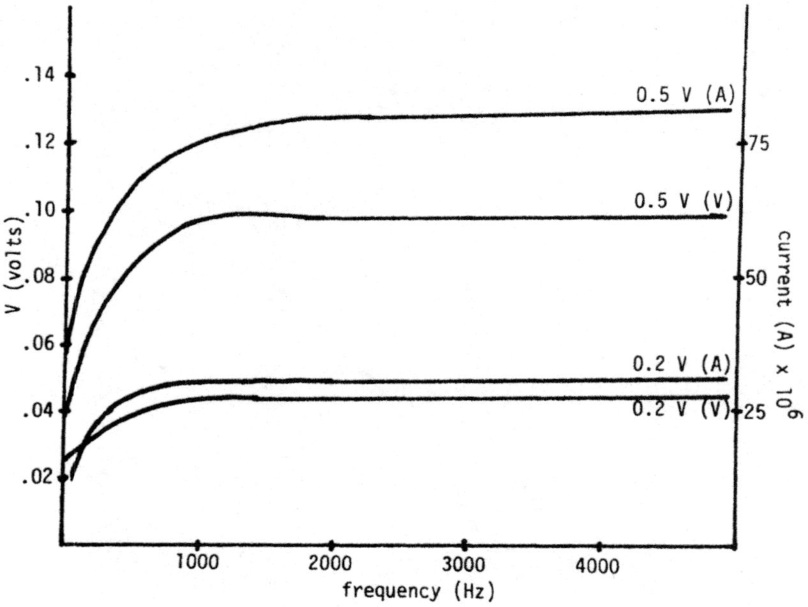

Fig. 2. Voltage and current dependence of inner electrodes on frequency at polarizing voltages of 0.2 and 0.5 V.

Problems are inherent when polarizing metal electrodes are used in fluids. Fig. 3 is a basic equivalent circuit of a conductance cell used for determining specific resistance. Z_1 and C_{DL1} refer to electrode 1 with R_S being the resistance to be measured. The faradaic impedance Z is high and of no concern if voltages are low

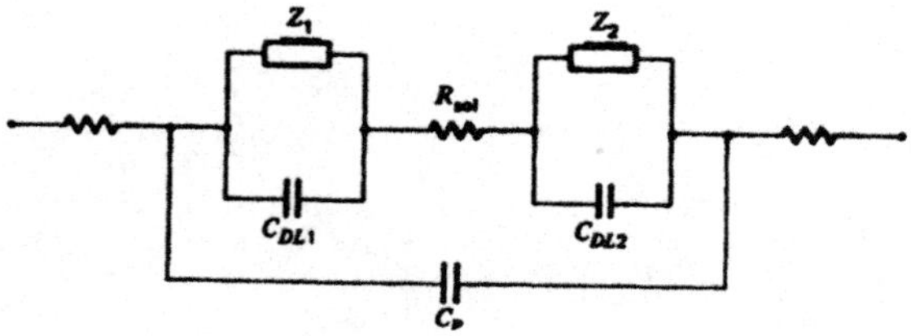

Fig. 3. Equivalent circuit for a conductivity cell.

and do not exceed E^0 of the metal electrodes that would produce undesirable chemical reaction. CdL1 represents the double layer capacitance and ideally should be large to prevent large capacitive impedance as determined by $Xc = 1/2\Pi fc$. Parallel capacitance of the tissue or solution, Cp, is negligible for solution resistances less than 10^4 ohms (Siskind, 1965).

Phase angle measurements at 1000 Hz with an oscilloscope determined the capacitive impedance and was compared to the specific resistance found with Wheatstone bridge measurements. The phase angle is less than 5 degrees which is 90 ohms capacitive reactance.

It is recommended that specific resistances be measured between 1000 and 5000 Hz at less than 0.7 V and a cell constant determined for each geometrical configuration to prevent capacitive, as well as other physical and chemical problems in determining the least possible electrical conductance parameters in stimulating tissues of this type.

REFERENCES

Sawyer, D.T., and J.L. Roberts, Jr. (1974). Experimental Electrochemistry for Chemists. J. Wiley and Sons, New York.

Siskind, C.S. (1965). Electrical Circuits, 2nd Ed., McGraw-Hill, New York.

EFFECT OF DIFFERENT ELECTROMAGNETIC PULSED STIMULATIONS ON THE SKELETAL GROWTH OF CHICK EMBRYO

S. Saha, A. Pal, G.N. Reddy and J.A. Albright
Biomechanics Laboratory
Department of Orthopaedic Surgery
L. S. U. Medical Center
Shreveport, Louisiana 71130

ABSTRACT

We studied the effect of pulsing electromagnetic fields of different pulse shapes and magnitudes on the skeletal as well as overall growth of chick embryos. We noted both accelerating and inhibitory effects of pulses of varying parameters. Our study indicates that embryonic growth can be modified by electromagnetic stimulation.

KEYWORDS

Electromagnetic Stimulation; chick embryo growth; skeletal growth; PEMF

INTRODUCTION

Successful clinical use of pulsing electromagnetic fields in the treatment of non-unions and pseudarthroses indicates a definite positive effect. This positive effect, however, cannot be quantified without a suitable control group. Also, little is known about the mechanisms involved in the influence of electrical stimulation on osteogenesis. We, therefore, used incubating fertilized chicken eggs in our evaluation of the effect of pulsing electromagnetic fields on bone growth. This model system was chosen due to its low cost and ease of operation, compared to an animal model. Our goal was to obtain a dose-response curve for pulsed electromagnetic fields based on the criterion of skeletal growth of chick embryo.

MATERIALS AND METHODS

Three tunable pulse generators were designed and fabricated. Pulsing electromagnetic field (PEMF) of several desired shapes, amplitudes, and repetition frequencies could be produced with these pulsers. These pulse generators were used to supply current to a rectangular coil (39 cm x 31 cm, 90 turns of 24 AWG coated copper wire). This coil was used for generating electromagnetic field inside an incubator where fertilized eggs were kept in trays. The magnetic field at the location of each egg was measured by a sensitive probe (Magnetek) which was calibrated at 60 Hz. The magnetic field was maximum close to the coils, near the edge of the tray and minimum at the center.

The fertilized chicken eggs (Indian River, high breed quality) were collected from a local hatchery in five different batches, each batch containing 240 eggs. Each time 240 eggs were divided equally into test and control grups and incubated in two different incubators (Lehay Manufacturing Co., Model 416E; and National Appliance) kept in adjacent rooms. The temperature was maintained approximately constant at 37°C and humidity at 80 to 85% in most of the experiments. Due to minor variations in the temperature and humidity of the two incubators, experiments with some pulse shapes were repeated by interchanging the test and control groups between the incubators.

The initial weights of the eggs were determined. The test group was then exposed to a particular pulsing electromagnetic field as described earlier and the control group was simutaneously incubated in a separate incubator. The eggs were placed approximately 30° inclined to vertical with their blunt face on top and for most groups, they were turned back and forth twice daily. Several embryos from both groups were examined on different days starting from the 7th day of incubation, until the day of hatching i.e. 21st day. After being removed from the shells, the embryos were washed with distilled water and their wet weights and lengths were measured. They were kept soaked in individual containers, first in 95% ethyl alcohol for 48 hours and then in potassium hydroxide solution for another 24 hours. Subsequently their skeleton was stained in Alizarin red-S for 24 hours. The stained tibiae and femora of the chicks were dissected under a microscope removing all soft tissues. Lengths of each tibia and femur were then measured within an accuracy of 0.5 mm using a large magnifying glass. The bones were then dried at 80°C for 24 hours in a vacuum oven (Fisher Isotemp, model 281) and their bending strength determined using a special rig (Saha and others, 1982).

The dried bone samples were then heated at 580°C for 24 hours using a muffle furnace and their percentage of ash content was determined.

RESULTS

Altogether five batches of eggs were examined, each batch being subdivided equally between a test and a control group. Each test group was exposed to an electromagnetic field of a particular pulse shape. The details of various types of pulses used in this study are shown in Fig. 1. The pulse no. 5 in fact consisted of a train of repetitive burst of pulses.

Embryos exposed to an electromagnetic stimulation with a pulse type No. 1 (Fig. 1) with a repetition rate of 200/sec., produced an increased rate of growth compared to the control group. This result was verified by repeating the experiment, interchanging the test and control groups between the incubators. Detailed results of this group have been reported before (Saha and co-workers, 1981, 1982 a,b) and thus will not be repeated here.

When fertilized eggs were exposed to pulsed electromagnetic fields as shown in No. 2, 4 and 5 (Fig. 1), the rate of growth of the chick embryos were less when compared to their individual control groups. An example of this decreased embryo weight for group B is shown in Figure 2. The long bones from these test groups also showed decreased length, weight and mechanical strength compared to the controls. The test group of embryos exposed to the pulse shape as shown in No. 3 (Fig. 1), showed mixed results with increased growth on some days and decreased growth on other days.

Our result clearly shows that the rate of embryo growth is affected by electromagnetic stimulation. However, before the effect of indidivual pulse shapes can be interpreted for clinical relevance, several factors should be considered. We have already pointed out that the magnitude of the stimulating field varied in each group depending on the location of the egg. Pulsed electromagnetic stimulation when used clinically, is generally applied only for 8 to 12 hours per day (Bassett and co-workers, 1977, 1981). In this study, the embryos were exposed to a continuous PEMF. We are continuing our study to examine the effect of PEMF when the fertilized eggs are stimulated by a PEMF only for a short time during their incubation period.

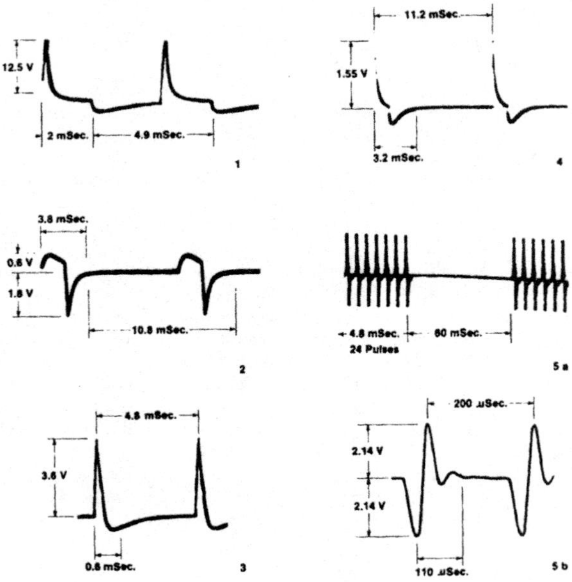

Fig. 1. Induced pulse shapes as recorded at the center of the tray containing the fertilized eggs.

ACKNOWLEDGEMENT

This research was supported by a USPHS Research Grant No. 5 R01 26020 and funds from Shriners Hospital for Crippled Children and by a RCDA Award No. 5 K04 AM 00756 from USPHS to Dr. Saha.

REFERENCES

Bassett, C.A.L., A. A. Pilla, and R. J. Pawluk (1977). Clinical Orthopaedics, 124: 128-143.

Bassett, C.A.L., S. N. Mitchell, and S. R. Gaston (1981). J. Bone and Joint Surg., 63-A: 511-523.

Saha, S., A. Pal, J. A. Albright, and R. P. Misra (1981). Trans. 1st Ann. Mtg. Bioelec. Repair and Growth Soc., 1: 76.

Saha, S., A. Pal, J. A. Albright, and R. P. Misra (1982a). Trans. 28th Ann. Mtg. of Orthop. Res. Soc.

Saha, S., A. Pal, S. Pal and J. A. Albright (1982b). Proc. 1st Southern Biomedical Eng. Conf. (ed. by S. Saha), Pergamon Press.

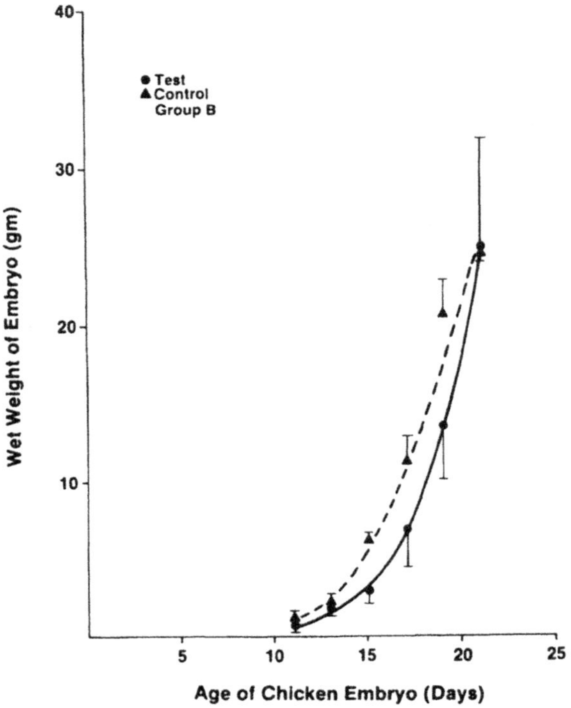

Fig. 2. Comparison of the wet weights of chick embryos from the test and control groups. The test group (B) was stimulated by pulse shape No. 2 as shown in Fig. 1.

ELECTROGENIC ION TRANSPORT AND BIOELECTRIC EXCITATION

J. Astbury* and B. Lowenhaupt**

*Dept. of Electrical Engineering, Univ. of Pittsburgh, PA

**Dept. of Biology, Edinboro State College, Edinboro, PA

***Authors appear in alphabetical order.

ABSTRACT

Equations of a new quantitative theory of bioelectric excitation are presented. Aspects of the theory are discussed, especially a positive feedback mechanism essential to an all-or-none action potential. In addition, relevance of the theory to cellular behavior is discussed, specifically growth and repair.

KEYWORDS

Bioelectricity; electrogenic ion transport; action potential; electrophysiology; bioelectrochemistry; nerve impulse; cellular growth and repair.

INTRODUCTION

A new theory of bioelectric excitation (Lowenhaupt, in preparation) is quantitatively consistent with electrophysiological observations. It is termed electrogenic because its focus is electrogenic and biochemical in contrast to the ionic, inorganic interpretation of the membrane theory. Differential equations relevant to this theory express metabolite concentrations and reaction densities in the cell membrane. Integration of these equations gives the metabolite concentrations, electron and ion currents, and the transmembrane potential in the steady and non-steady states.

There are two sets of differential equations, called the first generation and second generation equations. Starting with a primitive model of interacting biochemical reactions in two reaction sequences, the first generation equations were developed on the computer by trial and error, and simplified by using a thermodynamic assumption. While the first generation equations predicted many electrophysiological data, the thermodynamic assumption used to symplify the equations was found to be inappropriate over part of the range of an action potential. Then the assumption was dismissed, and the more sophisticated second generation equations were developed. These equations are able to predict results over the full range of an action potential.

THE FIRST GENERATION EQUATIONS

$$d(\beta)/dt = J_{GF}(\gamma)(\chi)^2\exp(Fs_p/RT) - J_{AG}(\beta) + J_{GA}$$

$$d(\gamma)/dt = J_{FE}|G - (\gamma)| - J_{GF}(\gamma)(\chi)^2\exp(Fs_p/RT)$$

$$d(\chi)/dt = J_{BC}(\beta) - J_{AB}(\chi)$$

$$p_{xm} = (RT/F)\ln(\chi) + s_p$$

THE SECOND GENERATION EQUATIONS

$$d(\beta)/dt = \{J_{GF}(\gamma) - J_{FG}(\beta)|G-(\gamma)|\}$$
$$\{(H\Psi)\exp(Fp_{xm}/RT)/|1+J_s(H\Psi)\exp(Fp_{xm}/RT|\} - J_{AG}(\beta) + J_{GA}$$
$$d(\gamma)/dt = J_{FE}|G-(\gamma)| - J_{EF}(\gamma) - \{J_{GF}(\gamma) - J_{FG}(\beta)|G-(\gamma)|\}$$
$$\{(H\Psi)\exp(Fp_{xm}/RT)/|1+J_s(H\Psi)\exp(Fp_{xm}/RT|\}$$
$$d(H\Psi)/dt = (\beta)\{J_{BC} - J_{CB}(H\Psi)^{(1+\vec{A}_{BCi})}\exp(\vec{A}_{BCi}Fp_{xm}/RT)\} - J_{CD}(H\Psi)$$
$$d(H\omega)/dt = J_{CD}(H\Psi) - J_{DE}(H\omega) + J_{ED}(H\omega)^{-(\vec{A}_{DEo})}\exp(-\vec{A}_{DEo}Fp_{xm}/RT)$$
$$s_i = -F\{(\beta)|J_{BC} - J_{CB}(H\Psi)^{(1+\vec{A}_{BCi})}\exp(\vec{A}_{BCi}Fp_{xm}/RT)|$$
$$-J_{DE}(H\omega) + J_{ED}(H\omega)^{-(\vec{A}_{DEo})}\exp(-\vec{A}_{DEo}Fp_{xm}/RT)\}$$
$$+(p_{xm} - p'_{K})|D + E_K(\beta)| + 0.6811(p_{xm} - p'_{Na})E_K(\beta)$$

Capitalized Roman letters represent constants (except H, which stands for hydrogen); s_p and s_i represent the stimulus potential and stimulus current, respectively; Greek letters in parentheses represent organic metabolite concentrations; p_{xm} represents the membrane potential. (The second generation has five equations; five variables, all of which are functions of time and s_i).

IMPORTANT ASPECTS OF THE THEORY

The electrogenic theory parallels Mitchell's chemiosmotic theory. Mitchell speaks of vectorial metabolism to refer to biochemical reactions which take place in such a way that material moves through a membrane. If this material is electrically charged, such as electrons, then this constitutes an electric current through the membrane. Mitchell's theory applies to ATP synthesis in the inner mitochondrial membrane, whereas our theory applies to bioelectric events in the cell membrane.

Figure 1 shows a U-shaped pathway of biochemical reactions in the cell membrane similar to the U-shaped pathways of Mitchell's chemiosmotic theory. Notice that

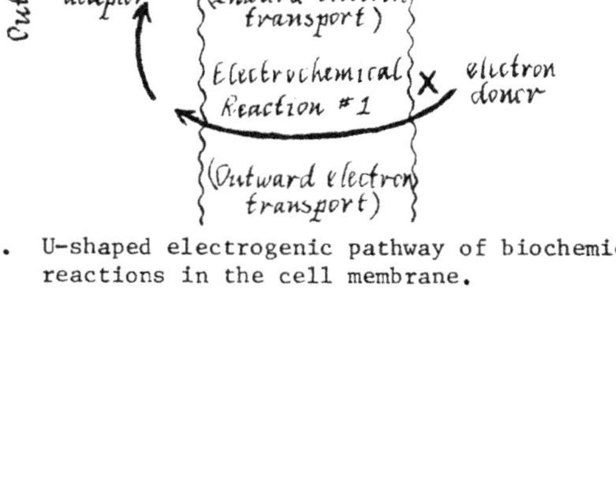

Fig. 1. U-shaped electrogenic pathway of biochemical reactions in the cell membrane.

electrochemical reaction 1 transports charge one direction through the membrane, and electrochemical reaction 2 transports charge in the opposite direction. Both electrochemical reactions are links in the same biochemical sequence, and therefore in steady state the two reactions must proceed equivalently.

The resting state is considered as the steady state, in which an electrochemical reaction step at X is rate-limiting for the entire sequence. However, non-steady state is also possible, in which the reaction at X is not rate-limiting, and the concentrations of reaction intermediaries of this sequence are changing. In this case, the two electrochemical reactions do not proceed equivalently.

Consider the status of the resting, steady state membrane, when the reaction at X rate-limits the entire sequence. Under this circumstance electrochemical reaction two proceeds almost reversibly, in other words, at quasi-equilibrium. This means that the transmembrane potential changes when the reaction at X is no longer rate-limiting.

The theory suggests that the enzyme for the reaction at X is turned off or on reversibly when calcium is attached to or released from a carrier molecule, respectively. We call this molecule beta. There is feedback between the calcium reactions and the electrochemical reactions because the electrochemical reactions generate an electric field which modulates the calcium reactions, and in turn, the calcium reactions modulate the electrochemical reactions. To clarify this point, let us present Fig. 2:

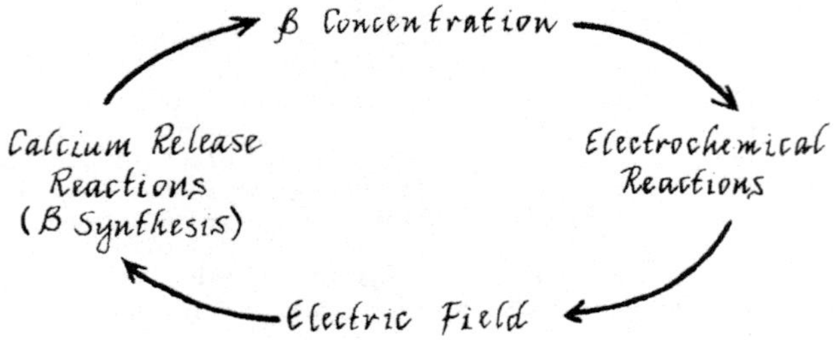

Fig. 2. Positive feedback mechanism of the electrogenic theory.

Notice that an electric stimulus can increase the electric field which would increase the rate of release of calcium from calcium-beta, and therefore would increase the concentration of beta without calcium. This in turn increases the density of the electrochemical reactions, which alters the electric field further, and so on. This provides rationale for the positive feedback of an all-or-none action potential, which the experimental data illustrates. However, the excited state does not persist, because the calcium release reaction uses up another metabolite, which we call gamma. As the gamma concentration becomes rate-limiting, the calcium re-attachment reaction begins to catch up with the release reaction. As a result, beta concentration declines, and the density of electrochemical reaction one decreases. And so the resting membrane potential is restored.

EXPERIMENTAL EVIDENCE

The equations describe this scheme quantitatively, and computer simulations generated by means of these equations agree with diverse electrophysiological observations. There is flexibility in the interpretations, therefore we cannot determine how many of our interpretations will indure further study. Nevertheless, the agreement of the equations with the facts of electrophysiology should be evident.

Computer simulations are quantitative for: resting and action potentials, rheobase and the all-or-none behavior of bioelectric excitation, summation and other subthreshold responses, the strength-duration relation, accommodation, refractory and partially refractory periods, voltage clamp currents, membrane rectification, calcium and drug effects. The action potential and the voltage clamp currents are shown in Figures 3 and 4, respectively.

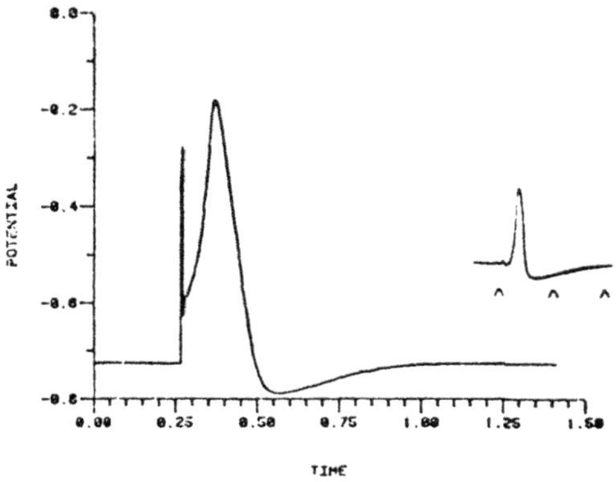

Fig. 3. Action Potential

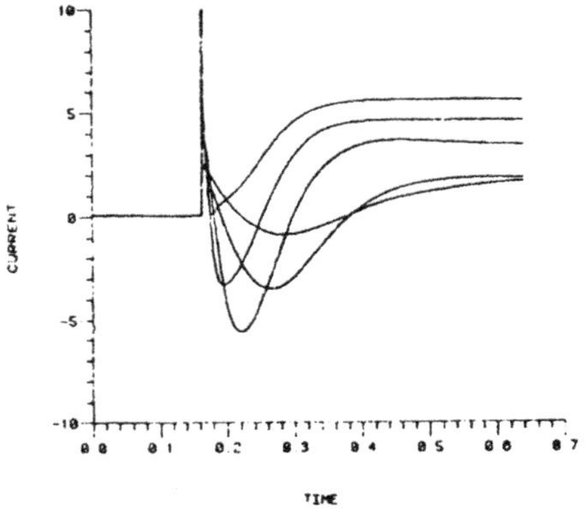

Fig. 4. Voltage Clamp Currents

CELLULAR GROWTH AND REPAIR

The electrogenic concept is organic and biochemical, just as a molecular biologist mist expect to regulate such cellular behavior. Consider the electric field at X in Fig. 2. This is a highly intense field across an interphase within a region of molecular dimensions, just such an electric field that would be necessary to modulate a biochemical reaction. If this biochemical reaction modulates growth and repair, we have a rationale for the role electricity plays therein.

The biochemical details of growth and repair have yet to be studied. We hope however, that this theory may bring forth new insights into this and other areas of cellular behavior.

ACKNOWLEDGMENT

We are grateful to the Hybrid Computer Laboratory at Pennsylvania State University where these figures were obtained. Also thanks to Marica Kapusta and Judy Jones.

REFERENCES

Cole, K.S. 1968. Membranes, ions and impulses. University of California Press, Berkeley.

B. Lowenhaupt, Bioelectricity and Cellular Behavior, Unpublished: Readers can obtain a copy of the typescript by sending $25.00 U.S. or the equivalent in other currency to the author.

Mitchell, P. 1966. Chemiosmotic coupling in oxidative and photosynthetic phosphorylation. Glynn Research, Ltd., Bodmin, Cornwall, England.

A DIFFERENTIAL METHOD FOR MEASURING ELECTRICAL CHARACTERISTICS OF BONE

G. N. Reddy and Subrata Saha
Biomechanics Laboratory
Department of Orthopaedic Surgery
LSU Medical Center
Shreveport, LA 71130

ABSTRACT

A differential method for measuring electrical impedance properties of bone is described. This method avoids stray capacitance effects and thus, measures true electrical characteristics of bone.

KEYWORDS

Electrical properties of bone, differential method, stray capacitance, impedance, capacitance, resistance of bone.

INTRODUCTION

Although orthopaedic surgeons are increasingly using electrical stimulation to treat non-unions and congenital pseudarthrosis (Brighton and others, 1979; Bassett, 1981), the mechanism of osteogenesis by bioelectricity is still unknown. For an analysis of the distribution of direct or induced current in bone, we need accurate data on the electrical properties of bone. Previously, we and others have attempted to measure the electrical impedance characteristics of bone (Durand and others, 1978; Saha and others, 1982). However, this technique did not allow separate measurements of the resistance and capacitance components of bone impedance. This paper describes a differential method for accurate determination of the specific resistance and capacitance properties of bone, avoiding errors due to cable and wiring capacitance.

MATERIALS AND METHODS

The block diagram for the measurement of bone impedance is shown in Fig. 1. The signal generator (Wavetek Model 180) was used to generate required variable frequency signals. The frequency counter (HP Model 5233L) was used for accurately measuring the frequency applied to a test specimen, thus avoiding any calibration errors on the signal generator. We observed calibration errors up to 2 to 3%. A gain-phase meter (HP Model 3575A) was used to measure 1) the ratio of voltage across the reference circuit (V_o) to input voltage (V_{in}) in db and 2) the phase (ϕ) of the output voltage.

Direct Method

In this method, the reference circuit was removed and a bone sample was introduced. As shown in Fig. 2, without the reference circuit, the bone impedance ($Z_s\angle\theta$) was computed as follows (with input voltage (V_{in}) as reference):

$$\frac{V_o \angle\phi}{V_{in}} = \frac{Z_s\angle\theta}{R_{sr} + Z_s\angle\theta} \qquad (1)$$

$$\therefore \quad Z_s\angle\theta = \frac{\left|\frac{V_o}{V_{in}}\right| \cdot \angle\phi}{1 - \left|\frac{V_o}{V_{in}}\right| \cdot \angle\phi} \qquad (2)$$

In equation (2), the ratio of V_o/V_{in} and phase angle (ϕ) are measured with the gain-phase meter. The ratio of V_o/V_{in} is, however, measured in db. Let,

$$x = \text{amphitude ratio of } V_o/V_{in} \text{ in db} = 20 \log_{10} V_o/V_{in} \qquad (3)$$

$$\therefore \quad \frac{V_o}{V_{in}} = \text{Antilog } (x/20) = A \qquad (4)$$

From equation (2), the bone impedance is then given by

$$Z_s\angle\theta = \frac{R_{sr} \cdot A \angle\phi}{1 - A \angle\phi} \qquad (5)$$

TABLE 1 Measurement of Bone Impedance by the Direct Method.

Freq. f (kHz)	Amp. Ratio x (db)	Phase Angle ϕ (deg.)	A = Antilog (x/20)	$Z_s\angle\phi = \frac{R_{sr}A\angle\phi}{(1 - A\angle\phi)}$ (kΩ)	$R_s = \frac{\lvert Z_s \rvert}{\cos\phi}$ (kΩ)	$C_s = \frac{-\sin\phi\theta}{\omega \lvert Z_s \rvert}$ (pf)
100	- 1.1	- 14.6	0.8810	32.22$\angle$-71.01°	99.32	46.52

Table 1 illustrates the calculation of bone impedance with the direct method. The impedance values are for axial direction for a compact bovine bone.

Differential Method

The electrical equivalent circuit for the differential method is shown in Fig. 2. Here a reference reading (the amphitude ratio of V_o/V_{in} in db (x) and output voltage phase angle (ϕ) are noted with a gain-phase meter to estimate the stray capacitance. Such reference reading is necessary at each frequency of interest. From the reference reading (x, ϕ), the reference impedance ($Z_r\angle\theta 1$) is calculated as

$$Z_r\angle\theta 1 = \frac{R_{sr} \cdot A\angle\phi}{1 - A\angle\phi} \tag{6}$$

where A = Antilog (x/20).

Once the reference impedance is calculated, the circuit's stray capacitance (C_x) is calculated as

$$C_x = C_r' - C_r$$

$$C_x = -\frac{\sin\ \theta 1}{\omega \cdot Z_r} - C_r \tag{7}$$

On computing $Z_r\angle\theta 1$, a second reading (x', ϕ') is noted, with bone sample across the reference circuit, from the gain-phase meter. The combined parallel impedance is then given by

$$Z'\angle\theta 2 = \frac{R_{sr} \cdot A'\angle\phi'}{(1 - A'\angle\phi')} \tag{8}$$

where A' = Antilog (x'/20).

From Equations (6) and (8), the sample impedance $Z_s\angle\theta$ is computed.

$$Z_s\angle\theta = \frac{Z_r\angle\theta 1 \cdot Z'\angle\theta 2}{(Z_r\angle\theta 1 - Z'\angle\theta 2} \tag{9}$$

The sample resistive (R_s) and capacitive (C_s) for a parallel RC circuit are then given by

$$R_s = \frac{|Z_s|}{\text{Cos}\ \theta} \quad \text{and} \quad C_s = \frac{-\text{Sin}\ \theta}{\omega|Z_s|} \tag{10}$$

Table 2(a) and Table 2(b) combined together illustrate the bone sample impedance at 100 kHz.

TABLE 2a Differential Method Using Computational Procedure 1

	Amp. Ratio x (db)	Phase Angle ϕ (deg.)	A = Antilog (x/20)	$Z_r\angle\theta 1 = \frac{R_{sr}A\angle\phi}{(1 - A\angle\phi)}$ (kΩ)	$Z'\angle\theta 2 = \frac{R_{sr}A'\angle\phi'}{1 - A\angle\phi'}$ (kΩ)
Without Sample	- 3.5	- 16.8	0.6683	15.99∠-45.00°	-
With Sample	- 4.0	- 16.3	0.6309	-	14.27∠-40.48°

TABLE 2b

$Z_s\angle = \dfrac{Z_R\angle\theta 1 \cdot Z'\angle\theta 2}{Z_R\angle\theta 1 - Z'\angle\theta 2}$ (kΩ)	$R_s = \dfrac{Z_s}{\cos\theta}$ (kΩ)	$C_s = \dfrac{-\sin\theta}{2\pi f Z_s}$ (pf)
109.02/-7.94°	110.07	2.01

Specific Impedance

The specific impedance is the impedance computed based on resistance per unit volume (cm^3) and capacitance per unit volume (cm^3). From the known geometry of the sample using its resistance and capacitance, Table 3 presents the procedure for obtaining the specific impedance. The values of R_s and C_s are taken from Table 2(b).

TABLE 3 Computation of Specific Impedance

Sample Resist. R_s (kΩ)	Sample Capac. C_s (pf)	Specific Resist. $R_{sp} = R(A_a/L_a)$ (kΩ)	Specific Capac. $C_{sp} = C(L_a/A_a)$ (pf)	Specific React. $X_{sp} = 1/\omega C_{sp}$ (kΩ)	Specific Imped. $Z_{sp} = \dfrac{R_{sp}X_{sp}}{(R_{sp}^2 + X_{sp}^2)^{1/2}}$ (kΩ)	Phase Angle $\theta_{sp} = \tan^{-1} - \dfrac{R_{sp}}{X_{sp}}$ (degree)	$Z_{sp}\angle\theta_{sp}$ (kΩ)
110.07	2.01	15.84	13.97	113.8	15.69	- 7.92	15.69 /-7.92°

$A_a = 1.1 \times 0.59 = 0.649\ cm^2$ $L_a = 4.51$ cms. $f = 100$ kHz

RESULTS

The results of measurements made with known impedance circuits, with the Direct and the Differential Methods, indicated that the impedance measurements were significantly influenced by the wiring and cable capacitance. The effect of stray capacitance was particularly pronounced in low capacitance (order of few pico-farards) circuits. The measurments made on several bone samples with the digital-LCR meter (HP Model 4262-A) show that the bone has capacitance in the order of ten to few hundred pico-farards depending on the sample size. Thus, due to the stray capacitance considerable errors are introduced if the bone impedance is measured by the Direct Method. The Differential Method presented here will minimize such errors. Figure 3 shows the differences in the impedance measured with the Direct and the Differential Method. The impedance measured with the Differential Method shows a much smaller decrease in impedance with frequency than with the Direct Method. In the case of the Direct Method, the effects of stray capacitance is responsible for rapid decrease in impedance. The comparatively smaller decrease in impedance with frequency using the Differential Method is due to two factors: 1) the stray capacitance effects are minimized and, 2) the frequency dependence of the sample capacitance (C_s) itself. Figure 4 shows the frequency dependence of a bone sample capacitance.

CONCLUSION

Our results indicate that by using the Differential technique, one can minimize the error due to stray capacitance, while measuring the dynamic impedance properties of bone.

ACKNOWLEDGEMENT

This research was supported by a USPHS research grant No. 26020.

REFERENCES

Brighton, C. T., J. Black, and S. Pollock (1979). Electrical Properties of Bone and Cartilage. Grune and Stratton.

Bassett, C. A. L., S.N. Mitchell, and S.R. Gaston (1981). J. Bone and Joint Surg., 63-A:511-523.

Durand, B., P. Christel, S. Stan, and T.C. Mulier (1978). Electric Stimulation of Bone Growth and Repair. Springer-Verlag, pp. 19-24.

Saha, S., M.V. Kamath, and J.A. Albright (1982). Trans. 1st Ann. BRAGS Meet., pp. 25.

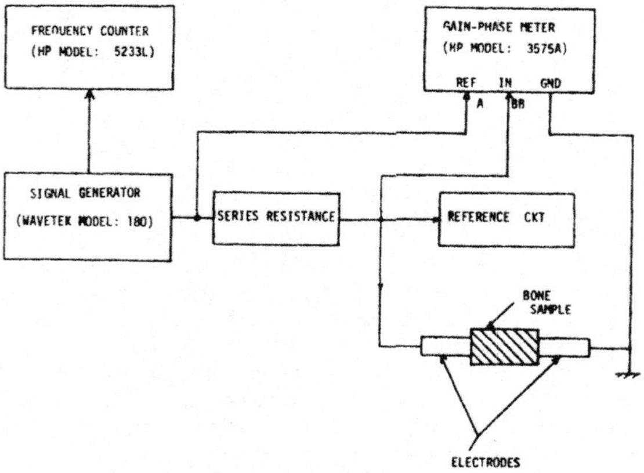

Figure 1: The block diagram for measurement of bone impedance by the differential method.

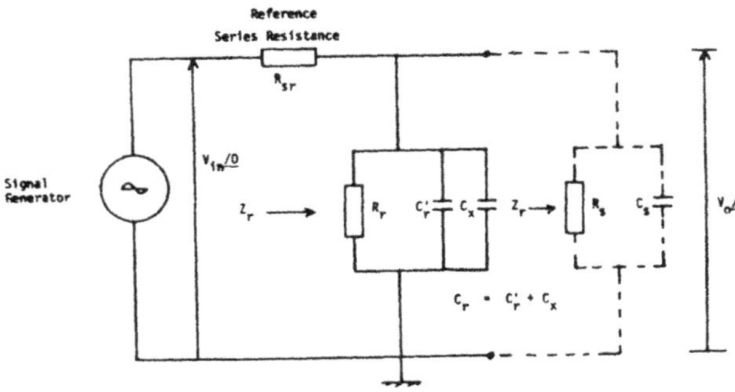

Figure 2: Electrical equivalent circuit for the differential method.

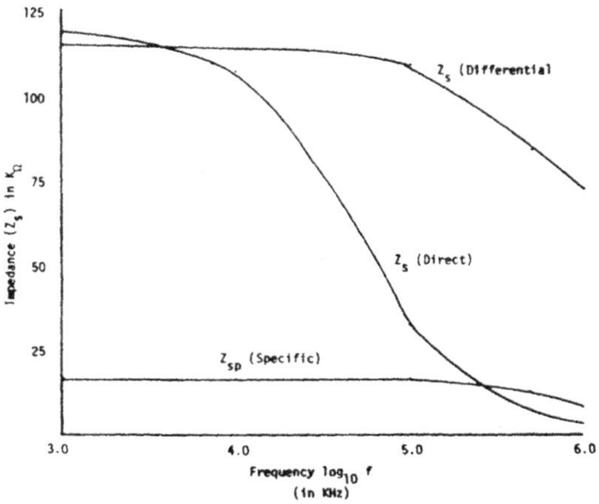

Figure 3: Impedance frequency dependence characteristic of a compact beef bone specimen (Axial Direction).

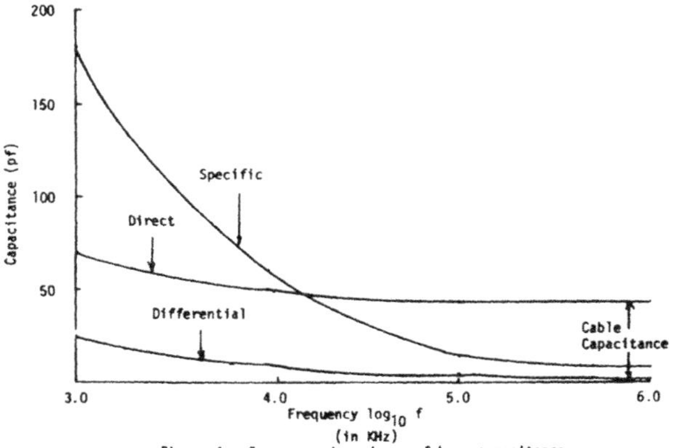

Figure 4: Frequency dependence of bone capacitance.

Session 7
ORTHOPAEDIC BIOMECHANICS

BIOMECHANICAL EVALUATION OF BONY DEFECTS REPAIRED WITH NORMAL, CARBON FIBER, AND WIRE REINFORCED BONE CEMENT

S. Saha, V. Saatdjian, S. Pal and J.A. Albright

Biomechanics Laboratory
Department of Orthopaedic Surgery
L. S. U. Medical Center
Shreveport, Louisiana 71130

ABSTRACT

Standardized defects were produced in-vitro in human and bovine long bones. These defects were repaired with normal, carbon fiber reinforced and wire reinforced bone cement. When tested mechanically, the wire reinforcement group sustained highest maximum load in bending.

KEYWORDS

Pathological fractures; bone cement; carbon fiber reinforced; bending strength; fracture energy.

INTRODUCTION

Patients suffering from bone metastases are often subject to pathological fractures. Intelligent treatment of such patients is hampered due to lack of any available information on how bony lesions affect the load carrying capacity of long bones. Therefore, one of our objectives was to determine the relationship between the nature and size of a bony defect and the corresponding reduction in the strength of a bone. Fixation of such fracture with a large part of the bony cortex missing is often a difficult and challenging procedure. This has led several authors to use bone cement as an adjunct in the fixation of pathological fractures (Ray et al, 1974 and Harrington et al, 1972, 1976). However, little information is available on the effectiveness or load bearing capacity of these procedures. Previously we have shown that wire and fiber reinforcement significantly improves the mechanical properties of bone cement (Taitsman and Saha, 1977; Saha and Warman, 1979; Pal and Saha, 1982). In this study we investigated the possible use of wire or fiber reinforced bone cement in improving the strength of such surgical constructs.

MATERIALS AND METHODS

Fresh bovine tibiae were collected from a local slaughter house and stored at $-20^{\circ}C$ until used since freezing does not affect the bone properties significantly. A rectangular osteotomy was performed in the mid-diaphysis of each tibia using a fine hand saw. To keep the amount of bone removed approximately the same, a rectangular template 75mm x 60mm was used for marking the outline on the curved surface of the bones. The lateral dimensions and cortical thickness of each tibia was then measured and recorded. The bone specimens were then divided into three groups so that their average dimensions (A-P & M-L) and cortical thickness in each group were approximately the same.

These defects in the first and second groups of bovine bone were repaired by normal bone cement and 1% (by weight) carbon fiber reinforced bone cement respectively. In the third group, 0.56mm diameter stainless steel wire (316L premium quality suture wire) was inserted through minute holes (1.5mm dia. and 10mm pitch) drilled along the periphery of the defect and it was covered with this interwoven network as shown in Fig. 1.

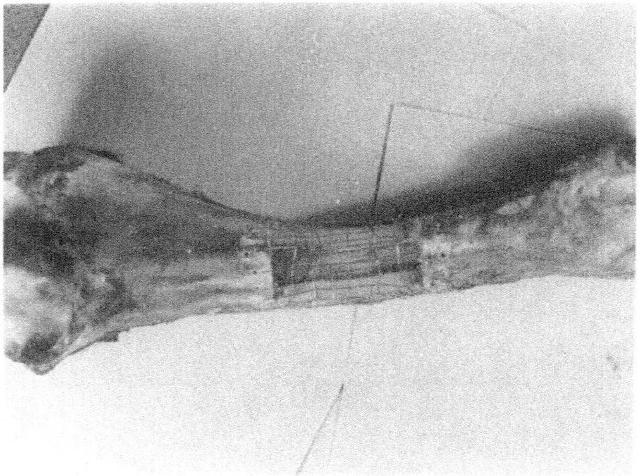

Figure 1. A bovine tibia with a rectangular cut area in it which is subsequently interwoven with stainless steel wire and finally filled with normal PMMA.

Subsequently the defect was filled with normal bone cement to the level of the original bone surface. For this study Surgical Simplex-P bone cement (Howmedica, Inc.) and for fiber reinforcement, chopped graphite fiber 7mm length and 8μm diameter (Hercules AS-type) were used.

Similarly a second study on the repair of bony defects by normal and carbon fiber reinforced bone cement was conducted using human humerus collected from adult embalmed cadavers. These bones were 29.7 to 31.5cm in length with a cortical thickness of 2 to 4.4mm. A diamond shaped area, 5cm x 2cm, was marked on the curved surface of the mid-diaphysis with the long axis along the length of the bone. A fine band saw was used to cut out the marked area and thus create a standardized defect. The marrow was removed from the vicinity of the defect and the bones were divided into three groups so that the approximate section modulus at the cut section was similar in each gruop. In one group, the defect was not repaired, and the two other groups were repaired with normal(N)and 1% by weight of carbon fiber reinforced bone cement respectively.

All specimens were tested under three point bending mode at a cross-head speed of 7.5mm/min, using an Instron servohydraulic mechanical testing machine (model 1320).

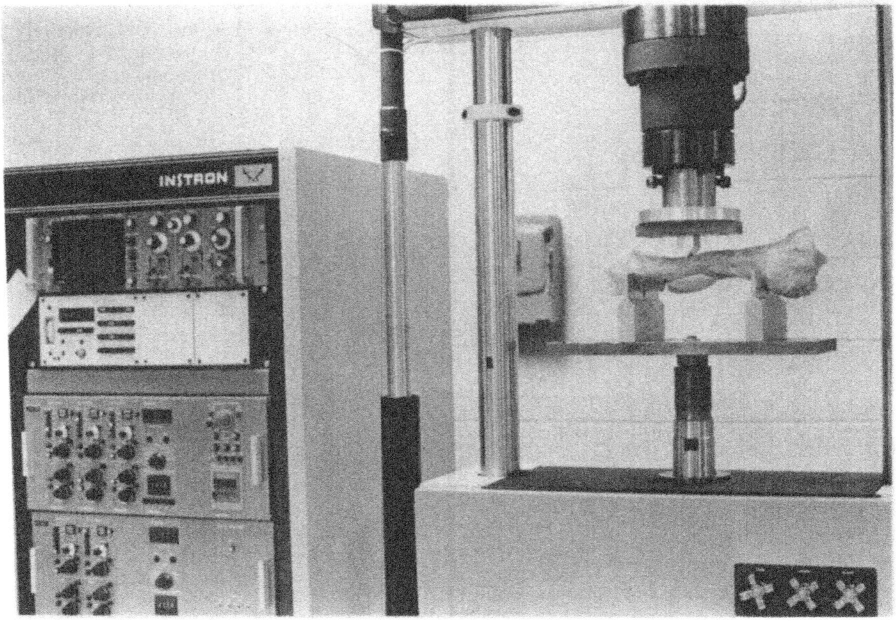

Figure 2. The three point bending test set up for testing the bovine tibia with the cut defect repaired with bone cement.

The bending load and deformation were automatically recorded in the built-in X-Y recorder of the Instron testing system. Those records were used to calculate the energy absorbed before the final fracture occurred.

RESULTS AND DISCUSSION

The results of the mechanical testing of the bovine tibia are shown in Table 1.

Table 1: Mean (1 standard deviation) values of the ultimate bending moment and energy absorbed at failure of bovine tibia with repaired bone defect. During testing the defect was placed on the tension side of the specimen.

Group	Type of Repair	Ult. Bending Mom., NM	Energy Absorbed, J
I	Normal PMMA	184. (41.4)	3.43 (1.28)
II	Carbon fiber - PMMA	199. (48.7)	
III	Stainless steel - PMMA	228 (69.8)	4.01 (2.58)

Table 1 shows that the wire-reinforcement improved the bending strength by 24% and the energy absorption also increased by 17% when compared to samples repaired with normal PMMA without any reinforcement. Carbon-fiber reinforcement did not improve the bending strength significantly. This is expected, because in group III, the wires allowed transfer of tensile load from the bone to the bone cement, making it a composite beam. On the other hand, for group I & II, there was no bond between the bone and PMMA and thus only the bone resisted most of the bending moment. However, if the cement mass was on the compression side, then the cement would resist compressive stress and thus the construct would be significantly stronger. This may be the case in many in-vivo situations. The test result of human long bones is tabulated below.

Table 2. Mean and (1 std. dev.) values of ultimate bending moment and maximum deflection at failure of human humerii repaired with normal and carbon fiber reinforced PMMA.

Type	Defect on Tension Side, Gr. I		Defect on Compr. Side, Gr. II	
of sample	Ult. bending moment, Nm	Max. Def. mm	Ult. bending moment, Nm	Max. Def. mm
Defect not repaired	29.96 (10.93)	3.78 (1.33)	--	--
Repaired with normal PMMA	31.72 (14.92)	3.29 (1.27)	48 (12.1)	5.11 (2.15)
Repaired with CF-PMMA (1 w/o)	33.08 (10.15)	3.89 (1.18)	51.16 (8.4)	4.94 (1.82)

It is evident from Table 2 that when the defect was placed on the tensile side, its repair with normal or carbon fiber reinforced bone cement did not improve the strength of the construct significantly. However, when the defect was on the compression side, the use of bone cement did improve the flexural strength considerably. In group I, the fracture initiated from bone in all the cases. In case of group II the fracture occurred in the bone cement when it was not reinforced. Fracture never initiated from CFR-PMMA mass in group III and it was bone which failed first, indicating the superior strength of CFR-PMMA in such applications.

REFERENCES

Harrington, K.D., Johnston, J.O., Turner, R.H., Green, D.L. (1972) J. Bone and Joint Surg., 54-A: 1665-1676.

Harrington, K.D., Sim, F.H., Enis, J.E., Johnston, J.O., Dick, H.M., Gristina, A.G., (1976) J. Bone and Joint Surg., 58-A, 1047-1055.

Pal, S. and Saha, S. (1982) Biomaterials, (in press).

Ray, A.K., Romine, J.S. and Pankovitch, A.M. (1974), Clin. Orthop., 202, 182-185.

Saha, S. and Warman, W. (1979) Orthop. Trans. 3, p. 169.

Taitsman, J.P. and Saha, S. (1977) J. Bone and Joint Surg., 59A (3), 419-425.

DEVELOPMENT OF MATHEMATICAL MODEL OF NEURO-MUSCULATURE SYSTEM IN THE FOREARM

Y. YOUM

DEPARTMENT OF MECHANICAL ENGINEERING
THE CATHOLIC UNIVERSITY OF AMERICA
WASHINGTON, D.C. 20064

ABSTRACT

This paper describes the development of a realistic mathematical model of the neuro-musculo skeletal system of the human forearm.

To facilitate an analytical model, ten cadevers were studied anayomically to define the points of origin and insertion of the elbow flexors and extensors and the kinematic equations are derived from the model. Dynamic behavior of skeletal muscle is described with a viscoelastic model consisting of parallel elastic element, damper and a passive elastic element.

The quantitative relationship between mechanical response and some appropriate measure of neural activation is obtained by integrated electromyogram. To determine the parameters involved in the analytical model, a series of experiments, isometric tension, static load, constant rotation and quick release experiment are performed and carefully studied.

KEYWORDS

Biomechanics; muscle model; neural control; kinematics of upper extremity.

INTRODUCTION

The understanding of the mechanical behavior of the skeletal muscle in a quantitative manner is of great interest not only from the fundamental viewpoint of anatomy and physiology but also it is of vital importance in many biomechanical and clinical applications such as design and evaluation of sophisticated modern limb prosthetic devices. Basic quantitative knowledge is equally important for designing in robot. The chain saw accidents were direct motive for this project to investigate human reflex mechanism in the upper extremeties. In the following development of an analytical model, Youm's anatomical and kinematic analyses of the human forearm and Soechting et al's. muscle and neural control model study on human skeletal muscle were applied.

SYSTEM MODEL OF THE FOREARM

In elbow flexion-extension, two musculoskeletal segments, forearm and upper arm, are involved. The forearm consists of the radius and ulna, whose main motions are supination-pronation between the radius and ulna and flexion-extension around the elbow joint which is formed by the proximal ulna and the distal humerus of the upper arm. The prime elbow flexor muscles are the brachialis, the biceps and the brachioradialis, and the prime extensor is the triceps. Even though the normal elbow in flexion has a 10-15 degrees of carrying angle, the elbow joint is basically a hinged type joint and thus only planar motion is considered.

To facilitate an analytical model, ten cadacers were carefully studied anatomically to obtain the points of origin and insertion of the muscles are chosen at approximately the mid point of the muscle-bone contacting area.

Knowing that the center of rotation in the elbow flexion-extension is located at the center of the trochlea Youm (1979) a ratio between the length of the humerus and the distance from the center of rotation to each mid point of muscle attachment is obtained from the average of ten cadaver studies (see Fig.1).

MUSCLE MODEL

Each muscle fiber contains several hundred to several thousand myofibilis. Each myofibilis in turn has, lying side by side, about 1500 myosin filaments and two times this many actin filaments which are responsible for muscle contraction. The muscle contraction occurs by a sliding filament mechanism. The sliding filament is caused by attractive forces that develop between the actin and myosin filaments. These forces are the result of mechanical interaction of the cross-bridges of the myosin filaments with the actin filaments.

To describe dynamic behavior of skeletal muscle a viscoelastic model consisting of parallel elastic element, neurally controlled contractile element and damper connected in series with a passive elastic element as shown in Fig.2 is examined. In the mathematical derivation, the following assumptions are made: (a) An individual muscle fiber has a rheological model as shown in Fig.2, (b) every muscle fiber in different muscle group has the same mechanical properties, and (c) therefore, the muscle forces of different muscle groups depend on the number of muscle fibers which are expressed as the cross sectional area of the muscle groups.

NEURAL CONTROL SYSTEM

The main muscle sensory receptors are the muscle spindles. They transmit impulses almost continuously through the posterio roots in the spinal cord, where they excite the anterior motor neurons, which in turn provide the necessary nerve stimuli for muscle. The spindle can be simulated in two different ways. First, by stretching the whole muscle, this lengthens the extrafusal fibers and therefore also stretches the spindle. Second, by contracting the intrafusal muscle fibers while the extrafusal fibers remain at the normal length. Therefore, in effect, the muscle spindle acts as a comparator of the lengths of the two types of muscle fibers. By applying single, equivalent control theory (Houk, 1967), the total effective control over the net moment produced at the elbow joint by contractile activity will be considered. Also introduced is the time-delay parameter. This is the transmission times between muscle stretches and contractile response.

EMG ACTIVITIES AND CONTRACTILE FORCE

The mechanical properties of muscle depend on its level of neural activation, therefore, it is quite desirable to find quantitative relationships between mechanical response and some appropriate measure of neural activation. The only readily available measure of neural activation is the electromyogram (EMG). It has been shown by several investigators that the resultant contractile force in the muscle is proportional to the integrated EMG. Experimentally, an averaged EMG signal from each pair of electrodes is amplified, full wave rectified and integrated. It is a function which increases monotonically, but sporadically, with time. Over appropriate time intervals, the average slope of this function can be measured and the slope is then taken as the average EMG amplitude over the interval in question.

GOVERNING EQUATIONS

The summary of the governing euations are listed as follows. In the following equations, subscript 1 stands for the biceps, 2 for the brachioradialis, 3 for the brachialis and 4 for the triceps.

A. System Equation

$$F_1a_1(\theta) + F_2a_2(\theta) + F_3a_3(\theta) - F_4a_4(\theta) - T\ell - W\ell_{c.g}\cos\theta = I\ddot{\theta}$$

B. Kinematic Equations

a. Muscle lengths

$$\ell_1 = (\ell/6)(37 + 12\cos\theta)^{1/2}$$
$$\ell_2 = (\ell/9)(53 + 28\cos\bar{\theta})^{1/2}$$
$$\ell_3 = (\ell/9)(10 + 6\cos\theta)^{1/2}$$
$$\ell_4 = \ell + 0.08\ell\theta \quad (\theta \text{ in radians})$$

b. Moment arms

$$a_1 = (\ell\sin\theta)/\ (37 + 12\cos\theta)^{1/2}$$
$$a_2 = (\ell\sin\theta)/9(53 + 28\cos\theta)^{1/2}$$
$$a_3 = (\ell\sin\theta)/3(10 + 6\cos\theta)^{1/2}$$
$$a_4 = 0.08\ell$$

c. Changes in muscle length

$$x_i = -a_i(\bar{\theta})(\theta - \bar{\theta}) \qquad i = 1,2,3$$
$$x_4 = a_4(\bar{\theta})(\theta - \bar{\theta})$$

where θ is the final forearm position and $\bar{\theta}$ is the initial position of the forearm.

d. Velocities of muscles

$$\dot{x}_i = -a_i(\theta)\dot{\theta} \qquad i = 1,2,3$$
$$\dot{x}_4 = 0.08\ell\dot{\theta}$$

C. Equations for Muscles

$$\dot{F}_i + A_iF_i = (K_{pi} + K_{si})x_i + A_iK_{pi}x_i + A_iP_i \qquad i = 1 \text{ to } 4;\ A = K_s/C$$

D. Contractile force and Integrated EMG

$$a_i(\theta)P_i = \kappa_i E_i \qquad i = 1 \text{ to } 4$$

where κ_i is the proportionality constants.

E. Neural Control System

$$A_1P_1a_1(\theta) + A_2P_2a_3(\theta) + A_3P_3a_3(\theta) - A_4P_4a_4(\theta) = -[\beta(\theta - \bar{\theta}) - \gamma(\dot{\theta} - \dot{\bar{\theta}})]$$

where β is control constant for change in length and γ is control constant for the rate of change in length.

To determine the parameters involved in the governing equations, a series of experiments, isometric tension, static load, constant rotation and quick release experiment are performed and studied.

ACKNOWLEDGEMENT

This work was supported by the U.S. National Bureau of Standard, Grant NB80NAHA1033.

REFERENCES

Bigland, B and Lippold, O.C.J. (1954) Jr. of Physiology, 123, 214-224.
Houk, J.C. and Henneman, E. Brain Research, 5, 433-451
Soechting, J.F., Stewart, P.A., Hawley, R.H., Paslay, P.R., and Duffy, J. (1971). Jr. of Dynamic Systems, Measurement, and Control, 93, 195-226.
Youm, Y and Flatt, A.E. (1979). Jr. of Biomechanics, 12, 245-255.
Zahalak, G.I. and Heyman, S.J. (1979). Jr. of Biomechanical Engineering, 101, 28-37.

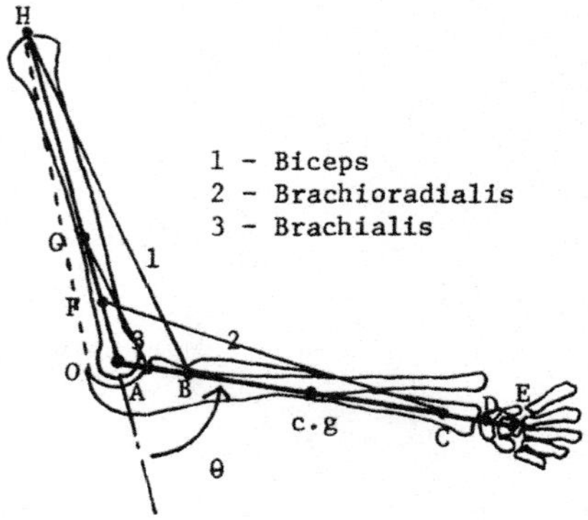

OH = OE = ℓ   c.g = 8ℓ/15
OA = ℓ/9   OF = 2ℓ/9
OB = ℓ/6   OG = ℓ/3
OC = 7ℓ/9
Weight of forearm = 0.03 W_{body}

Fig.1 System model of the forearm

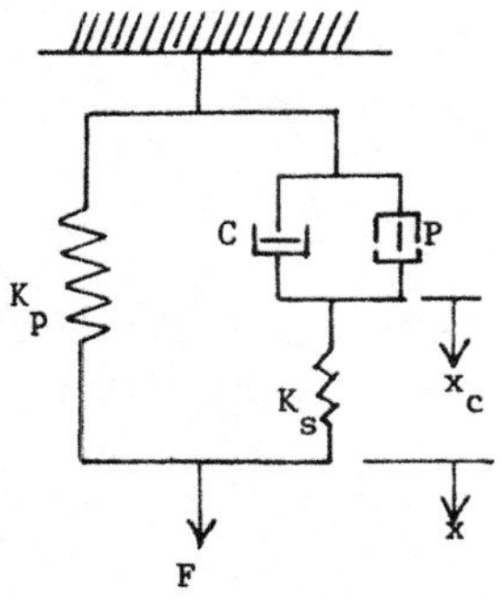

F : force in muscle
K_p : parallel spring constant
K_s : series spring constant
C : Damping coefficient
P : neurally controlled contractile force
x : displacement of muscle measured from the resting position
x_c : displacement of contractile element due to neural stimulation

Fig.2 Four element model of muscle

THE SCOPE OF FINITE ELEMENT MODELS IN PROVIDING ANSWERS TO CLINICAL ORTHOPAEDIC PROBLEMS

Sarma S Vangala , Dhanjoo N Ghista , R U Repo

Departments of Surgery, Medicine and Biomedical Engineering, McMaster University and Chedoke-McMaster Rehabilitation Centre, Box 2000, Station A, Hamilton. Ontario. L8N 3Z5 . CANADA .

ABSTRACT

A review of some of the problems analysed by Finite Element Techniques is made. In spite of the elegance of some of the formulations, they have not yielded any biomechanically significant information that the clinician could benefit by. Moreover, there is usually a conflict between the results obtained using analytic tools and those obtained by experimental means. An example is cited of a model which uses simplistic mechanics which may be employed to yield useful surgical guidelines. It is proposed that Finite Element Models will be most gainfully employed if some form of biologic response mechanism is incorporated into the formulation. This latter step will depend upon a knowledge of bone remodelling coefficients which can only be obtained by experimental means.

KEYWORDS

Orthopaedic Biomechanics; Finite Element Models; *In vivo* Experimental Models; Bone Remodelling.

INTRODUCTION

The methodology by which structure stiffness coefficients may be calculated and how they may be used in the analysis of complex structures by digital computers was first laid down in 1956 (Turner and his co-workers, 1956). This technique was not given a name until 1960 (Clough, 1960) when the term the Finite Element Method (FE Method) was coined. Since then, the FE Method has burgeoned in use (mainly due to a parallel improvement in computer size and speed) in most areas of engineering.

Orthopaedic Biomechanicians have not been slow in appreciating the power of the method and, since 1970, the activity in this area of application of the FE Method has seen a steady increase.

AIMS AND OBJECTIVES

The purpose of the present paper is to attempt to provide answers to the following three questions with respect to FE Models in Orthopaedic Biomechanics:

i What have FE Models achieved ?

ii What is being done at the present time in this area ? and
iii Where is the FE Method headed to in the future ?

FINITE ELEMENT MODELS IN ORTHOPAEDIC BIOMECHANICS - A Review

Three areas of Orthopaedics have received more attention from the FE Modeller than any other areas, *viz.*,

i internal fracture fixation plates,
ii thermal and mechanical aspects of total joint replacement arthroplasty components (usually those of the Hip) using polymethylmethacrylate bone cement, and
iii modelling of intact bones.

Internal Fracture Fixation Plates

Most of the studies concerning these devices,which use FE Models, usually make a comparison between the results of the model and those from *iv vitro* and *in vivo* experiments (Rybicki, Simonen and Weiss, 1972; Plant and Bartel, 1974; Simon and his colleagues, 1977). Especial emphasis has been placed on determining the stresses in the host bone in the proximity of the device. Based on the magnitudes of these stress levels, conclusions were drawn on the remodelling processes as a consequence of Wolff's Law. The usual conclusion was that existing plates are too stiff and there is a need for plates of lower stiffness. Time only will tell if the results (clinical and radiographic) from the newer, lower stiffness plates will be better than those using conventional plates. However, the Orthopaedic Surgeon, who knew all along — based on radiographic evidence — that the plates were too stiff, was compensating for the effects of Wolff's Law by either removing the plates as early as clinically advisable or was instructing the patient in the correct use of his limb after plate removal by the use of aids or by a strict *regimen* of physiotherapy. If the conclusions arrived at by the models are first experimentally verified *in vivo*, then the Orthopaedic Community and the patient, more importantly, will be greatly benifited from the results. The need to follow such a procedure is aptly justified in the light of a recent experiment. Using an experimental model, Szivek and his colleagues (1981), concluded that the bone remodelling under a plate is influenced more by the bond between the plate and bone than by the stiffness of the plate.Faced with these conflicting results, the Orthopaedic Surgeon is going to continue to use conventional plates and to take the precautions indicated earlier.

Thermal and Mechanical Analyses of THR Devices

Jefferiss, Lee and Ling (1975) were among the first workers to formulate a thermodynamic model for polymerising PMMA bone cement. Anxiety had been expressed by many Surgeons on the temperature of the polymerising mass and the possible deleterious effects on the surrounding tissue. It had been proposed that the fibrous tissue layer that forms around the cement was a consequence of the high temperatures. Using a simple Finite Difference Model, the above-mentioned authors concluded that the temperatures could not lead to tissue necrosis and,hence,to the fibrous layer. Huiskes and his co-workers (1977,1978), developed a FE Model incorporating vascular flow, a realistic heat source and interface resistance. Based thereon, they concluded that tissue necrosis was inevitable and recommended certain remedial procedures. An elegant experimental study by Linder(1977) clearly demonstrated that the tissue was responding not to the temperature at the interface or even to the monomer (as some had suggested) but to the surgical insult occasioned by the procedure. Thus far, the FE Modeller has not been able to explain the cause of the fibrous layer in a convincing manner and, in the face of the overwhelming experimental evidence pointing to surgical preparation of the cement bed as the causal factor (Miller and his co-workers, 1977; Halawa and his colleagues, 1978; Lee,Ling and Vangala, 1978), it is most unlikely he will succeed. Thus, the Surgeon is going to put credence on the experimental findings in explaining the fibrous layer.

The above applies to the design of total joint replacement devices. The CAD (Computer Assisted Design) Stem, which was based entirely on an FE Model has shewn, in *in vitro* experimental tests (Weightman, 1976), that it may lead to bone resorption in the *calcar femorale*. Clinical experience to date with this particular device at the authors' centre has proven satisfactory.

Modelling of Intact Bones

The experience to date in this area of FE Method application has been most gainful. Both predictions (based on theory) and actual bones seem to be in remarkable agreement with regard to deflexions and trajectorial architecture (Hayes, Svenson and Schurman, 1978; Woo and his co-authors, 1979). These have been rather recent publications indicating that the FE Modeller has become aware of newer formulations as well as of limitations of the modelling technique.

FINITE ELEMENT MODELS - Current Status

After more than two decades in use, the FE Method is able to model the bone as an inhomogeneous, anisotropic continuum (Brown, May and Ferguson, 1980) and the cement-device interface as a non-linear junction (Hampton and Andriacchi, 1980). Furthermore, awareness is growing as to the limits imposed on the model as a result of the assumptions made. There is also an increasing trend in verifying results obtained through FE Models by experimental means.

FINITE ELEMENT MODELS - *Quo vadis* ?

From the three areas considered, certain conclusions may be drawn :

i In the case of plate design, FE Models have not told the Surgeon any more than he already knew and also no more than experimental models could have yielded. However, FE Models have a lot to contribute to this area of research if, based on experimental data, they incorporate proper boundary conditions <u>and</u> take into account the biologic response of tissue. For the latter, controlled experiments are needed to determine the strain remodelling coefficients.

ii FE Models have not been able to explain the the fibrous layer on the basis of thermodynamics because it is very probable that the layer is a consequence of acrylic cement bed preparation. FE Models should aim at determining the fate of the bone in the cement as well as how a gradual transition in strength from cement to bone (owing to the interstitial cement and the resulting increase in interface shear strength) affects implant life. For such an analysis, a knowledge of bone remodelling would be required. Another application would be in laying down specifications for implants made of novel materials based on stress analysis incorporating bone remodelling.

iii Armed with a knowledge of bone remodelling, the vexing question of collar or no collar for implants cemented to the femur can be answered. The FE Modeller could quite easily determine if collar-calcar-contact, shewn to be favourable in *in vitro* testing (Oh,1977), can be maintained (even <u>if</u> initially obtained). Further, the fate of the bone in apposition with the device <u>can</u> be determined. A recent paper, (Crowninshield and co-workers, 1981), using the FE Method indicated that collar-calcar-contact can be favourable in certain situations. However, an implant devoid of any collar seems to have had no deleterious effects on the *calcar femorale* (Black and his co-authors, 1979).

It is important, for establishing the professional role and usefulness of Biomechanicians, at the outset to carefully formulate the problem from the viewpoint of its clinical relevancy. This has to be done in conjunction with the associated Surgeon. A solution which will lead to a tangible application and implementation is then derived. The cases of elaborate FE Modelling of spinal and head injury applications do not measure up to the extent of biomechanical modelling effort. On the other hand, the failure of FE Models of joint replacement devices in explaining failures (both fatigue and aseptic loosening) may be attributed to their not taking into account appropriate biologic response mechanisms. The latter can only be determined from *a priori in vivo* experimentations.

One cannot discount how adroitly even simplistic mechanics can be employed to yield useful surgical guidelines. Matthew,Sonstegard and Dumbleton (1981) studied the repair of intertrochanteric fractures using a sliding nail device. They considered several failures and, based on very basic mechanics, compared the biomechanics of several clinical alternatives. Based thereon, they were able to outline guidelines on the choice of treatment. Here,the problem was formulated to answer well-defined surgical queries. Furthermore, the clinical picture was not lost sight of.

CONCLUSION

FE Models are only now coming into their own in solving Orthopaedic problems. However, to be able to give meaningful results, they will have to incorporate bone remodelling into their formulations. For the latter, *in vivo* experimental data will need to be generated. Moreover, the Orthopaedic Bioengineer must attempt to see if simplistic models based on basic mechanics can be fruitfully employed prior to embarking upon complex Finite Element formulations.

REFERENCES

Black J, AG Sew Hoy, AJC Lee, RSM Ling and SS Vangala: (1979).
J Bone and Joint Surgery, (61 B) p515

Brown TD, EW Margaret, FH Fu and AB Ferguson : (1980).
Proc Int Conf Finite Elements in Biomechanics, Tucson,AZ. Feb 18-20.

Clough RW : (1960)
Proc 2nd Conf Electronic Computation , ASCE, Pittsburgh, PA. Sept 8-9.

Crowninshield RD, RA Brand, RC Johnston and DR Pedersen : (1981)
Clin Orth, 158 , p270.

Halawa M, AJC Lee, RSM Ling and SS Vangala : (1978).
Arch Orth Trauma Surg, 92, p19.

Hampton SJ and TP Andriacchi : (1980).
Proc Int Conf Finite Elements in Biomechanics, Tucson, AZ. Feb 18-20

Hayes WC, LM Svenson and DJ Schurman : (1978).
J Biomechanics, 11, p407.

Huiskes R, JR de Wijn, TJJH Slooff and ThJG van Rens : (1977).
Proc 1st Con Evaluation of Biomaterials, Strasbourg, France. Sept 26-29.

Huiskes R and JR de Wijn : (1978).
Proc 24th Annual Meeting ORS, Dallas, TX.Feb 11.

Lee AJC, RSM Ling and SS Vangala : (1978).
Arch Orth Trauma Surg, 92, p1.

Linder L : (1977).
J Bone and Joint Surgery, (59 A), p82.

Matthews LS, DA Sonstegard, JH Dumbleton : (1981).
In Clinical Biomechanics : A Case History Approach (edited by) J Black and JH Dumbleton, Churchill Livingstone, p116.

Miller J, DL Burke, W Krause, A Ahmed, L Kelebay and G Tremblay : (1977).
Proc 23rd Annual Meeting ORS, Las Vegas, NE.

Oh I : (1977)
Proc 5th Open Scientific Meeting of the Hip Society, CV Mosby Co., St Louis.

Plant RE and DL Bartel : (1974).
Advances in Bioengineering, Proc ASME (WAM), p85.

Rybicki EF, FA Simonen and EB Weiss : (1972)
J Biomechanics, 5, p203.

Simon BR, SLY Woo, GM Stanley, SR Olmstead, MP McCarthy, GF Hemmott and WH Akeson : (1977).
J Biomechanics, 10, p79.

Szivek JA, GC Weatherly, RM Pilliar and HU Cameron : (1981).
J Biomedical Materials Research, 15, p853.

Turner MJ, RW Clough, HC Martin and LC Topp : (1956).
J Aero Sci , 23, p218.

Weightman B : (1976).
In Advances in Artificial Hip and Knee Joint Technology (edited by) M Schaldach and D Hohman, Springer Verlag, New York, Heidelberg and Berlin.

Woo SLY, BR Simon, WH Akeson and MP McCarthy : (1977).
J Biomechanics, 10 , p87.

A STUDY ON THE KINEMATIC PARAMETERS OF NORMAL HUMAN LOCOMOTION

DL Venkateswara Rao, B Venkatappaiah and N Ramanathan
Biophysics Lab., Cent. Leath. Res. Inst., Adyar,
Madras 600 020, India.

ABSTRACT

Using a two dimensional cinematographic technique the kinematic parameters (both linear and angular) of the lower limb for ten normal children during barefoot walking were obtained to suggest a standard pattern for normal human locomotion.

KEYWORDS

Horizontal component, Vertical component, Velocity, Acceleration, Leg marker, Knee joint, Metatarsophalangeal joint, Ankle joint, Heel.

INTRODUCTION

Analysis of human locomotion provides information required for the development of prosthesis and aids for the handicapped and also to evaluate the success of such designs. Studies have been made earlier on the kinematic analysis of locomotion for normal and abnormal subjects by Winter and colleagues (1972, 1974), Sutherland and Hagy (1972), Foley and co-workers (1979), Sutherland and others (1981) using two dimensional and three dimensional motion analysis techniques. In order to compare the data acquired on the pathological gaits, especially for children, a standard data for normal human barefoot walking is essential. So, using a cinematographic technique the kinematic parameters viz. horizontal and vertical components of velocity and acceleration of the various joints namely, knee joint, metatarsophalangeal joint (MTP joint), ankle joint and heel, the angular velocity and acceleration of the segments joining at the ankle are measured, computed and reported.

MATERIALS AND METHODS

Five boys and five girls whose ages and body weights range between 8 - 12 years and 24 - 32 kg respectively, were chosen as the test subjects. Their feet have well developed arches and they have no abnormality in their lower limbs.

Black markers (1 cm X 1 cm) were applied to the skin (see Fig. 1) to help identify the joint levels, one over each of the following sites: lateral femoral condyle, lateral malleolus and lateral prominence of the fifth metatarsophalangeal joint. Then the subjects were asked to practise walking at their normal speed (100 steps/min.). The length of the walkway chosen for photographing was three metres.

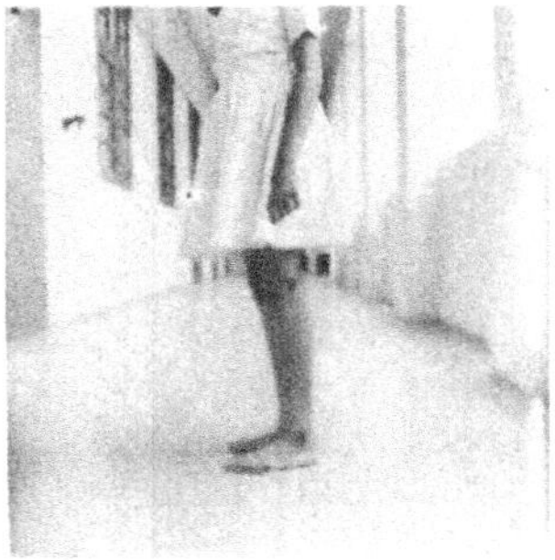

Fig. 1. Leg markers affixed to the lower limb.

A 35 mm cine camera which photographed at a speed of 25 frames/sec. was placed at a distance of five metres perpendicular to the direction of walking. This position allowed a continuous recording of five strides for the study.

The processed film was displayed on the viewing screen of a microfilm reader. The film could be viewed frame by frame. Taking the first frame, in which the subject stood on both feet before starting to walk, as a reference frame the X and Y co-ordinates of each marker were traced frame by frame. Joining these points provided the displacement patterns for each of the leg markers as shown in Fig. 2. The method of computing suggested by Grieve and others (1975) was used with the help of an IBM 370 computer, for obtaining the linear parameters. The angles between the segments joining at the ankle were measured in each frame and the angular velocity and acceleration calculated using the time interval between each frame.

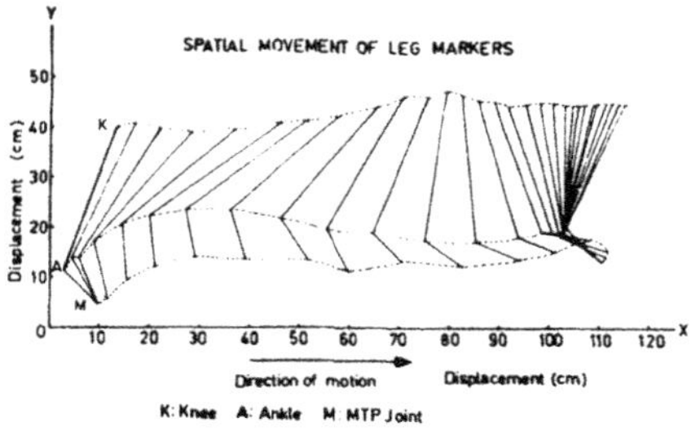

Fig. 2. Displacement patterns of the leg markers.

RESULTS AND DISCUSSION

Figures 3 and 5 illustrate the horizontal components of velocity and acceleration respectively of the leg markers and the heel. The vertical components of the same are plotted in Fig. 4 and Fig. 6. These linear parameters are plotted as a function of time. These curves are for one single stride.

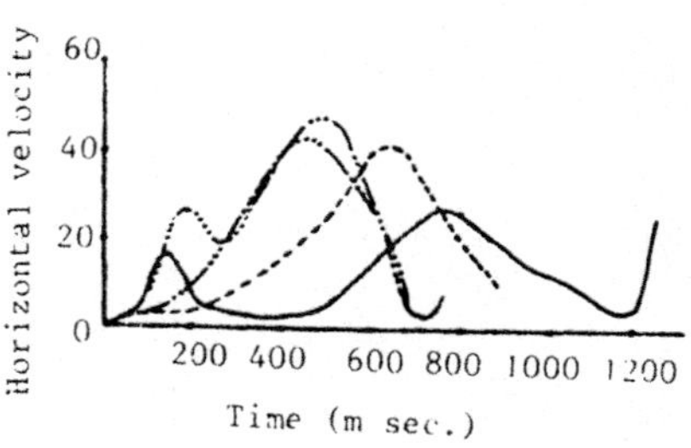

Fig. 3.

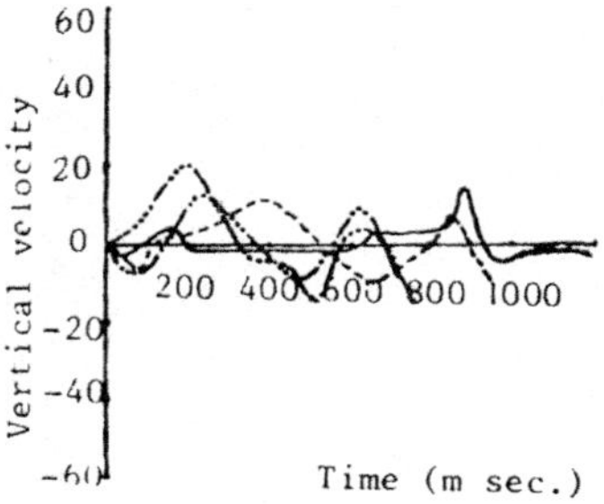

Fig. 4.

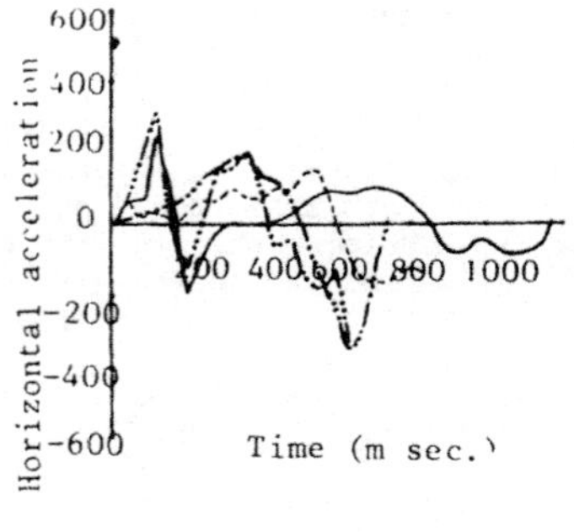

Fig. 5.

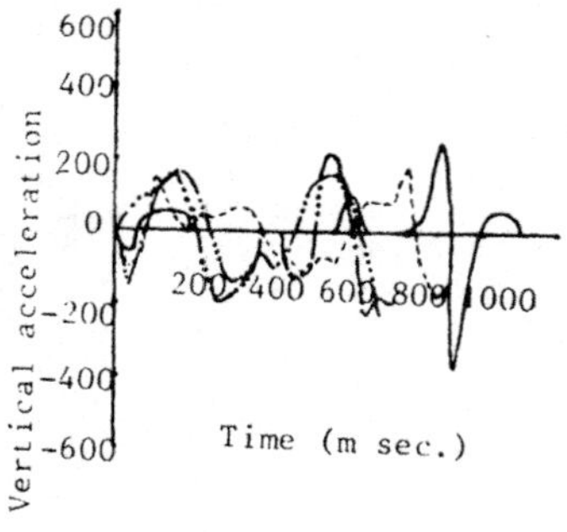

Fig. 6.

(———) knee joint
(-----) ankle joint
(-····-) MTP joint
(-·····-) heel

All Y values to be multiplied 5 times for exact value.
Units: Velocity - cm/sec.
Acceleration - cm/sec^2.

The angular parameters are tabulated as shown below:

Table. Angular Parameters

S.No.	Description	Mean	Std.Dev.
1.	Max.angle (Dorsi.)	27°	7
2.	Max.angle (Plantar.)	12°	2
3.	Max.ang.vel. (Dorsi.)	$332^o/sec$	131
4.	Max.ang.vel. (Plantar.)	$386^o/sec$	149
5.	Max.ang.accl. (Dorsi.)	$5446^o/sec^2$	1002
6.	Max.ang.accl. (Plantar.)	$5446^o/sec^2$	1125

The horizontal component of velocity appears to become maximum during a period ranging between 58% and 72% of the total movement time for the leg markers and the heel. This shows that the lower limb as a whole attains a peak velocity in the forward direction after about 2/3 rds of the journey time. The occurrence of the peak horizontal and vertical components of velocity of the heel at 29% indicates that at this time the heel takes off from the ground at a faster rate than at the heel-rise time.

The vertical component of velocity for the MTP joint is maximum in the upward direction at 36% of its movement time i.e. in swing phase. The second peak in the upward component of velocity in the case of MTP joint at 89% may be due to the impact of the heel in heel-strike. The downward velocity is maximum in the lateral-swing phase occuring at 64% of its total cycle time. Similarly, in the case of the ankle joint the maximum upward component occurs at 44% and 94% of its total cycle time i.e. mid-swing and after heel-strike respectively. The downward component is maximum at 72% i.e. lateral-swing phase. But in the case of the knee joint there is only one peak at 82% of its total cycle time i.e. just before heel-strike.

Retardation in the movement patterns of the heel, the MTP joint and the knee joint starts after 71% of their total cycle times i.e. lateral-swing phase. The maximum downward acceleration in all cases occurs after 64% of the time i.e. the lower limb starts approaching the heel-strike after about 64% of the total cycle time. There are many peaks and troughs in the acceleration components for the ankle joint showing that the ankle movement is somewhat jerky.

The heel takes off from the ground after a lapse of about 30% of the time and then the foot starts accelerating till it reaches the mid-swing phase whereafter the lower limb retards in the lateral-swing phase till it approaches the next heel-strike. At and after the heel-strike, the knee joint, the MTP joint and the ankle joint show a peak upward velocity which is required in preparing the heel for the next heel-strike.

REFERENCES

Foley, C.D., A.O. Quanbury and T. Steinke (1979). Kinematics of normal child locomotion - A statistical study based on TV data. J. Biomechanics, 12, 1-8.

Grieve, D.W., Doris Miller, M. Mitchelson, J. Paul and A.J. Smith (1975). Photographic analysis. In H.T.A. Whiting (Ed.), Techniques for the Analysis of Human Movement 1st ed. Unwin Bros., Woking, Surrey, pp. 15-20.

Sutherland, D.H. and J.L. Hagy (1972). Measurement of gait movements from motion picture film. J. Bone Jnt. Surg. 54-A, 787-797.

Sutherland, D.H., Richard Olshen, Les Cooper, Marilynn Wyatt, Judy Leach, Scott Mubarak and Paul Schultz (1981). The pathomechanics of gait on duchenne dystrophy. Develop. Med. Child Neurol., 23, 3-22.

Winter, D.A., A.O. Quanbury, D.A. Hobson, H.G. Sidwall, Garry Reimer, Brain G. Trenholm, T. Steinke and H. Shlosser (1974). Kinematics of normal locomotion - A statistical study based on TV data. J. Biomechanics, 7, 479-486.

Session 8

SOFT TISSUE MECHANICS

DETERMINATION OF TRUE POINT-BY-POINT HEART WALL STIFFNESS

D.B. Rivers*, W.F. Ranson* and J.B. Caulfield**

*College of Engineering
**School of Medicine
University of South Carolina, Columbia, S.C.

ABSTRACT

A measurement system capable of determining the instantaneous changes in heart wall geometry through the use of laser speckle interferometry and acoustical holography is described. These data aid in developing improved heat models with nonstandard geometry and variable wall compliance.

KEYWORDS

Measurement; heart geometry; laser speckle; acoustical holography; heart wall compliance.

SUMMARY

In previous investigations to date of the dynamic geometry of the heart (Glantz, 1980; Mirsky and Pasipoularides, 1980; Janz and Grimm, 1973; Ghista, Brady and Radhakrishnan, 1973) assumptions have had to be made as to the general shape of the heart chambers, typically a sphere or ellipsoid. In addition, the assumption of constant wall stiffness has always been used to further simplify the model. These models have then been used to justify and interpret empirical physiologic data of both diseased and healthy hearts. However, the ventricle is neither a sphere nor an ellipsoid but a complexly shaped three dimensional structure which follows no linear shape change. Furthermore, the point-by-point wall stiffness around the ventricles is obviously neither linear nor constant, particularly in diseased hearts.

The instantaneous true geometry of the heart wall can be determined during diastole and passive filling by combining the techniques of laser speckle interferometry and ultrasonic holography. Both of these techniques have extensive applications to non-destructive testing but neither has yet been applied to medical imaging.

The use of laser speckle interferometry to obtain surface strains of opaque solids has been documented by Ranson and Moslehy (1979). When a diffuse surface of a structure is illuminated by coherent light, a grainy speckle effect is imaged by the eye or camera due to the random interference of light reflected from the surface. When the speckle pattern of an undeformed surface is combined with the pattern of the surface after deformation, a complex diffraction halo modulated by light and dark bars of light is produced where the distance between the bars is inversely proportional to the distance between the undeformed and deformed speckle; thus the surface strains can be measured.

This technique is only directly applicable to plane surfaces normal to the coherent light source. However, if the contour of a surface is known, then the laser speckle data can be modified using simple trigonometric analysis so that the surface strains of the contoured surface may be determined. This contour can be determined using acoustical holography.

The principle of acoustical holography is mathematically identical to that of light holography (Hildebrand and Brenden, 1972). Two underwater ultrasonic beams are both aimed at the same area of water surface. When only one of these beams passes through some solid object to be imaged, a ripple pattern is formed on the surface of the water which is the interference pattern of the acoustical hologram. By use of a collimated coherent light source, this ripple pattern produces an image of the internal voids of the imaged object. If an acoustical lens is placed between the object and the liquid surface so that both are located at twice the lens' focal distance on opposite sides of the lens, then the reconstructed light image contains only that spatial information from that part of the object which is exactly twice the focal distance from the lens. Spatial information from the object from either side of this holographic imaging plane is diffused and, therefore, not imaged. Consequently, by moving the object back and forth through this plane, one can view successive two-dimensional "cuts" of the solid object. These sectional views can be imaged directly by a high resolution digitizing camera so that computer image edge enhancement can be performed.

After digitizing the serial heart sections, a three-dimensional computer model of the heart can be generated by connecting the sections together. In this way the exact internal and external volumes can be calculated for a disatolic heart. This same technique allows for the determination of point-by-point external surface contours so that the external heart wall strains can be calculated. Furthermore, by using a pulsed set of ultrasonic beams and coupling the phases of the pulses to that of an infusion pump, a complete set of serial sections of the heart can be recorded for various pressure increments, thereby allowing for an exact determination of the pressure-volume relationship for any heart. Also, this same data set, after edge enhancement, allows for calculation of the point-by-point wall thickness changes as a function of pressure and, therefore, determination of the instantaneous stiffness at each point on the heart walls.

This technique has been successfully demonstrated by digitizing the acoustical holographic images of diastolic rat hearts and is currently being applied with passive filling to generate true pressure volume relationship.

ACKNOWLEDGEMENT

This work is funded by National Heart, Lung, and Blood Institute grant No. 1RO1HL27533-01.

REFERENCES

Ghista, D. N., A. J. Brady, and S. Radhakrishnan (1973). A three dimensional analytical (rheological) model of the human left ventricle in passive-active states. Biophys. J., 13: 832-854.

Glantz, S. A. (1980). Computing indices of diastolic stiffness has been counterproductive. Federation Proc., 29: 162.

Hildebrand, B. P., and B. B. Brenden (1972). An Introduction to Acoustical Holography. Plenum Press.

Janz, R. F., and A. F. Grimm (1973). Deformation of the diastolic left ventricle. I. Nonlinear elastic effects. Biophys. J., 13: 689-704.

Mirsky, I., and A. Pasipoularides (1980). Elastic properties of normal and hypertrophied cardiac muscle. Federation Proc., 39: 156-161.

Ranson, W. F., and F. A. Moslehy (1979). Laser speckle measurements of surface displacements and derivations of a deformed solid. Proceedings First Cairo University Conference on Mechanical Design and Production, Cairo, Egypt.

REGIONAL AND GLOBAL LEFT VENTRICULAR DEFORMATION: PRESSURE-STRAIN ANALYSIS

J.Y. Kresh and S.K. Brockman
Division of Cardiothoracic Surgery, Department of Surgery
Jefferson Medical College, Thomas Jefferson University
Philadelphia, PA

ABSTRACT

In non-beating isolated canine left ventricles, the dynamic relationship between intraventricular pressure resulting from a periodic oscillation of chamber volume, including diameter, wall thickness, and regional segment dimensions were plotted and fitted to an exponential pressure-strain model, assuming a quasi-static large deformation. The observed difference between global and regional pressure-dimension strain stiffness coefficients can be attributed to basic constitutive and geometric considerations and not necessarily to the complex anisotropic or heterogeneous nature of cardiac muscle properties. Thus, providing indices which appropriately characterize the regional and global left ventricular chamber stiffness.

KEYWORDS

Pressure-strain analysis; viscoelastic properties; stiffness coefficient.

INTRODUCTION

The characterization of regional and global ventricular elastic, viscous, and inertial properties is necessary for the understanding of both normal ventricular chamber mechanics and disease states such as infarct, hypertrophy and cardiomyopathy which contribute to left ventricular dysfunction. In this study, non-linear elastic theory and quasi-static large deformation analysis were applied to the description of mechanical properties of the myocardium.

MATERIALS AND METHODS

Dimension sonomicrometry. While the heart was still beating, three pairs of miniature ultrasonic (5MHz) transducers were implanted and used for measuring instantaneous chamber dimensions. One pair of crystals (5mm. in diameter) was placed near the minor equator, measuring antero-posterior endocardial diameter (DIA). The second pair of crystals (2 and 5mm. in diameter) was implanted across the left ventricular free-wall near the endocardium and epicardium respectively. The third pair (2mm. in diameter) was implanted in the circumferential plane mid-wall to the endocardium. The transit-time of a 5MHz soundwave traveling between each crystal pair was detected and converted to distance using known velocity of sound in blood-tissue medium (1580m/sec.). The measurement system also included a pulsatile blood pump communicating with the isolated left ventricular chamber, generating sinusoidal waveforms of selectable frequencies (0-100s/min.). The volume displaced by the pump was measured using an LVDT transducer. Left ventricular pressure was measured by means of a catheter-tip micromanometer. The aorta was cross-clamped at its root and the heart arrested with saturated KCl solution, left ventricular weight was 130-200gm., shell volume, 122-188ml. The right ventricle was dissected free so it does not exert a restraining force. The left atrium was clamped at the level of the atrioventricular groove without distortion to ventricular geometry. All analog signals were digitized using a 12-bit analog-digital converter and analyzed using a dedicated graphics computer (Tektronix-4052).

Theory. The instantaneous relationship between measured pressure and volume, including diameter and regional segment dimensions were plotted and fitted to an exponential pressure-strain model for the left ventricle assuming quasi-static large deformation.

$$P = P_o \cdot (e^{\kappa\varepsilon} - 1) \tag{1}$$

In Equation (1) κ and P_o are non-linear elastic coefficients, ε is the natural strain defined for volume, diameter, wall thickness and segment length as $\varepsilon_v = \ln(V/V_o)$, $\varepsilon_d = \ln(D/D_o)$, $\varepsilon_w = \ln(W/W_o)$, $\varepsilon_s = \ln(S/S_o)$ where V_o, D_o, W_o, and S_o are the zero-stress global and regional geometric dimensions. This analytic representation is analogous to isolated unstimulated muscle mechanics models where the elastic stiffness was found to be proportional to stress (σ): $d\sigma/d\varepsilon = \sigma\kappa_m + C$ where κ_m is an elastic stiffness and C a constant. Applying Equation (1) and strain definition it can be mathematically demonstrated that κ_v is equal to $dP'/P'/dV/V$ and $\kappa_d = dP'/P'/dD/D...$, where $P' = P - P_o$ and $d\varepsilon_v = dV/V$, $d\varepsilon_d = dD/D...$, which yields an analogous pressure-strain relationship: $dP'/d\varepsilon = \kappa P'$. This definition of stiffness provides an excellent normalization property for comparing ventricles of different shapes and sizes.

RESULTS AND DISCUSSION

Figure 2A. shows a typical experimental result of pressure-dimension cyclic loading and unloading for measurements made in Fig. 1. The outstanding feature to note is the similar exponential relationship between pressure and chamber dimensions including global chamber volume. In addition, at a relatively low sinusoidal volume pump rate (25s/min.) there was a noticable hysteresis between the phase of loading and unloading in both the regional and global cyclic pressure-dimension plot. Specifically, the regression analysis for the full cycle (including negative pressure) in the above experiment resulted in $\kappa_v = 3.32$ and a $P_o = 4.37$ ($r = 0.997$). The regression analysis for $P > 0$: $\kappa_v = 4.12$ and $P_o = 2.61$ ($r = 0.998$); $\kappa_d = 16.18$ and $P_o = 2.79$ ($r = 0.981$); $\kappa_w = 19.06$ and $P_o = 2.02$ ($r = 0.990$). When the harmonic oscillation rate is decreased to 10s/min., the resulting $\kappa_v = 3.76$ and $P_o = 3.06$ ($r = 0.991$); $\kappa_d = 15.08$ and $P_o = 3.22$ ($r = 0.982$); and $\kappa_w = 12.88$ and $P_o = 3.9$ ($r = 0.959$). The difference in these results suggest possible contribution by viscous and/or inertial properties to the effective stiffness coefficient observed at higher strain rates. Figure 3 exhibits data from an experiment where segment length was measured. In addition, equatorial circumferential mid-wall stress (σ) was computed based upon an ellipsoid of revolution model for the left ventricular chamber (Mirsky, 1979). The regression analysis based upon a composite data (loading and unloading) give rise to $\kappa_v = 1.85$ and $P_o = 3.29$ ($r = 0.991$); $\kappa_d = 9.89$ and $P_o = 3.16$ ($r = 0.989$); $\kappa_s = 6.62$ and $P_o = 4.7$ ($r = 0.989$) and $\kappa_w = 11.35$ and $P_o = 3.00$ ($r = 0.985$). The stiffness coefficients derived using the constitutive circumferential stress-

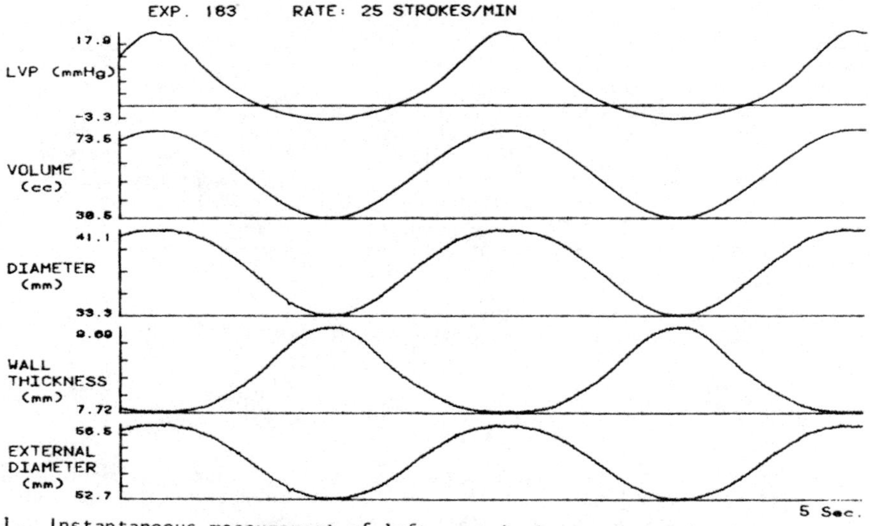

Fig. 1. Instantaneous measurement of left ventricular pressure (LVP) chamber volume, equatorial diameter, wall thickness, and external diameter for pump rate of 25s/min. with 43ml. stroke.

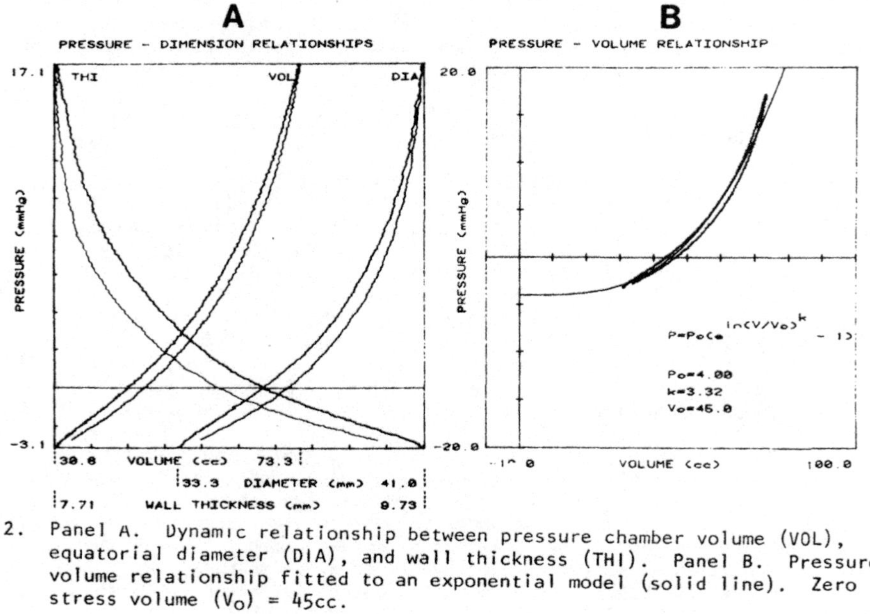

Fig. 2. Panel A. Dynamic relationship between pressure chamber volume (VOL), equatorial diameter (DIA), and wall thickness (THI). Panel B. Pressure-volume relationship fitted to an exponential model (solid line). Zero stress volume (V_o) = 45cc.

strain relation were $\kappa_v = 2.36$, $\sigma_o = 4.32$ ($r = 0.993$); $\kappa_d = 12.81$ and $\sigma_o = 3.97$ ($r = 0.991$) and $\kappa_s = 9.01$, $\sigma_o = 5.8$ ($r = 0.992$). Consistently higher values for the stiffness coefficient derived from the computed circumferential stress-strain analysis are principally due to the higher peak stress (mid-wall region) when compared to intraventricular pressure-strain analysis. Moreover, both results indicated similar direction of change. Combined results based upon fifteen experiments (at low strain rate, less than 10% $sec.^{-1}$), the pressure-volume strain stiffness κ_v was 3.18 ± 0.86 and P_o was 3.03 ± 1.13. The pressure minor axis strain stiffness constant κ_d was 15.73 ± 5.32 with P_o of 2.91 ± 1.11. The wall thickness strain stiffness constant κ_w was 22.72 ± 8.72 with P_o of 3.55 ± 1.21. Pressure segment length strain elastic coefficients κ_s was 22.32 ± 9.85 with P_o of 3.67 ± 1.46.

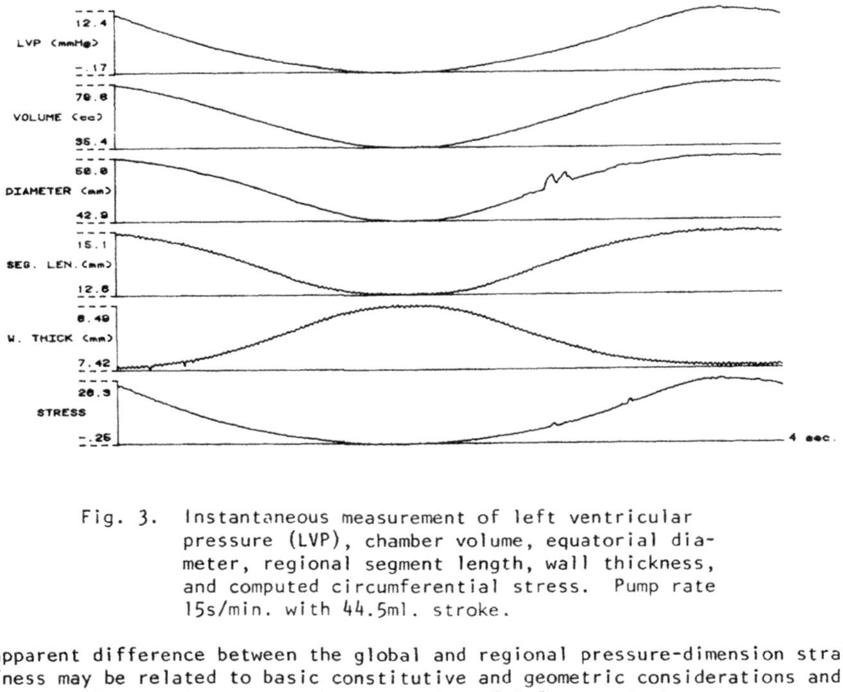

Fig. 3. Instantaneous measurement of left ventricular pressure (LVP), chamber volume, equatorial diameter, regional segment length, wall thickness, and computed circumferential stress. Pump rate 15s/min. with 44.5ml. stroke.

The apparent difference between the global and regional pressure-dimension strain stiffness may be related to basic constitutive and geometric considerations and not necessarily to anisotropy or heterogeneity of left ventricular structure. In addition, the deformation in a particular dimension (region) can be attributed to the relative number of fibers oriented in each axis, (Streeter, 1969), the largest strain occurring in the plane of the greatest number of myocardial fibers (circumferential direction). This in part explains the relatively lower κ_s when compared to κ_w. The wall stiffness coefficient κ_w lumps fibers in the epicardial and endocardial regions where muscle fibers are predominantly oriented parallel to these boundary surfaces. During low volume infusion or inflation, the ventricle can be characterized by an ellipsoidal geometry, and as the volume is increased it assumes a more spheroidal geometry with shortening of the long axis and a substantial increase in the external diameter (preserving a constant shell volume) with lesser change in wall thickness, as was demonstrated by a $\kappa_w > \kappa_d$. It is therfore suggested that the volume or bulk stiffness κ_v can be substantially lower than the elastic coefficient based upon regional dimensions and equatorial diameter strain. It can be demonstrated analytically that if the left ventricle is assumed to be a sphere, volume stiffness $\kappa_v = 1/3\ \kappa_d$. At higher strain rates (20% $sec.^{-1}$) pressure-dimension plot exhibit noticable hysteresis between trajectory of cyclic loading and unloading suggesting the presence of inertial property , viscous elasticity and/or stress relaxation. These effects will give rise to a more complex (frequency dependent) viscoelastic modulus which may better describe myocardial wall properties and cardiac filling phase. For a given frequency, this effect can be simulated by adding a viscous component to the model described by Equation (1). A hysteresis loop will result when the strain direction (negative strain rate) is reversed during the transition from loading to unloading phase (see Fig. 4).

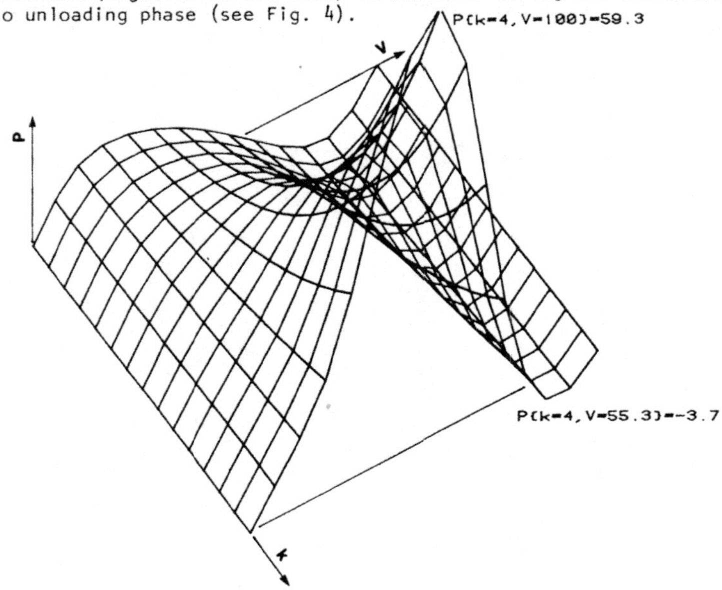

Fig. 4. The three-dimensional plot illustrates the effect of added viscosity on generated pressures by adding a strain rate term to Equation (1). For a given value of stiffness (κ), the attained pressure at a negative volume-strain rate is considerably lower when compared to that at positive strain rate (for identical chamber volume). The unloading trajectory can give rise to negative pressures for volumes greater than the elastic equilibrium value.

REFERENCES

Mirsky, I., Rankin, J.S. (1979). The effects of geometry, elasticity, and external pressure on the diastolic pressure-volume stiffness-strain relations. Circulation Research, 44:601-611.

Streeter, D.D.Jr., Spotnitz, H.M., Patel, D.J., Ross, J.Jr., Sonneblick, E.H. (1969). Fiber orientation in a canine ventricle during diastole and systole. Circulation Research, 24:339-347.

ACKNOWLEDGEMENT

We would like to acknowledge Jeffrey Sims and Steven Rappaport and thank Marian Donato for preparing this manuscript.

ANISOTROPY OF THE AORTIC TISSUE IN THE AXIAL AND TRANSVERSE DIRECTIONS

Jafar Vossoughi*, Ramesh N. Vaishnav*, and Dali J. Patel**
*School of Engineering and Architecture
The Catholic University of America, Washington, D. C. 20064, and
**College of Medicine, Department of Physiology and Biophysics
Howard University, Washington, D. C. 20059

ABSTRACT

Mechanical properties of the porcine aortic tissue were studied by subjecting circumferential and longitudinal strips of the tissue to uniaxial states of stress. Stress-strain curves and transverse strain vs. axial strain curves were obtained for strips in the circumferential (θ) and longitudinal (z) directions. The fact that these curves were distinct suggested anisotropic behavior of the tissue. To quantify the degree of separation of these curves directly, indexes of anisotropy were computed as ratios of stresses and strains in the θ and z directions at equal values of axial strains or stresses. It is shown that various stress and strain ratios, which would have a uniform value of unity if the tissue were isotropic, are far from having a constant value of one.

KEYWORDS

Mechanical properties, transverse strain, anisotropy, aorta, elasticity.

INTRODUCTION

Considerable work has been done toward the study of the mechanical properties of the arterial tissue (Dobrin, 1978; Patel and Vaishnav, 1980). Most of the studies treat the tissue as isotropic in behavior. However, from considerations of the structure of the vascular tissue, it appears more likely that the tissue is anisotropic. It is also clear from the experiments of Vaishnav et al (1972), Young et al (1977), Cox (1975) and Dobrin and Doyle (1970) that the mechanical properties of the tissue in the circumferential and longitudinal directions are not identical. In this paper we present further direct evidence of this aniostropic nature of the arterial tissue on the basis of uniaxial tests on porcine thoracic aortic strips. We also show how the deformations in the two transverse directions compare with each other in the circumferential and longitudinal uniaxial tests.

METHODS

In vitro uniaxial tests were performed on strips from 7 porcine aortic segments from 7 animals (average weight, 180 lbs and average age, 6 months) using a test system described elsewhere in detail (Vossoughi and Vaishnav, 1980). In a typical experiment the aortic segment was cut open longitudinally in the dorsal aspect and flattened onto a plane surface so as not to stretch it in any direction.

Rectangular specimens (5 mm x 30 mm) were carefully cut out through the tissue thickness in the circumferential and longitudinal directions. Fifteen to 30 strips were cut out from each segment depending on the size of the segment. The thickness of each strip was measured at several locations using a thickness gage and the average value used as the specimen thickness. Four gage marks in the form of dots (Fig. 1) were imprinted on the intimal surface of the segment. The axial engineering stress (σ) in the specimen computed by dividing the force value by the cross-sectional area of the specimen. The strain (ε) in the axial and transverse directions was determined by photographing the gage marks pattern at any desired instant and measuring the dimensions on the photographic negative using a traveling microscope. The second transverse strain could be calculated by assuming the tissue to be incompressible (Carew et al, 1968). Before carrying out an actual test, the specimen was preconditioned by subjecting it to three cycles of loading and unloading up to a maximum strain level of 0.5 approximately.

RESULTS

Figure 2 shows curves of engineering stress versus engineering strain for the circumferential and longitudinal directions. The curves represent averages for 7 segments and have been obtained by averaging strains corresponding to various stress levels. Each point on the curve for the circumferential direction is based on 80 strips and that for the longitudinal direction is based on 105 strips. By observing that a given strain level corresponds to a higher stress in the circumferential direction than in the longitudinal direction, we can conclude that the tissue is stiffer in the circumferential direction than in the longitudinal direction. This phenomenon was observed in all of the seven segments.

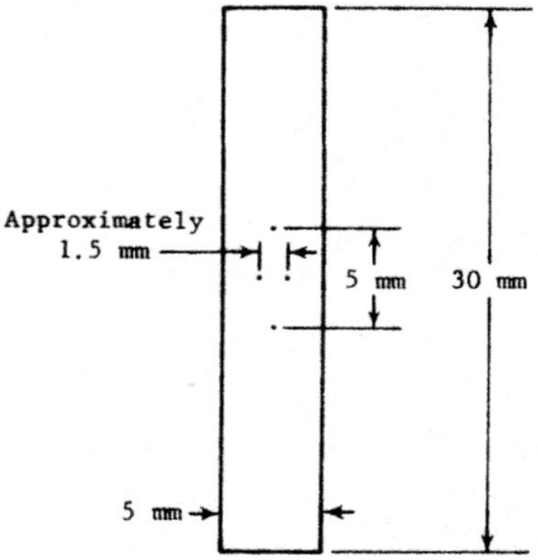

Fig. 1. Geometry of uniaxial test specimen with gage marks.

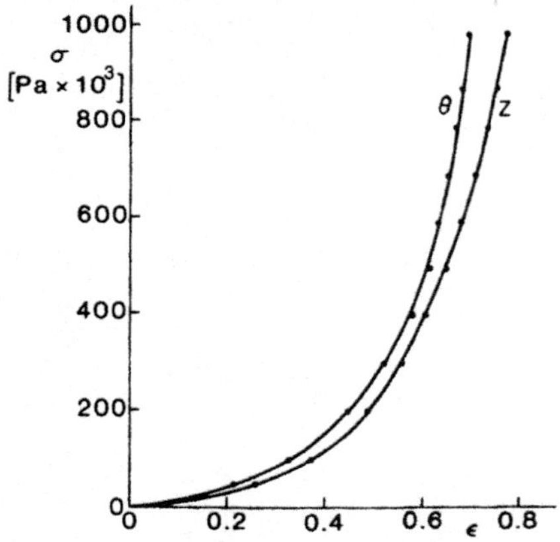

Fig. 2. Average curves of stress (σ) vs. strain (ε) for circumferential (θ) and longitudinal (z) strips.

Figure 3 shows a plot of the ratio $\varepsilon_z/\varepsilon_\theta$ (which may be called the strain index of anisotropy) versus stress from the two curves at corresponding levels of stress, and Fig. 4 shows a plot of the ratio σ_θ/σ_z (which may be called the stress index of anisotropy) versus strain at the corresponding levels of strain. The ratio $\varepsilon_z/\varepsilon_\theta$ varies from 1.05 to 1.20 and ratio σ_θ/σ_z varies from 1.20 to about 1.75. For an isotropic material, both of these curves would have been straight lines parallel to the axis of abscissa and having an ordinate value of unity.

To quantify the degree of anisotropy in the transverse directions, several additional relationships among strains were studied. Figures 5 and 6 show the transverse strains associated with the axial strains in the uniaxial tests on the

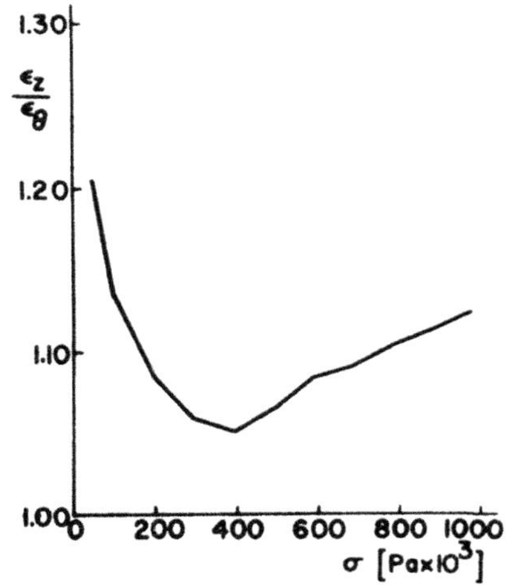

Fig. 3. Variation of the strain index of anisotropy $\varepsilon_z/\varepsilon_\theta$ vs. σ.

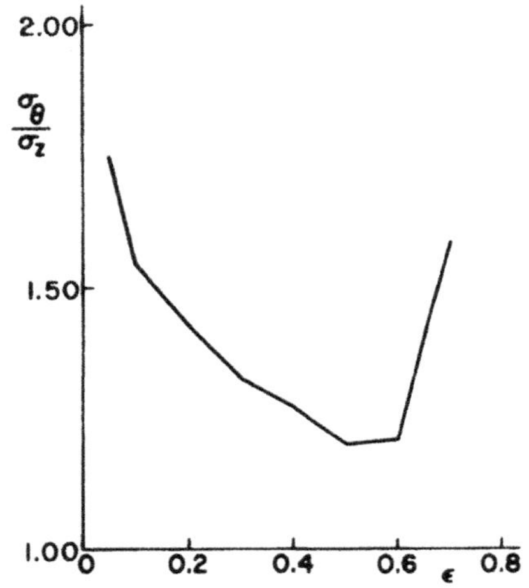

Fig. 4. Variation of the stress index of anisotropy σ_θ/σ_z vs. ε.

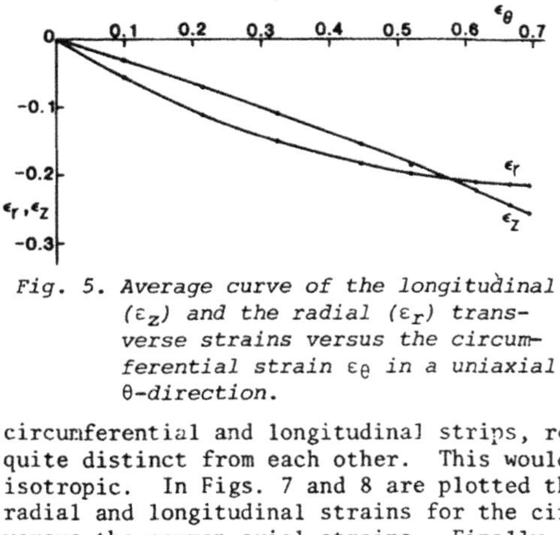

Fig. 5. Average curve of the longitudinal (ε_z) and the radial (ε_r) transverse strains versus the circumferential strain ε_θ in a uniaxial θ-direction.

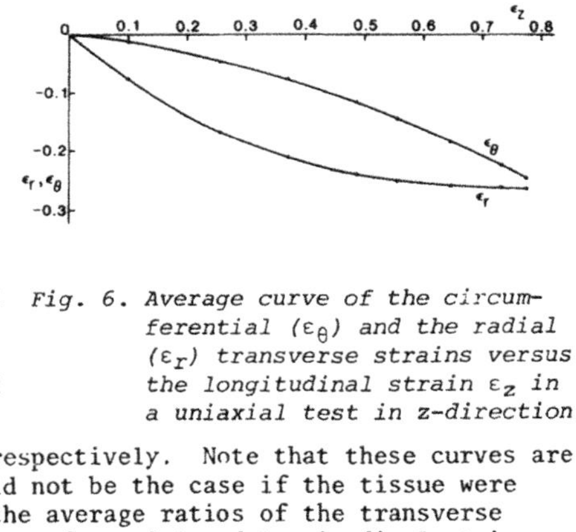

Fig. 6. Average curve of the circumferential (ε_θ) and the radial (ε_r) transverse strains versus the longitudinal strain ε_z in a uniaxial test in z-direction.

circumferential and longitudinal strips, respectively. Note that these curves are quite distinct from each other. This would not be the case if the tissue were isotropic. In Figs. 7 and 8 are plotted the average ratios of the transverse radial and longitudinal strains for the circumferential and longitudinal strips versus the common axial strains. Finally, in Figs. 9 and 10 are plotted various

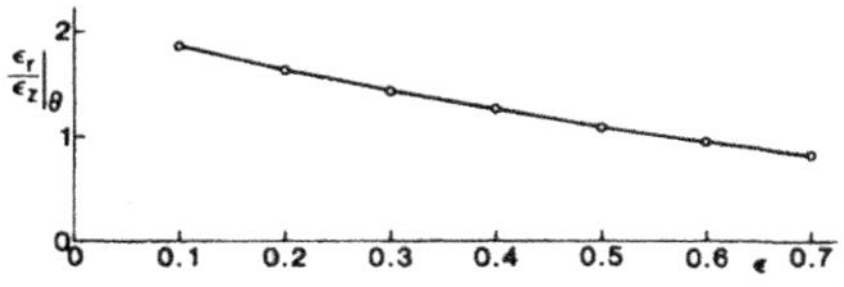

Fig. 7. Average ratio of the transverse radial and longitudinal strains $\varepsilon_r/\varepsilon_z$ for the circumferential strips versus the common axial strain.

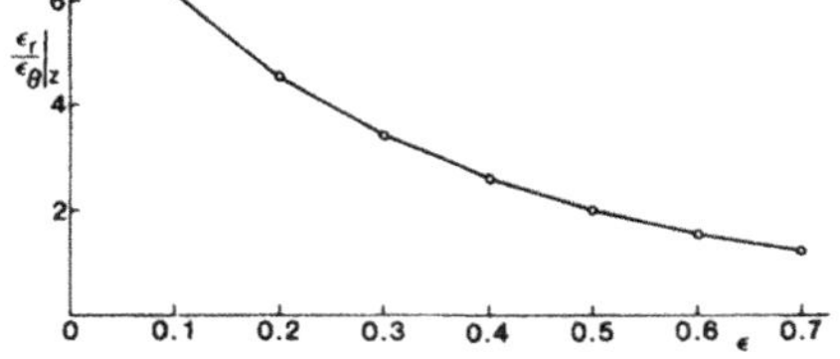

Fig. 8. Average ratio of the transverse radial and circumferential strain $\varepsilon_r/\varepsilon_\theta$ for the longitudinal strips versus the common axial strain.

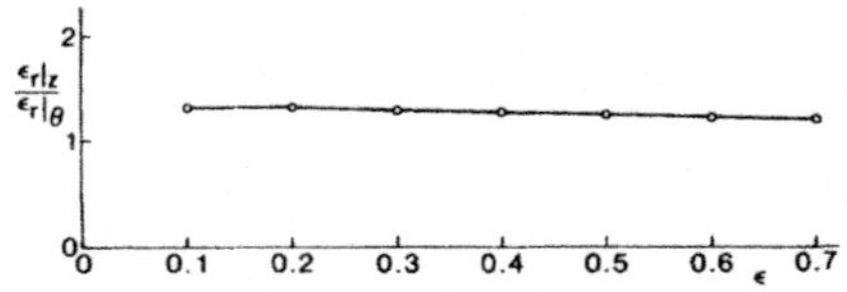

Fig. 9. Average ratio of the transverse radial strains from the longitudinal $\varepsilon_r|_z$ and circumferential $\varepsilon_r|_\theta$ tests at equal axial strain.

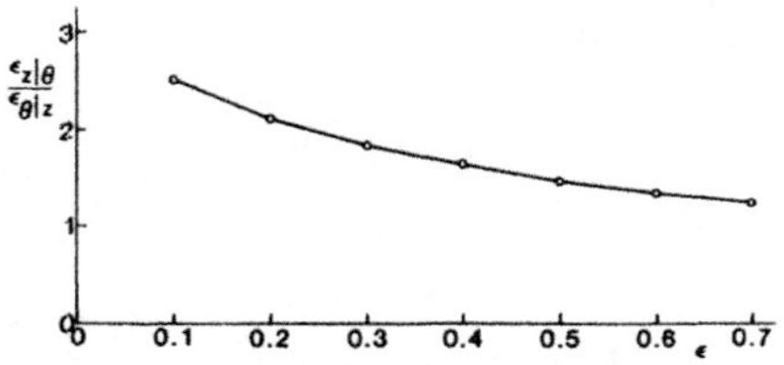

Fig. 10. Average ratio of the transverse longitudinal strain in the circumferential test $\varepsilon_z|_\theta$ and the transverse circumferential strain in the longitudinal test $\varepsilon_\theta|_z$ at equal axial strain.

ratios of transverse strains from the circumferential and longitudinal tests at equal axial strains. Again, the curves are distinct, which is another indication of anisotropy.

It is thus clear that the tissue behavior is nonlinear and anisotropic. Not only are the axial strain responses of circumferential and longitudinal strips quantitatively different, but the corresponding transverse strain responses are markedly different also. The tissue is distinctly stiffer in the circumferential direction than in the longitudinal direction. This corroborates our previous findings based on inflation-extension experiments on tubular specimens (Vaishnav et al, 1972 and Young et al, 1977). This investigation also provides quantitative information on the degree of error one would make in considering the tissue to be isotropic.

ACKNOWLEDGMENTS

We gratefully acknowledge the support of NSF grant CME8006338.

REFERENCES

Carew, T. E., R. N. Vaishnav, D. J. Patel. (1968). Circ. Res., 23, 61-68.
Cox, R. H. (1975). J. Biomech., 293-300.
Dobrin, P. B. (1978). Physiol. Reviews, 58, No. 2, 397-460.
Dobrin, P. B. and J. M. Doyle (1970). Circ. Res., 23, 105-119.
Patel, D. J. and R. N. Vaishnav (1980). Basic Hemodynamics and Its Role in Disease Processes. Univ. Park Press, 504 p.
Vaishnav, R. N., J. T. Young, J. S. Janicki, and D. J. Patel (1972). Biophys. J., 12, 1008-1027.
Vossoughi, J. and R. N. Vaishnav (1980). In 1980 Advances in Bioengineering, V. C. Mow (Ed.), ASME NY, 313-316.
Young, J. T., R. N. Vaishnav, and D. J. Patel (1977). J. Biomech., 10, 549-559.

THE ELASTICITY OF THE URINARY BLADDER FOR LARGE DEFORMATIONS

C. Regnier*, H. Kolsky**, P. Richardson***, J. Susset* and G. Ghoniem*

*Department of Surgery, Roger Williams General Hospital, Providence, RI
**Division of Applied Mathematics, Brown University, Providence, RI
***Division of Engineering, Brown University, Providence, RI

ABSTRACT

The purpose of this paper is to investigate the theoretical basis of a new clinical test called "Rapid Injection Cystometry" (RIC), aimed at quantifying the elastic properties of the urinary bladder wall. The theory of large elastic deformations is applied to the inflation of a spherical rubber balloon, taken as a first approximation model, and during the stretching phase of the RIC test. The Treloar form of strain energy function, which gives a good representation of vulcanized rubber , does not have enough terms to account for all the data in the case of the urinary bladder. A series expansion of the former is suggested, i.e., a polynomial type. Some clinical examples are given.

KEYWORDS

Urinary bladder; cystometry; elastic properties; rubber; strain energy function.

INTRODUCTION

In order to analyse the elastic response of various hollow viscera such as the urinary bladder, Osborne performed in 1909 a number of experiments on bladders and rubber balloons. He found that these two types of material behaved differently, but he assumed that they obeyed Hooke's Law perfectly, i.e., have a linear stress-strain relationship for infinitesimal displacements. This assumption is clearly invalid for this problem since the strains are finite and the stress-strain response is highly non-linear. The alternative approach to the problem of finite deformation is in terms of strain energy function. This was first used successfully by Mooney (1940); later, Rivlin (1948) and his co-workers applied this concept to the finite three dimensional deformation of many structures composed of rubber-like materials.

In this paper, we first apply the theory of large elastic deformations to the inflation of a spherical rubber balloon and then see if the same theory can be applied to the urinary bladder. We describe an experimental way to measure the elastic response of the bladder without effective relaxation and compare it with that of vulcanized rubber.

THEORY

Let us consider the inflation of a thin spherical balloon from an initial radius R to a final radius r. We assume that the material is isotropic, homogeneous and incompressible. Mooney (1940), and Rivlin (1948) have emphasized that if the material is elastic in its undeformed state, the energy stored in the deformed state is defined as the strain energy function W for the material. The work $\Delta P dV$ done to increase the enclosed volume V is equal to the change of energy of the balloon, where ΔP is the transmural pressure difference. Therefore $\Delta P dV = 4\pi R^2 dW$, which gives

$$\Delta P = (1/R\lambda^2)(dW/d\lambda) \tag{1}$$

We want to express the pressure in terms of λ. Let us now consider the balloon as a thick spherical shell initially of internal radius R_i and external radius R_o under internal and external pressures P_i and P_o, respectively. We split it into a series of nested membranes, like onion layers, each of initial thickness dR. The pressure drop across one such membrane can be expressed as

$$-dP = (1/R\lambda^2)((dW(\lambda,\lambda,\lambda^{-2})/d\lambda)dR) \tag{2}$$

The material being incompressible, its volume remains constant which gives $r^3 = R^3 - R_i^3 + r_i^3$. By using $r = \lambda R$ in equation (2) and by replacing dR in equation (1), one can write as described by Pipkin (1970)

$$P_i - P_o = P(\lambda_i) - P(\lambda_o) \tag{3}$$

with

$$P(\lambda) = \int_1^{\lambda} \frac{1}{\lambda^3 - 1} \frac{d}{d\lambda} W(\lambda,\lambda,\lambda^{-2})d\lambda \tag{4}$$

The problem of formulating the mechanical properties of our material is thus resolved into the problem of determining the form of strain energy function for a pure homogeneous deformation. It was pointed out by Mooney (1940) that the strain energy function for an incompressible material can be expanded as an infinite series in terms of the first and second strain invariants I_1 and I_2 involving powers of (I_1-3) and (I_2-3). Treloar (1943) suggested the simple approximation $W = (G/2)(I_1-3)$, where G represents the shear stress modulus. Equation (3) becomes

$$(P_i - P_o)/G = 2(1/\lambda_o - 1/\lambda_i) - 0.5(1/\lambda_i^4 - 1/\lambda_o^4) \tag{5}$$

MATERIAL AND METHODS

Eighteen patients underwent a rapid injection cystometry test as part of a complete urodynamic investigation. We seek an experimental condition where the bladder can reasonably be approximated to an elastic body, to characterize the elastic part of our stress-strain relationship. This is done by fast stretching; the stretching time is short compared with the effective relaxation time. The condition of fast stretching can be satisfied by injecting a fluid at fast speed into the urinary bladder. The technique consists of injecting 100ml of saline solution at a rate of 33ml/s into the bladder, and recording the pressure continuously. The subjects were classified into three categories (see Table 1). Two catheters were used; one was a #F16 silicone Foley for injection and the other was a #F6 ureteral type for pressure recording. The latter was connected to a pressure gauge transducer which was sent to an amplifier and to a recorder. All injections were followed by, at least, a 3 minute observation period to allow the pressure to stabilize. The initial volume was about 4ml for a bladder thickness estimated at 3mm. The calculated shear stress modulus represents the elasticity, i.e., stiffness of the bladder wall at a given stress, and is measured in dyn/cm^2.

RESULTS

Equation (5) is plotted on Fig. 1; an experiment performed on a rubber balloon shows the same pattern. The pressure rises very quickly to a maximum and then, on continued inflation, falls slowly. For asymptomatic subjects, there is a constant rise in the average peak pressure for 5 consecutive and cumulative injections (see Table 1). By plotting the shear stress modulus G as a function of the internal extension ratio λ_i, it is found that G is not constant but increases relatively sharply (Fig. 2). For patients with neurogenic bladders, the average peak pressure at each injection is significantly higher ($p<0.05$) than normal for both males and females (Table 1). For acontractile bladders, there is little or no rise in pressure as the bladder is filled (Table 1). The bladder can almost be approximated to an "empty bag".

Table 1 Inflation Pressures for five Consecutive and Cumulative Injections

Type of Bladder	Sex	P_1^*	s.d.	P_2^*	s.d.	P_3^*	s.d.	P_4^*	s.d.	P_5^*	s.d.
Normal	M	4.4	1.3	5.6	0.5	8.6	0.9	11.5	3.2	17.8	6.6
(n = 11)	F	5.9	1.6	7.3	1.6	9.5	2.6	10.9	4.4	12.7	4.9
Neurogenic	M	82.7	37.8	100.5	31.2	-	-	-	-	-	-
(n = 4)	F	20.9	10.8	43.8	34.4	51.0	36.6	73.1	56.1	-	-
Acontractile	M	1.5	-	1.5	-	2.0	-	3.0	-	3.0	-
(n = 3)	F	1.5	0.7	2.9	1.4	4.9	1.4	6.4	0.7	7.8	1.4

*Pressures are expressed x10^{-3} in dyn/cm^2.

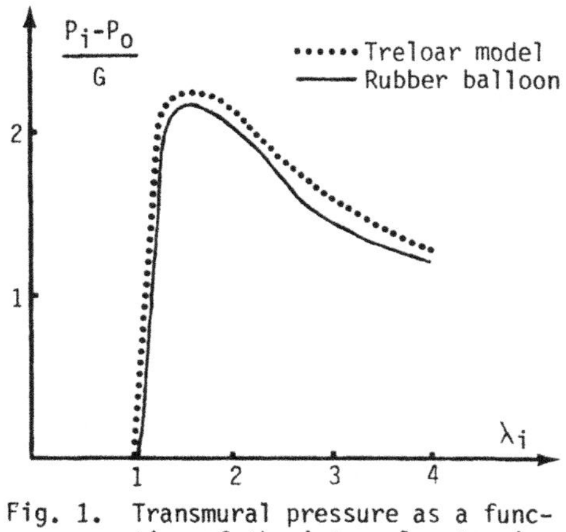

Fig. 1. Transmural pressure as a function of the internal extension ratio λ_i.

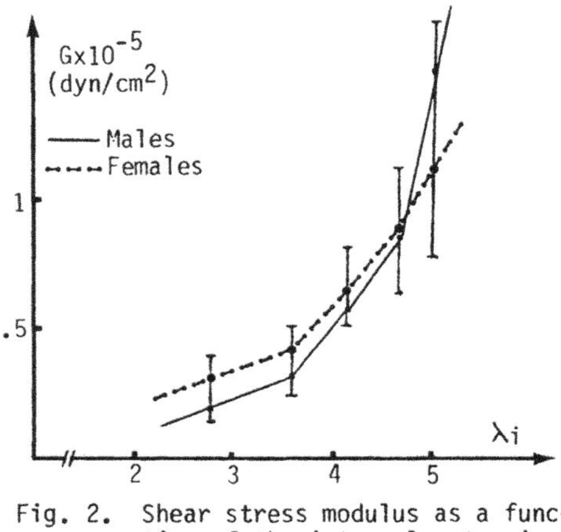

Fig. 2. Shear stress modulus as a function of the internal extension ratio λ_i.

DISCUSSION

Even when finite strain is considered, the elastic response of a normal bladder is entirely different from that of a rubber balloon; where the pressure within a balloon first exhibits a crest upon inflation and then decreases, that of a normal bladder rises almost monotonically. For urinary bladders, the shear stress modulus can be assumed to be an exponential function of the internal extension ratio λ_i (Regnier, 1981). For a rubber balloon, it is necessary to introduce a more general expression for the strain energy function of the form $W = (G/2)(I_1-3) + \Phi(I_2-3)$ to account for all the data due to the fact that the shear stress modulus increases with the internal extension ratio especially above 2.5. Fig. 2 (shear stress modulus vs extension ratio for a human bladder) shows that the increase in the shear modulus with the extension ratio is higher in magnitude from that for a rubber balloon, previously described. In such a case, the term we have introduced, $\Phi(I_2-3)$, is not sufficient to explain this behavior because it cannot account for such increases. If we recall the general form of strain energy function proposed by Mooney (1940) for vulcanized rubber, we may introduce another expression by expanding the former into a series of terms involving powers of (I_1-3) and (I_2-3). Two other major issues should be resolved before we can achieve our goal; the measurement of bladder thickness and bladder irritation. If we recall equation (5), we see that the shear stress modulus depends on internal and external radii. No direct means is presently available for their measurement. Lisco (1949) caculated the weight of the urinary bladder as 0.21% of the body weight and assumed it represented the weight of the bladder muscle but this value was not very well documented and concerned healthy subjects. X-rays provide some useful information about bladder thickness only when the muscle is hypertrophied and therefore cannot be a reliable tool for all patients. Ultrasound should probably be investigated but the resolution between bladder material and surrounding structures may be so small that it will be part of the error involved in the measurement. The other issue, bladder irritation, is suggested by the degree of inflammation of the bladder wall observed on patients having an indwelling catheter. The RIC test cannot appreciate quantitatively the "true" muscle properties due to the bladder hyperactivity. In the same way, if the RIC test is done immediately after emptying an overdistented bladder, one will only measure the very low response of a "flat bag" inflated by sudden 100ml increments of saline. We have a major artifact between the clinical implication of the patient's condition (e.g. insertion of catheter, bladder stone,...) and the pure passive mechanical properties of his bladder wall. In this sense, the RIC test is somewhat limited to an "instantaneous screening" of the bladder condition.

ACKNOWLEDGEMENT

The authors would like to express their gratitude to the National Institute of Arthritis, Metabolism and Digestive Diseases which supported this work under Grant No AM 26893.

REFERENCES

Lisco, L. (1949). Scan. J. Urol. Nephrol., Suppl. 6, 11-33.
Mooney, M. (1940). J. Appl. Phys., 11, 582-592.
Osborne, W. A. (1909). Proc. R. Soc. (London), B81, 485-499.
Pipkin, A. C. (1970). In Tremonese (Ed.), Non-linear Continuum Theories in Mechanics and Physics and their Applications. Roma. pp 57-67.
Regnier, C. H. (1981). M.Sc. Thesis, Brown University.
Rivlin, R. S. (1948). Phil. Trans. R. Soc. (London) A, 240, 459-508.
Treloar, L. R. (1943). Nature, 151, 616.

INTERCONVERSIONS OF VARIOUS TYPES OF SECANT AND INCREMENTAL MODULI FOR SOFT BIOLOGICAL TISSUE

Jafar Vossoughi and Ramesh N. Vaishnav
School of Engineering and Architecture
The Catholic University of America
Washington, D. C. 20064

ABSTRACT

Elastic behavior of most biological tissue is nonlinear in terms of the usual definitions of stress and strain. Because of the large geometric changes involved, two different types of stress (engineering stress and true stress) and several different types of strain (for example, engineering strain, Green-St. Venant strain and natural strain) can be used to describe uniaxial test data for such materials, with each choice of a stress-strain pair yielding a different curve. Linearizations using secant or incremental moduli of elasticity can be carried out in terms of many alternate definitions. For a choice of two stress and three strain descriptions we may define six sets of secant moduli and nine sets of incremental moduli. Here presented are explicit relationships in tabular form among the different types of secant and incremental moduli at arbitrary levels of deformation. It is hoped that the tables will serve as ready tools to compare and interconvert uniaxial test data presented by different investigators.

KEYWORDS

Incremental moduli, secant moduli, modulus of elasticity, soft tissue.

INTRODUCTION

The elastic stress-strain response of the vascular tissue is nonlinear in terms of the usual definitions of stress and strain (Patel and Vaishnav, 1980). Modern nonlinear theories of mechanics can be used to describe some essential aspects of this response. However, it is often convenient, particularly in uniaxial stress tests, to linearize the response, either by considering the stress at a certain level of strain as being proportional to the strain at that level with the constant of proportionality depending on the level of strain, or by considering increments of stress as being proportional to increments of strain, again with the constant of proportionality depending on the level of strain. The former idea leads to the definition of a secant modulus of elasticity, and the latter to that of an incremental modulus of elasticity (Vaishnav et al, 1981). Even at a specific level of strain, the values of secant and incremental moduli depend on the choice of the measures of stress and strain. Analytical relationships among the incremental moduli based on a few different important definitions of stress and strain have been reported elsewhere (Vaishnav et al, 1981). Here we give relations for interconversions among various measures of secant and incremental moduli of elasticity based on two definitions of stress and three of strain. With availability of these interconversion formulas one can readily obtain any one type of secant or incremental modulus at a chosen level of strain from the knowledge of any other type along with the value of the extension ratio at that level of strain.

STRESSES AND STRAINS

Consider a strip of a soft biological tissue with dimensions ℓ_o, b_o, h_o in the undeformed, unstressed state. Assume that when the strip is subjected to a uniaxial tensile stress in the direction along the dimension ℓ_o, the new dimensions of the strip become ℓ, b and h, respectively. If F denotes the force responsible for this extension, we may define the engineering stress S_i and the true stress S_f as

$$S_i = \frac{F}{b_o h_o} \; ; \quad S_f = \frac{F}{bh}$$

We can also define, in terms of the extension ratio $\lambda = \ell/\ell_o$, the following three types of strain:

$$\varepsilon = \lambda - 1; \quad \gamma = \tfrac{1}{2}(\lambda^2 - 1); \quad \nu = \ell n \lambda$$

which are the engineering strain, the Green-St. Venant strain, and the natural strain, respectively.

It is generally assumed that soft biological tissue is incompressible (Carew et al, 1968; Vossoughi et al, 1980). For such a material, $\ell bh = \ell_o b_o h_o$ which implies that $S_f = \lambda S_i$. In the following, we assume this relationship to hold.

Either of the two stresses S_i or S_f can be plotted against any one of the 3 strains ε, γ, or ν. We can thus have at least 6 distinct stress-strain curves for each uniaxial tensile test of a sfot biological tissue capable of undergoing large deformations. In the following we will designate the stresses S_i and S_f as the stresses of the first and second kind, respectively, and the strains ε, γ, and ν as the strains of the first, second, and third kind, respectively.

SECANT MODULI OF ELASTICITY

The ratio of a stress to a strain is called a secant modulus of elasticity. For the two types of stress and three types of strain we can have six definitions of secant moduli. For example, we can define the secant modulus of type 23 as

$$\overset{23}{E_s} = \frac{S_f}{\nu}$$

where the superscripts 23 denote the use of the second type of stress and the third type of strain. Other secant moduli can be defined in a similar manner. Table 1 gives interconversion multipliers for each type of secant modulus.

INCREMENTAL STRESSES AND STRAINS

Consider now the case where the dimensions ℓ, b and h are further changed to $\ell + d\ell$, $b + db$ and $h + dh$ by application of a small additional force dF in the direction parallel to the dimension ℓ. In this case it is possible to define three types of incremental stresses dS_i, dS_f and $d\tau$, and three types of incremental strains $d\varepsilon$, $d\gamma$ and $d\nu$, as

$$dS_i = \frac{dF}{b_o h_o} \; ; \quad dS_f = d\left(\frac{F}{bh}\right) = dS_i + S_i d\lambda; \quad d\tau = \frac{dF}{bh} = \lambda dS_i$$

$$d\varepsilon = d\lambda; \quad d\gamma = \lambda d\lambda; \quad d\nu = \frac{d\lambda}{\lambda}$$

For the significance of these incremental quantities, see (Vaishnav et al, 1981). We shall refer to dS_i, dS_f and $d\tau$ as the first, second and third incremental stress, and $d\varepsilon$, $d\gamma$ and $d\nu$, as the first, second and third incremental strain, respectively.

INCREMENTAL MODULI

The ratio of an incremental stress to an incremental strain gives rise to an incremental modulus. Since there are 3 incremental stresses and 3 incremental strains defined above, there are 9 possible incremental moduli in a uniaxial state of stress. For example, the incremental modulus of type 23 is given by

$$\overset{23}{E}_i = \frac{dS_f}{d\nu}$$

where the superscripts 23 denote the use of the second type of stress and the third type of strain in computing the modulus. Table 2 gives all possible relations among the incremental moduli of various types. It is hoped that Tables 1 and 2 will serve as a ready tool to interconvert and compare uniaxial test data presented by different investigators.

ACKNOWLEDGMENTS

We gratefully acknowledge the support of NSF grant CME8006338.

REFERENCES

Carew, T. E., R. N. Vaishnav, and D. J. Patel (1968). Circ. Res., 23, 61-68.

Patel, D. J. and R. N. Vaishnav (1980). Basic Hemodynamics and Its Role in Disease Processes, Univ. Park Press, 504 p.

Vaishnav, R. N., J. Vossoughi and D. J. Patel (1981). Proceedings of 34th ACEMB.

Vaishnav, R. N., J. Vossoughi and D. J. Patel (1981). In Advances in Bioengineering, D. C. Viano(Ed.), ASME NY, 151-153.

TABLE 1

	$\overset{11}{E}_s$	$\overset{12}{E}_s$	$\overset{13}{E}_s$	$\overset{21}{E}_s$	$\overset{22}{E}_s$	$\overset{23}{E}_s$
$\overset{11}{E}_s$	1	$\frac{(\lambda + 1)}{2}$	$\frac{\ell n\lambda}{\lambda - 1}$	$\frac{1}{\lambda}$	$\frac{(\lambda + 1)}{2\lambda}$	$\frac{\ell n\lambda}{\lambda(\lambda - 1)}$
$\overset{12}{E}_s$	$\frac{2}{(\lambda + 1)}$	1	$\frac{2\ell n\lambda}{(\lambda^2 - 1)}$	$\frac{2}{\lambda(\lambda + 1)}$	$\frac{1}{\lambda}$	$\frac{2\ell n\lambda}{\lambda(\lambda^2 - 1)}$
$\overset{13}{E}_s$	$\frac{\lambda - 1}{\ell n\lambda}$	$\frac{(\lambda^2 - 1)}{2\ell n\lambda}$	1	$\frac{\lambda - 1}{\lambda\ell n\lambda}$	$\frac{(\lambda^2 - 1)}{2\lambda\ell n\lambda}$	$\frac{1}{\lambda}$
$\overset{21}{E}_s$	λ	$\frac{\lambda(\lambda + 1)}{2}$	$\frac{\lambda\ell n\lambda}{\lambda - 1}$	1	$\frac{(\lambda + 1)}{2}$	$\frac{\ell n\lambda}{\lambda - 1}$
$\overset{22}{E}_s$	$\frac{2\lambda}{(\lambda + 1)}$	λ	$\frac{2\lambda\ell n\lambda}{(\lambda^2 - 1)}$	$\frac{2}{(\lambda + 1)}$	1	$\frac{2\ell n\lambda}{(\lambda^2 - 1)}$
$\overset{23}{E}_s$	$\frac{\lambda(\lambda - 1)}{\ell n\lambda}$	$\frac{\lambda(\lambda^2 - 1)}{2\ell n\lambda}$	λ	$\frac{\lambda - 1}{\ell n\lambda}$	$\frac{(\lambda^2 - 1)}{2\ell n\lambda}$	1

TABLE 2

	E_i^{11}	E_i^{12}	E_i^{13}	E_i^{21}	E_i^{22}	E_i^{23}	E_i^{31}	E_i^{32}	E_i^{33}
E_i^{11}	$E_i^{11} = E_i^{11}$	$E_i^{11} = \lambda E_i^{12}$	$E_i^{11} = \frac{1}{\lambda} E_i^{13}$	$E_i^{11} = \frac{1}{\lambda}(E_i^{21} - S_i)$	$E_i^{11} = E_i^{22} - \frac{S_i}{\lambda}$	$E_i^{11} = \frac{1}{\lambda^2}(E_i^{23} - \lambda S_i)$	$E_i^{11} = \frac{1}{\lambda} E_i^{31}$	$E_i^{11} = E_i^{32}$	$E_i^{11} = \frac{1}{\lambda^2} E_i^{33}$
E_i^{12}	$E_i^{12} = \frac{1}{\lambda} E_i^{11}$	$E_i^{12} = E_i^{12}$	$E_i^{12} = \frac{1}{\lambda^2} E_i^{13}$	$E_i^{12} = \frac{1}{\lambda^2}(E^{21} - S_i)$	$E_i^{12} = \frac{1}{\lambda}(E^{22} - \frac{S_i}{\lambda})$	$E_i^{12} = \frac{1}{\lambda^3}(E^{23} - \lambda S_i)$	$E_i^{12} = \frac{1}{\lambda^2} E_i^{31}$	$E_i^{12} = \frac{1}{\lambda} E_i^{32}$	$E_i^{12} = \frac{1}{\lambda^3} E_i^{33}$
E_i^{13}	$E_i^{13} = \lambda E_i^{11}$	$E_i^{13} = \lambda^2 E_i^{12}$	$E_i^{13} = E_i^{13}$	$E_i^{13} = E_i^{21} - S_i$	$E_i^{13} = \lambda E_i^{22} - S_i$	$E_i^{13} = \frac{1}{\lambda} E_i^{23} - S_i$	$E_i^{13} = E_i^{31}$	$E_i^{13} = \lambda E^{32}$	$E_i^{13} = \frac{1}{\lambda} E_i^{33}$
E_i^{21}	$E_i^{21} = \lambda E_i^{11} + S_i$	$E_i^{21} = \lambda^2 E_i^{12} + S_i$	$E_i^{21} = E_i^{13} + S_i$	$E_i^{21} = E_i^{21}$	$E_i^{21} = \lambda E_i^{22}$	$E_i^{21} = \frac{1}{\lambda} E_i^{23}$	$E_i^{21} = E_i^{31} + S_i$	$E_i^{21} = \lambda E_i^{32} + S_i$	$E_i^{21} = \frac{1}{\lambda} E_i^{33} + S_i$
E_i^{22}	$E_i^{22} = E_i^{11} + \frac{S_i}{\lambda}$	$E_i^{22} = \lambda E_i^{12} + \frac{S_i}{\lambda}$	$E_i^{22} = \frac{1}{\lambda}(E_i^{13} + S_i)$	$E_i^{22} = \frac{1}{\lambda} E_i^{21}$	$E_i^{22} = E_i^{22}$	$E_i^{22} = \frac{1}{\lambda^2} E_i^{23}$	$E_i^{22} = \frac{1}{\lambda}(E_i^{31} + S_i)$	$E_i^{22} = E_i^{32} + \frac{S_i}{\lambda}$	$E_i^{22} = \frac{1}{\lambda^2} E_i^{33} + \frac{S_i}{\lambda}$
E_i^{23}	$E_i^{23} = \lambda^2 E_i^{11} + \lambda S_i$	$E_i^{23} = \lambda(\lambda^2 E_i^{12} + S_i)$	$E_i^{23} = \lambda(E_i^{13} + S_i)$	$E_i^{23} = \lambda E_i^{21}$	$E_i^{23} = \lambda^2 E_i^{22}$	$E_i^{23} = E_i^{23}$	$E_i^{23} = \lambda(E_i^{31} + S_i)$	$E_i^{23} = \lambda(\lambda E_i^{32} + S_i)$	$E_i^{23} = E_i^{33} + \lambda S_i$
E_i^{31}	$E_i^{31} = \lambda E_i^{11}$	$E_i^{31} = \lambda^2 E_i^{12}$	$E_i^{31} = E_i^{13}$	$E_i^{31} = E_i^{21} - S_i$	$E_i^{31} = \lambda E_i^{22} - S_i$	$E_i^{31} = \frac{1}{\lambda} E_i^{23} - S_i$	$E_i^{31} = E_i^{31}$	$E_i^{31} = \lambda E_i^{32}$	$E_i^{31} = \frac{1}{\lambda} E_i^{33}$
E_i^{32}	$E_i^{32} = E_i^{11}$	$E_i^{32} = \lambda E_i^{12}$	$E_i^{32} = \frac{1}{\lambda} E_i^{13}$	$E_i^{32} = \frac{1}{\lambda}(E_i^{21} - S_i)$	$E_i^{32} = E_i^{22} - \frac{S_i}{\lambda}$	$E_i^{32} = \frac{1}{\lambda^2}(E_i^{23} - \lambda S_i)$	$E_i^{32} = \frac{1}{\lambda} E_i^{31}$	$E_i^{32} = E_i^{32}$	$E_i^{32} = \frac{1}{\lambda^2} E_i^{33}$
E_i^{33}	$E_i^{33} = \lambda^2 E_i^{11}$	$E_i^{33} = \lambda^3 E_i^{12}$	$E_i^{33} = \lambda E_i^{13}$	$E_i^{33} = \lambda(E_i^{21} - S_i)$	$E_i^{33} = \lambda(\lambda E^{22} - S_i)$	$E_i^{33} = E_i^{23} - \lambda S_i$	$E_i^{33} = \lambda E_i^{31}$	$E_i^{33} = \lambda^2 E_i^{32}$	$E_i^{33} = E_i^{33}$

THE TENSILE PROPERTIES OF RAY FISH SKIN

A. Rajaram and N. Ramanathan

Biophysics Lab., Central Leather Res. Inst.,
Adyar, Madras 600 020, INDIA

ABSTRACT

This paper deals with the mechanical properties of the dorsal and ventral skin of a scaleless ray fish, measured parallel and perpendicular to the body axis. The tensile properties have been related to the arrangement of collagen in the dermis. The mechanical function of the skin obtained from the pectoral fins is also discussed in terms of the structure and properties.

KEYWORDS

Skin mechanical properties; ray fish skin; collagen structure; tensile properties; connective tissues; structure and function.

INTRODUCTION

The mechanical properties of skin have been the subject of study of many investigators over the years (Harkness, 1971; Viidik, 1976). A considerable amount of work has been done mainly on mammalian skin and efforts have been made to explain the properties from a theoretical point of view (Lanir, 1979). Even two dimensional analyses have been performed (Lanir and Fung, 1974; Cook and Alexander, 1976). However, there is comparatively little information about the mechanical properties of fish skins. Only the tensile strength seems to have been measured (Yamaguchi, Lavety and Love, 1976). In the present study, the mechanical properties of ray fish skin have been investigated and an attempt is made to correlate the properties with the structure and function of the tissue.

MATERIALS AND METHODS

Samples were obtained from the dorsal and ventral surfaces of the pectoral fins of the scaleless ray fish, Mobula diabolus, called the lesser devil ray. The approximate diameter of each fish was 50 cm. The ventral surface was white while the dorsal surface was pigmented black. The ventral skin was also thicker than the dorsal skin and was found to wrinkle up on removal from the body in a direction parallel to the body axis. After removing the flesh, samples of gauge length 1.2 cm. and width 4 mm were punched out with a die (Ridge and Wright, 1964) from an area 10 cm x 10 cm on the ray fish pectoral fins. Samples were taken in parallel and perpendicular directions with respect to the body axis and were stored in polythene bags at -20°C. A few drops of chloroform were sprayed into the bags to prevent bacterial attack. Altogether about thirty samples were obtained in each case and they were tested within three days of procurement. The tensile testing was performed with an Instron tensile testing machine at an extension rate of 0.5 cm/min. The tensile stress-strain characteristics were drawn by the averaging procedure suggested by Meredith (1945) after finding out the cross-sectional area from the geometrical measurements.

The dermal region of the samples was examined with a Cambridge Stereoscan SEM after removing the epidermis and suitable coating with gold in a sputtering unit. Sections (10-15 μ thick) were prepared for optical microscopy by the procedure of Gamori (1952). The sections were viewed in polarised light to differentiate collagen which exhibits birefringence. Histochemical staining was resorted to for identifying elastin (Fullmer, 1965).

RESULTS AND DISCUSSION

The stress-strain characteristics in tension for ray fish skin from the ventral and dorsal surfaces tested parallel and perpendicular to the body axis are shown in Fig. 1. Due to the crimped nature of ventral skin samples in the direction perpendicular to the body axis, the sample extends highly in this direction before

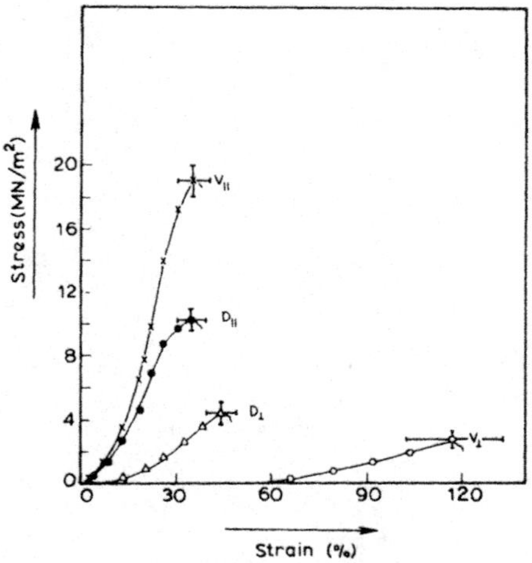

Fig. 1. Stress-strain characteristics of ray fish skin. V - ventral skin; D - dorsal skin. (∥) - parallel to body axis; (⊥) - perpendicular to body axis.

rupture. The extensions at break obtained here in the different cases is reasonable considering that for mammalian skin various values ranging from 15 percent to more than 100 percent have been noted (Lanir and Fung, 1974; Yamada, 1970). The tensile strength of skin also varies considerably. Veronda and Westman (1970) obtained a value of 12.5 MN/m^2 for cat skin. Yamaguchi, Lavety and Love (1976) have obtained values for fish skins ranging from 2.7 to 4.6 MN/m^2. The strength is higher in a direction parallel to the backbone which corresponds to the direction of Langer's lines and is lowest in the direction perpendicular to it (Harkness, 1972). The stress-strain characteristics can be explained in terms of the structure of the ray fish skin. The structure as observed in the SEM is shown in Fig. 2. It is evident that in the dorsal skin the fibres run in orthogonal directions with respect to the body axis. Considering the ventral skin, the fibres are predominantly uni-directional along the body axis. This accounts for the higher strength of the ventral skin in the direction of the body axis and lower strength in the orthogonal direction. The observations made in the SEM were

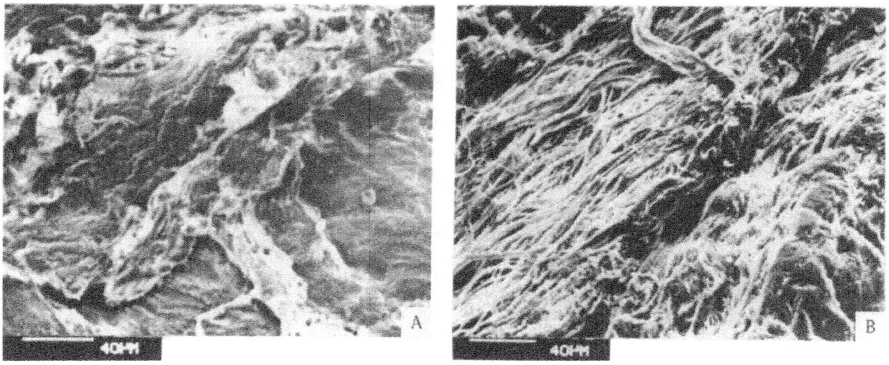

Fig. 2. Structure of ray fish skin dermis viewed in the SEM. (A) - Dorsal skin; (B) - Ventral skin.

supported by the optical studies of the sections made in polarised light. For cross-sections parallel to the body axis a high degree of orientation was observed for ventral skin; it was low for samples perpendicular to this direction and intermediate for the dorsal skin in both directions. The staining for elastin was more in the ventral skin samples perpendicular to the body axis, for which maximum breaking extension was observed.

The arrangement of the collagen fibres in the pectoral fin of the ray fish seems to be the result of the functional demands on the skin. Since the fish has a disc-like structure, the ventral skin has to bear the body weight and when the skin is removed, it contracts in a direction parallel to the body axis. The same phenomenon is observed in the belly skin of the hippopotamus which contracts on removal from the body (Harkness, 1972). This effect has been attributed to the elastin content. A circular hole punched on the ventral skin of the ray fish becomes oblong along the body axis showing that the Langer's lines are along the body axis.

Another aspect to be considered is the form of locomotion of the ray fish. It swims by the transverse motion of the pectoral fins with the waves increasing in amplitude towards the tail (Alexander, 1974). So the stress is more along the body axis and is obviously borne by the collagen fibres along this direction. The difference in arrangement of collagen in ray fish skins is very much in contrast to the orthogonal array of collagen in the skin of other types of fish which move by the longitudinal motion of their body (Nadol, Gibbins and Porter, 1969). Thus the mechanical properties and arrangement of collagen in ray fish seems to be a result of the functional demands on the skin.

REFERENCES

Alexander, R. McN. (1974). Functional Design in Fishes. B.I. Publications, Bombay.

Cook, T., H. Alexander, and M. Cohen (1971). Experimental method for determining the two dimensional properties of living human skin. Med. Biol. Eng. Comput., 15, 381-390.

Fullmer, H.M. (1965). In D.A. Hall (Ed.), Int. Rev. Conn. Tiss. Res., Vol. III Academic Press, New York and London. Chap. 1, pp 1-65

Gamori, G. (1952). Microscopical Histochemistry. Chicago University Press, Chicago.

Harkness, R.D. (1972). In H.R. Elden (Ed.), Biophysical Properties of Skin, Wiley Interscience, New York. Chap. 11, pp. 393-437.

Lanir, Y., and Y.C. Fung (1974). Two dimensional mechanical properties of rabbit skin - II Experimental results. J. Biomechanics, 7, 171-182.

Lanir, Y. (1979). A structural theory for the homogeneous biaxial stress-strain relationships in flat collagenous tissues. J. Biomechanics, 12, 423-436.

Meredith, J. (1945). The tensile behaviour of raw cotton and other textile fibres J. Text. Inst., 36, T107-T130.

Nadol, J.B., J.R. Gibbins, and K.R. Porter (1969). Collagen structure in basal lamella of fish skin. Devel. Biol., 20, 304-331.

Ridge, M.D., and V. Wright (1964). The description of skin stiffness. Biorheology, 2, 67-74.

Veronda, D.R., and R.A. Westman (1970). Mechanical characterisation of skin - finite deformation. J. Biomechanics, 3, 111-124.

Viidik, A. (1976). In D.A. Hall and D.S. Jackson (Ed.s), Int. Rev. Conn. Tiss. Res., Vol. VI. Academic Press, New York and London. Chap. 4, pp. 127-215.

Yamada, H. (1970). Strength of Biological Materials. William and Wilkins, Baltimore.

Yamaguchi, K., M. Lavety, and R.M. Love (1976). The connective tissues of fish - VIII Comparative studies on hake, cod and catfish collagens. J. Fd. Technol., 11, 389-399.

Session 9
MATHEMATICAL MODELLING

A SIMPLIFIED MODEL FOR PREDICTING MYOCARDIAL PO2 HISTOGRAMS

Roy W. Schubert*, John E. Fletcher** and D. D. Reneau*

*Department of Biomedical Engineering, Louisiana Tech University
Ruston, Louisiana 71272
**Laboratory of Applied Studies
Division of Computer Research and Technology
National Institutes of Health, Bethesda, Maryland 20205

ABSTRACT

A simplified mathematical model for oxygen transport to tissue was developed by replacing radial diffusion in the tissue with a finite capillary wall permeability and an infinite radial diffusion constant. The effect of this assumption is assessed by comparisons to a fully distributed model. The assumption is justified for parameters indicative of the cat heart. It is concluded that the simplified model adequately represents the transport of oxygen to heart tissue.

KEYWORDS

oxygen; microcirculation; mathematical ; heart; histograms; Krogh cylinder; model; transport; capillary; PO2

INTRODUCTION

The tissue level distribution of oxygen (O2) in the hemoglobinless perfused heart has been measured under controlled conditions with a Whalen PO2 microelectrode (Schubert, Whalen and Nair, 1978). In an attempt to understand tissue level details of O2 transport we compared that PO2 distribution to those predicted by mathematical models. Models incorporating only radial transport in the tissue compared poorly with the measured histograms. Blum's (1960) model with radial and axial diffusion was found to be in error (Schubert, 1976), and a correct solution had not been published. A simpler model was derived by modifying the radial transport phenomena so that axial diffusional transport could be included. This model displays trends seen in the experimental data. Unfortunately the best match was obtained by increasing the value of the axial diffusion coefficient, D, considerably beyond that accepted in the literature. This led to questions about the appropriatness of the radial transport assumptions. This could be answered only by seeking the correct solution to the problem originally posed by Blum. Salathe and Wang (1980) obtained the general mathematical solution, but not the correct specific solution. Fletcher and Schubert (1981) have presented the correct specific solution, although the computation is not trivial and is computer intensive. In this paper we outline the simplified model and compare its predictions with the complex model.

METHODS

Mass conservation principles applied to a Krogh cylinder arrangement of capillaries produced an equation for oxygen transport in the capillary:

$$dCc/dz=2*B1*(Ct-Cc) \quad (1)$$

where $B1=P/(Rc*v)$. This assumes axial convective transport and radial diffusion out of the capillary. P is capillary wall permeability; Rc, capillary radius; v, convective velocity; and Cc and Ct, oxygen concentration in the capillary and tissue respectively. The equation for transport in the tissue was:

$$D*(d^2Ct\ /dz^2)+B2*(Cc-Ct)=M \quad (2)$$

where $B2=(2*Rc*P)/((Rt^2)-(Rc^2))$; M, oxygen consumption; and Rt, tissue cylinder radius. Their is no radial flux to adjacent structures. Combining the above equations results in a third order ordinary differential equation. The solution is of the form:

$$Ct=C1+C2*z+C3*e^{\lambda_1*z}+C4*e^{\lambda_2*z} \quad (3)$$

where $\lambda_1, \lambda_2 = -B1 \pm \sqrt{B1^2+B2}$. The constants are found by applying the boundary condition that Cc is known at the exit and no axial diffusion flux exists at either end of the cylinder.

The more exact complex model was derived by using mass balance principles but radial transport was distributed. The details of this solution can be found elsewhere (Fletcher and Schubert,1981; 1982). It should be noted that the partial differential equation solution is in the form of an infinite set of modified Bessel functions. Problems arise when this slowly converging series is truncated at N. A Gibb's phenomena becomes apparent and can be removed. Nonphysical oscillations in the solution, produced by truncation can be smoothed by using a Cesaro' summation procedure. The results of this model show a plateau effect at the venous end of the tissue, reduction of the radial gradient at the venous end and enhancement of the radial gradient at the inlet.

RESULTS

Model parameter for cat heart (Fletcher and Schubert, 1982) were used. P was calculated to produce a radial flux equal to that from a space averaged tissue PO2 (Krogh cylinder). Fig. 1 shows details of the profile of Ct for the simplified (lower solid line) and complex (broken line) model as well as that predicted by the classic Krogh cylinder (upper solid line). Fig. 2 is the distribution of oxygen predicted by the models for a normal axial D. Fig. 3 shows results when the axial D is increased to twice its normal value. Fig. 4 represents the goal of our model matching in the form of measured PO2 distributions.

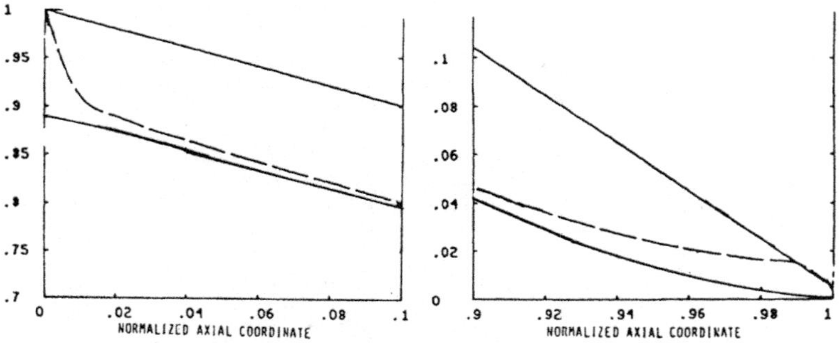

Fig. 1. Exit and entrance effects of axial diffusion on Ct.

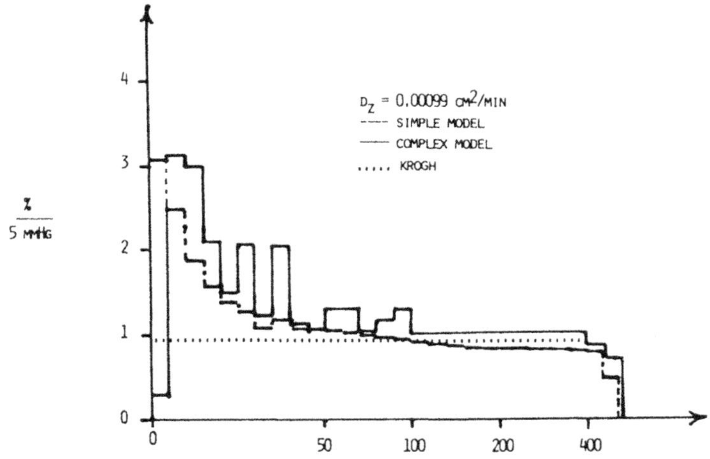

FIG. 2. PO_2 HISTOGRAM FOR NORMAL AXIAL DIFFUSION COEFFICIENT

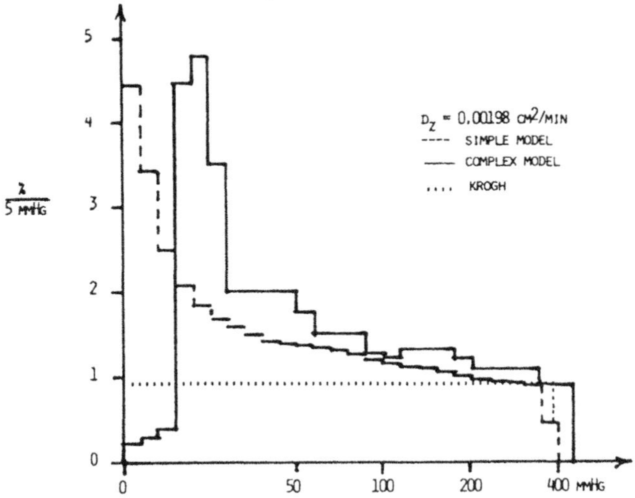

FIG. 3 PO_2 HISTOGRAM FOR LARGE AXIAL DIFFUSION COEFFICIENT

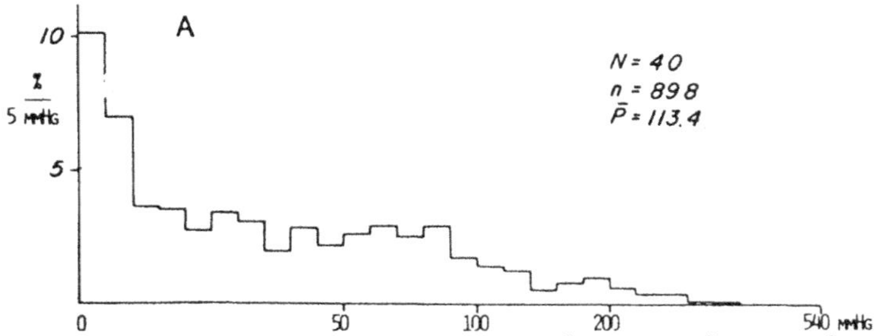

Fig. 4. Experimentally measured PO2 histograms from cat heart.

DISCUSSION

The left panel of Fig. 1 indicates that the model does not reflect entrance affects. This is a direct result of the radial modeling assumption. It is clear from the right panel that the model demonstrates trends seen in the complex model. The sudden turn to a low PO2 value at the venous end shown by the complex model has been shown to be an artifact generated by the series solution (Fletcher and Schubert, 1981). If one conceptually removes this artifact from the results then the models compare well and the distributions shown in Fig. 2 and Fig. 3 will be left shifted so that the peak will be against the ordinate. It is obvious from Fig. 2 and Fig. 4 that the classic Krogh cylinder falls far short of predicting the measured results. The inclusion of axial diffusion in both models shows the left shift seen in the experimental results. This is especially true when the value of the axial D is increased, Fig. 3. The rough appearance of the histogram from the complex model is a result of finite N and the physical cell size used to compute the histogram. The smooth prediction from the simple model would be preferred in any parameter fitting technique. Both models respond favorably to increased values of the axial D, and both need an axial diffusion coefficient in excess of the value accepted in the literature in order to mimic the experimentally measured distribution.

CONCLUSION

It can be concluded that the simple model of oxygen transport to heart tissue is as good as or perhaps better than the complex model in reflecting oxygen histograms at low PO2. It is disturbing that the model does not match the experimental data, except when the axial D is increased. Transport phenomena that might improve the match include axial diffusion in the capillary, Michaelis-Menten kinetics, and facilitated diffusion by myoglobin. On the other hand it might be reasonable to entertain the idea that the apparent axial D could be elevated in contracting heart. The complex model is the only model to reflect the steep radial gradients at the capillary inlet.

ACKNOWLEDGEMENT

This work was sponsored by Grant HL-22864 and NS-08802 from the USPHS.

REFERENCES

Blum, J.J. (1960). Concentration Profiles in and around Capillaries. Am. J. Physiol., 198, 991-998.

Fletcher, J.E. and Roy W. Schubert (1981), On the Computation of Substrate Levels in Cell Free Perfused Tissues. Submitted to Math. Biosci.

Fletcher, J.E., and Roy W. Schubert (1982). The Theoretical Prediction of Substrate Levels and Their Histograms in Cell Free Perfused Tissues. In Press, 1981 ISOTT Proceedings.

Schubert, Roy W. (1976). A Physiological and Mathematical Study of Oxygen Distribution in the Autoregulating Heart. Ph.D. Thesis, Case Western Reserve University.

Schubert, Roy. W., W.J. Whalen and P. Nair (1978). Myocardial PO2 Distribution: Relationship to Coronary Autoregulation. Am. J. Physiol., 234, H361-H370.

Salathe, E. and T. Wang (1980). Substrate Concentrations in Tissue Surrounding Single Capillaries. Math. Biosci., 49, 235-247.

MICHAELIS MENTEN KINETICS AS A MODELING ASSUMPTION IN A MODEL OF OXYGEN TRANSPORT IN HEART

Stan A. Napper and Roy W. Schubert

Department of Biomedical Engineering
Louisiana Tech University
Ruston, Louisiana 71272

ABSTRACT

The Michaelis-Menten kinetics model oxygen consumption rate is investigated as a modeling assumption in a mathematical study of oxygen transport to heart tissue. The model is derived from the classic Krogh capillary tissue cylinder geometry and based on physiological experiments in isolated cell-free perfused autoregulating cat hearts. This model is compared to a basic simplified model and to real data of oxygen distribution. Michaelis-Menten kinetics is found to be an improvement over the basic model for matching histograms of real data. However, further modeling assumptions will need to be studied because this effect falls short of completely matching the real data.

KEYWORDS

Michaelis-Menten kinetics, oxygen, heart, math model, histogram, microcirculation, transport, Krogh cylinder,cat

INTRODUCTION

A simple model for the steady state distribution of oxygen in tissue of heart was first derived by Schubert (1976) as the mathematical portion of a synergistic physiological/mathematical investigation into the mechanisms of local blood flow control in the heart. The laboratory experiments were conducted by recording oxygen tension via microelectrode from an isolated, isovolumic autoregulating cat heart perfused with cell-free Krebs-Heinsleit solution (Schubert, Whalen, and Nair; 1978). The basic model assumptions used to simulate the oxygen transport in this preparation include radially well-mixed capillary and tissue compartments, axial diffusion in the tissue,and constant oxygen consumption rate (Schubert, Fletcher and Reneau; 1982). These assumptions result from realization that heart capillaries are relatively long and thin. Comparisons between model and experiment are made through an oxygen distribution or PO2 histogram. Schubert discovered that experimental tissue PO2 histograms could be matched with this simplified model if the value of the axial diffusion coefficient was raised several times beyond reported values (Fletcher and Schubert, 1982). This large effective diffusion coefficient indicates that some other phenomena may be occuring which have the effect of increased diffusion rate. One phenomena which may help explain this effect is more accurate consumption model through nonlinear kinetics. The choice of nonlinear kinetics as a modeling assumption is logical from the understanding of how Michaelis-Menten kinetics actually affects the oxygen consumption. Consumption rate is lower at lower values of PO2 so that the gradient for axial diffusion is decreased towards the end of the cylinder where PO2 values are low. This effect might be mimiced as an increased apparent diffusion coefficient.

METHODS AND PROCEDURES

Derivation of the model equations follows classic deterministic techniques. Those equations are presented here.

Capillary:

$$dCc/dz = 2*B1*(Ct-Cc)$$

Tissue:

$$D *d(dCt/dz2)+B2*(Ct-Cc) = M*Ct/(Ct+Km)$$

where $B1=P/(r_c*v)$, $B2=(2*P*r_t)/(r_t*r_t-r_c*r_c)$, P is capillary wall permeability, r_c is capillary radius, r_t is tissue radius, M is maximum oxygen consumption, and Km is the Michaelis-Menten constant. The values for D , r ,L and M are appropriate cat heart values (Schubert, Whalen and Nair,1978; Fletcher and Schubert, 1982). P is determined from a radial space-average in the tissue which calculates P to account for the radial diffusion. Axial perfusate velocity,v ,and tissue radius are adjusted to a per capillary basis through the capillary density and whole organ flow. Km values are in the physiological range for oxygen consumption and are varied to establish a trend. The equations were normalized with respect to capillary entrance concentration and to length.

Keeping the experimental data in mind, the lowest value of PO2, which corresponds to venous concentration, is consistently between 0 and 5 mmHg. Therefore, one boundary condition is a value of 1.0 mmHg for Cc at the end of the cylinder. The common no-flux tissue conditions have been chosen for the other two constraints. An Adams-Bashforth-Moulton numerical integration scheme was employed for the third order ordinary nonlinear differential equation. A shooting technique was used to establish the initial conditions which produced the desired final values. A Simplex minimization was employed to obtain these required initial values.

RESULTS

Figure 1 shows an experimental PO2 histogram from the cat heart. Figure 2 showsthe low end (0 to 50 mmHg) of a PO2 histogram from the Michaelis-Menten kinetics model for Km values of 0,1, and 2 mmHg. A Km of 0 reduces to the basic model using constant consumption. Figures 3 and 4 are expanded views of the first 5% and last 10%, respectively, of the concentration profile down the Krogh cylinder length, in normalized variables. These figures show the relative effects of the different Km values.

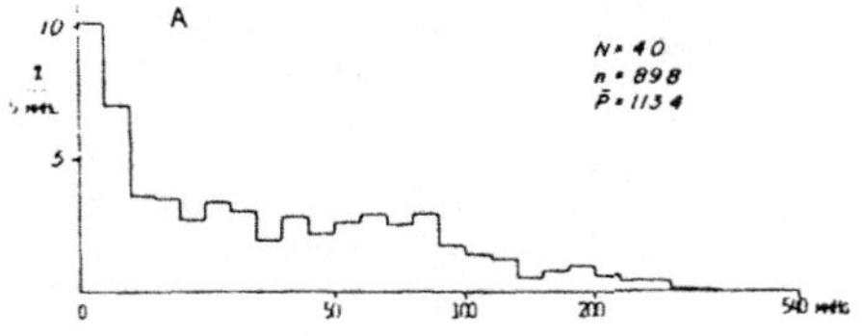

Fig. 1. Experimental PO2 histogram

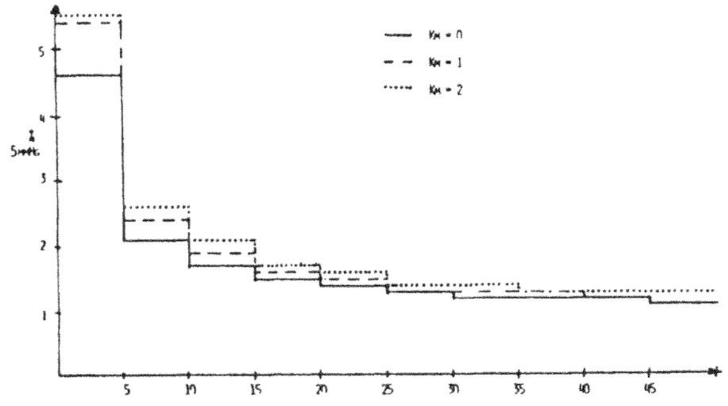

Fig. 2. Model PO2 histogram

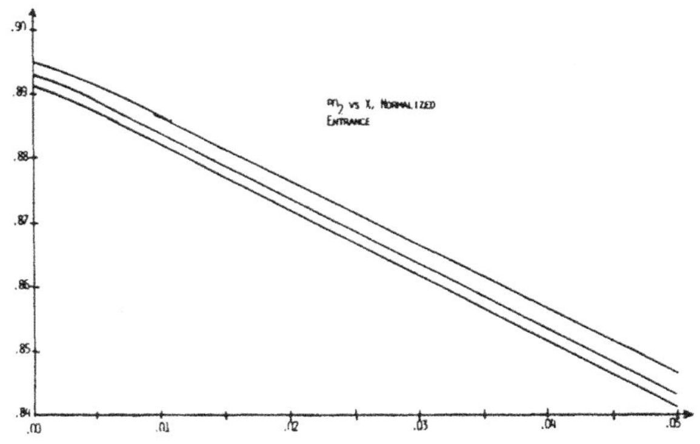

Fig. 3. PO2 profile in entrance region with normalized variables

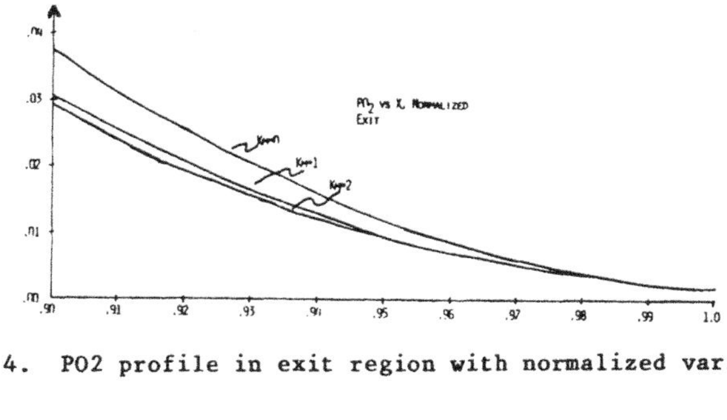

Fig. 4. PO2 profile in exit region with normalized variables

DISCUSSION

Figure 1 presents the ideal or the goal of the modeling work, that is, the physiological results which an accurate model must simulate. Figure 2 indicates that the Michaelis-Menten kinetics model does shift the histograms leftward as in the real oxygen histograms. Figure 3 merely indicates the different starting values required to produce final tissue PO2 concentration of 1 mmHg. The model proved to be very sensitive to the initial values, though this is expected from the exponential nature of the solution. Figure 4 presents further evidence that Michaelis-Menten kinetics has the effect of producing lower PO2 values as found experimentally. The figures also indicate that this modeling consideration is not enough to account for the high diffusion rates predicted by the basic model.

CONCLUSION

The effect of this modeling assumption on the ability of the model to predict oxygen tension histograms is clearly an improvement over the basic model with normal diffusion coefficient. More physiological consumption rates may then be part of the reason that a high diffusion coefficient is found. However, Michaelis-Menten kinetics falls short of actually matching the real data. Therefore, other modeling assumptions such as axial diffusion in the capillary and the oxygen-binding capacity of myoglobin should be investigated to further explain the high diffusion coefficient. The model may also be indicating that the values reported as normal diffusion coefficient for heart are not valid in isolated, autoregulating hearts where a certain degree of mixing may be possible because of the contractions.

ACKNOWLEDGEMENT

This work was supported by grant HL-22864 from the USPHS

REFERENCES

Fletcher, J.E., and Roy W. Schubert (1982). The Theoretical Prediction of Substrate Levels and Their Histograms in Cell Free Perfused Tissues. In Press, 1981 ISOTT Proceedings.

Schubert, Roy W. (1976). A Physiological and Mathematical Study of Oxygen Distribution in the Autoregulating Heart. Ph.D. Thesis, Case Western Reserve University.

Schubert, Roy. W., W.J. Whalen and P. Nair (1978). Myocardial PO2 Distribution: Relationship to Coronary Autoregulation. Am. J. Physiol., 234, H361-H370.

Schubert, Roy W., J.E. Fletcher and D.D. Reneau (1982). A Simplified Model for Predicting PO2 Histograms. 1st Southern Bioengineering Conference. June 1982.

Transport Limitation of Oxygen in Multicell Spheroids as Determined by Mathematical Modeling

R.S. Artigue and D.E. Mead

Rose-Hulman Institute of Technology
Terre Haute, IN 47803

ABSTRACT

A mathematical model has been developed to investigate the transport limitation of oxygen in tumor spheroids. The effects of zero order and Michaelis Menten kinetics for spheroids with and without necrotic cores have been studied. Results indicate elevated concentration profiles for cases employing Michaelis Menten kinetics compared to zero order, especially for those cases where oxygen supply is limited. Also elevated profiles were found as the necrotic core increases in size, indicating that oxygen supply is not limited and would not be responsible for the continued growth of the necrotic core.

KEYWORDS

Mathematical modeling; tumor spheroids; Michaelis Menten kinetics; necrotic core.

INTRODUCTION

Typically in vivo cancerous tumors are characterized by reduced vascularity and perhaps compression of the vasculature and subsequent reduction of blood flow by the expanding cellular mass (Rubin and Casarett, 1966). Because of the complex nature and wide variations of tumor vascularity in vivo, theoretical and experimental studies have used multicell spheroids as physiological models of tumors. Grown in vitro in oxygenated and nutritional medium, these cancer spheroids exhibit most of the characteristics of their in vivo counterparts (Sutherland and Durand, 1973; Franko and Sutherland, 1979).

Histological studies of multicell cancer tumors have shown that degenerative and necrotic cores progressively develop as the tumors grow past a certain diameter (Folkman and Hochberg, 1973). These cores consist of a center of dead tissue surrounded by a layer of non-cycling cells which may be aerobic, hypoxic or both. Cycling cells are found in the outer region of the tumors. The hypoxic fraction of the cells is often functioning in some altered, slowly metabolizing state and is the most resistant to irradiation (Durand and Sutherland, 1973; Hilland Pallavicini, 1981).

Since it has been postulated that the degenerative cores arise in spheroids from the transport limitations of oxygen, theoretical oxygen diffusion distances are investigated by mathematical modeling. Both zero order and Michaelis Menten kinetics have been studied with and without necrotic cores.

OXYGEN TRANSPORT MODEL

Assuming steady-state, spherical geometry and the homogenous distribution of metabolism and constant physical properties within indentifiable regions, the mathematical description of the model becomes

$$D \frac{1}{r'^2} \frac{d}{dr'} (r'^2 \frac{dP'}{dr'}) = M/\alpha \tag{1}$$

Here P' is the oxygen partial pressure for any radial position r' within the spheroid, D is the oxygen diffusion coefficient, α the oxygen solubility coefficient and M the metabolic rate. This equation is de-demensionalized with the following equations.

Let $P = \frac{P'}{P_o'}$ and $r = \frac{r'}{r_o'}$ where P_o' is the partial pressure of oxygen at the outer boundary of the sphere and r_o' is the radius of the sphere. Then by substitution and rearranging equation (1) becomes

$$\frac{1}{r^2} \frac{d}{dr} (r^2 \frac{dP}{dr}) = K \quad \text{where} \quad K = \frac{Mr_o^2}{D\alpha P_o} . \tag{2}$$

The boundary conditions that complete the mathematical description of the model depend on the case being studied. Another parameter introduced, besides K, concerns the relative thickness of the necrotic region (r_*).

For the case of no necrosis: $r_* = 0$ and for $1 \leq r \leq 0$

$$P|_{r=1} = 1 \quad \text{and} \quad \frac{dP}{dr}\Big|_{r=0} = 0 \tag{3}$$

The demensionless parameter (K) is constant for zero order kinetics and

$$K = \frac{Mr_o^2}{D\alpha P_o} \left(\frac{P}{K_m'+P}\right) \text{ for Michaelis Menten kinetics} \tag{4}$$

such that $K_m' = K_m/P_o$.

For the case where necrosis is present: $1 > r_* > 0$ and for $1 \geq r \geq r_*$

$$P|_{r=1} = 1 \quad \frac{dP}{dr}\Big|_{r=r_*} = 0 \tag{5}$$

Note that in the necrotic region ($r_* \geq r \geq 0$) the oxygen consumption rate is zero ($K = 0$).

Solutions have been obtained for constant metabolism with and without necrosis. The solution technique used in this case is straightforward integration. Also solutions have been obtained using Michaelis Menten kinetics for cases with and without a necrotic core. The Crank-Nicolson finite difference scheme is employed to solve these cases. The solutions for the various cases studied are plotted as a family of curves with K, r_*, and K_m' as parameters.

RESULTS

In Fig. 1 the maximum diffusion distance of oxygen in the spheroid is defined where a specific set of parameters result in K=1. Also the sensitivity of the position of the profiles to K is shown. Figure 2 demonstrates the effect of varying K_m' and in Fig. 3 the effect of the size of the necrotic region is displayed.

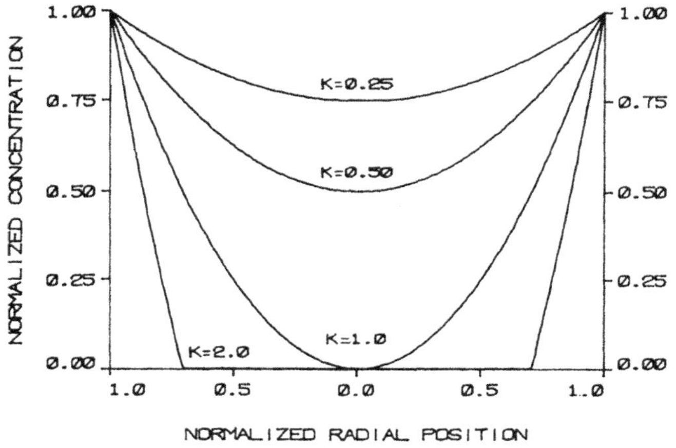

Fig. 1. Effect of K on normalized concentration in spheroid (zero order kinetics).

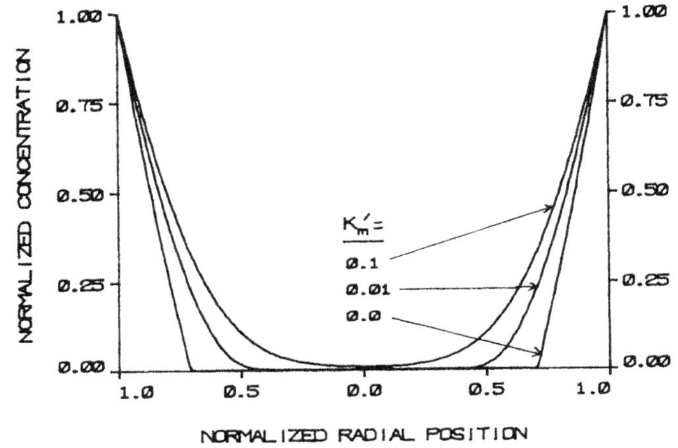

Fig. 2. Effect of K_m' on normalized concentration in spheroid with K = 2 (Michaelis Menten kinetics).

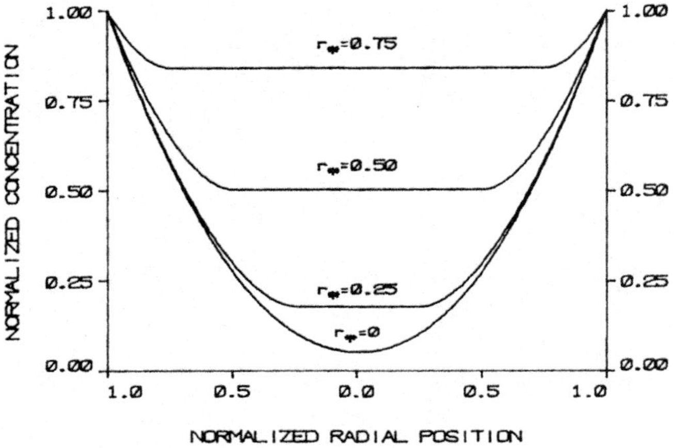

Fig. 3. Effect of r_* on normalized concentration in spheroid with K = 1 and $K_m{}'$ = 0.01 (Michaelis Menten kinetics).

CONCLUSIONS

The results from this modeling effort indicate that as expected, for the cases using Michaelis Menten kinetics, the oxygen partial pressure profiles are somewhat elevated relative to those in which a constant metabolic rate was used. Because of experimental evidence which indicates an increasing partial pressure of oxygen in the center of the spheroids as they grow past approximately 200 microns in diameter (Carlsson and others, 1979), it seems illogical that the continued growth of the necrotic region is the result of low oxygen levels. In light of past work and these results, the oxygen transport limitation would seem to induce the morphological changes that give rise to the identified three regions within large spheroids. And it is more likely that the transport limitation of another metabolite, whether into or out of the spheroid, is responsible for the continued growth of the necrotic core and non-cycling ring. Further work using multi-component mathematical models with various metabolic schemes will test this hypothesis.

REFERENCES

Carlsson, Jörgen, Carl-Göran Stalnacke, Helmuth Acker, Mohammed Haji-Karim, Sverker Nilsson and Börje Larsson (1979). Int. J. Radiation Oncology Biol. Phys., 5, 2011-2020.
Durand, R.E. and R.M. Sutherland (1973). Cancer Research, 33, 213-219.
Folkman, J. and M. Hochberg (1973). Journal of Exp. Medicine, 138, 745-753.
Franko, A.J. and R.M. Sutherland (1979). Radiation Research, 79, 439-453.
Hill, R.P. and M.G. Pallavicini ISOTT Presentation, Aug. 1981, (Detroit, MI).
Rubin, P. and G. Casarett (1966). Clin. Radiol., 17, 220-229.
Sutherland, R.M. and R.E. Durand (1973). Int. J. Radiat. Biol., 23, 235-246.

The Transport of Oxygen, Glucose, Carbon Dioxide and Lactic Acid in the Human Brain : Mathematical Models

R.S. Artigue and D.F. Bruley

Rose-Hulman Institute of Technology
Terre Haute, IN 47803
Louisiana Tech University
Ruston, LA 71272

ABSTRACT

Mathematical models of the transport of oxygen, glucose, carbon dioxide and lactic acid in the human brain have been developed and solved for both steady-state and dynamic cases which include normal conditions, arterial and venous hypoxia, reduced flow and hematocrit, hypocapnia, reduced glucose and transient arterial upsets. Results indicate that local concentration deficits of oxygen and/or glucose and excesses of lactic acid can exist within the system described by the models and therefore possibly could exist in the tissues of the human brain under similar conditions.

KEYWORDS

Mathematical modeling; Krogh Cylinder ; distributed parameter; Michaelis Menten kinetics; human brain.

INTRODUCTION

Mathematical modeling of oxygen transport had its beginning with the work of Krogh (1919a, 1919b) and more recently by others (Opitz and Schneider, 1950; Kety, 1957; Reneau and others, 1969; McCraken and others, 1972; Fletcher, 1975). These modeling efforts have either built upon the work of others by adding more detail in describing the transport processes and geometrical considerations, or studied more completely a certain aspect of the transport mechanisms.

The present study incorporates many of the important complex interactions of chemical components related through cellular metabolism and other important physiological phenomena. A few of these that are essential to the understanding of the cellular environment are the effect of carbon dioxide in shifting the oxygen dissociation curve (Bohr effect), the effect of oxygen release from hemoglobin on the carbon dioxide dissociation curve (Haldane effect), the formation of lactic acid when oxygen becomes limiting in oxidative metabolism, and the variation of blood flow due to upsets in arterial component concentrations.

The importance of developing distributed parameter models, even though the solution complexity is much greater than with a lumped approach, is that lumped models can only provide information on average values of component concentration and can not demonstrate the local concentration differences and gradients that arise in convective, reactive and diffusive systems.

TRANSPORT MODELS

Using the geometry of the Krogh Tissue Cylinder and assuming homogenous distribution of cellular metabolism and constant physical properties, the mathematical description of the models can be written as follows.

Capillary Equations

The capillary equation for each component can be written

$$\alpha_i\left(\frac{\delta C_i}{\delta t} + v\,\frac{\delta C_i}{\delta z}\right) + \beta_i + J_i = D_i\,\frac{\delta^2 C_i}{\delta z^2} + K_i \Delta C_i \qquad (1)$$

where C_i is the ith specie concentration, t is time, z is the axial position, D_i is the diffusion coefficient. The variables α_i and β_i account for the Bohr and Haldane effects for oxygen and carbon dioxide. J_i represents the mass transfer out of or into the capillary and K_i represents the interphase transport between the red blood cell and plasma.

Tissue Equations

The tissue equations can be written as

$$\frac{\delta C_i}{\delta t} = \nabla^2 C_i + M_i$$

where M_i is the metabolic rate for ith component. For oxygen the metabolism follows Michaelis Menten kinetics and for carbon dioxide, glucose and lactic acid, the metabolism is stoichiometrically related to that of oxygen.

Solution Techniques

With the appropriate boundary and initial conditions, the Price, Varga, Warren and Crank-Nicolson finite difference techniques were used to solve for the various cases studied.

RESULTS

In Fig. 1 the solution for the steady-state distribution of oxygen and carbon dioxide in the capillary are shown. Figure 2 demonstrates the oxygen tissue profiles at the venous end of the capillary for various blood flow rates. And in Figure 3 the lactic acid concentrations in the capillary are shown for various conditions.

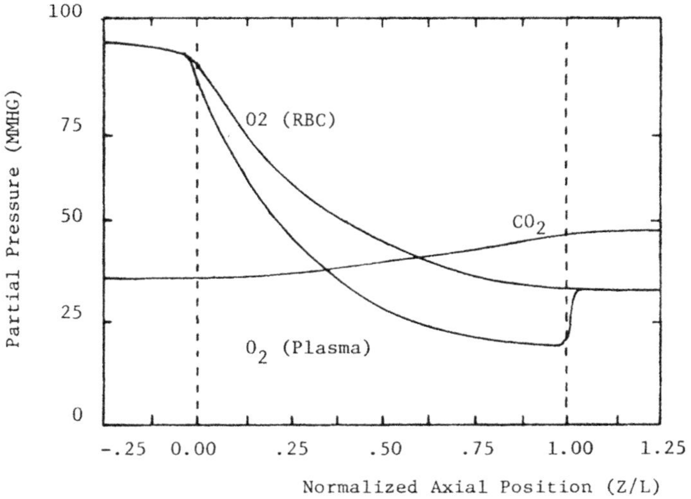

Fig. 1 Capillary oxygen and carbondioxide partial pressure profiles. (Normal conditions)

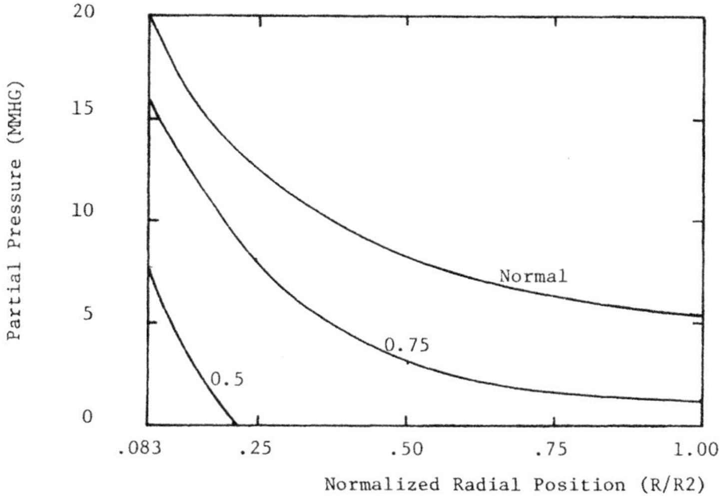

Fig. 2 Steady-state oxygen partial pressure profiles in the tissue at the venous end of the capillary. (Velocity as the parameter)

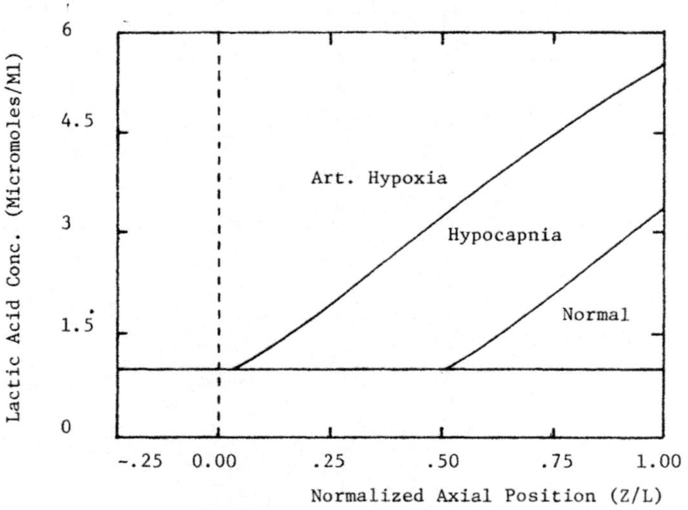

Fig. 4 Capillary lactic acid concentration profiles. (Conditions as indicated)

DISCUSSION

The models developed in the study have shown that under conditions of reduced flow and/or reduced arterial concentrations of carbon dioxide, an anoxic region forms in the tissue near the venous end of the capillary with a concomittant elevation in the lactic acid concentration.

REFERENCES

Fletcher, J.E. (1973). Adv. Exp. Med. Biol., Plenum Press, (Bruley and Bicher, eds.)
Kety, S.S. (1957). Fed. Proc., 16, 666-670.
Krogh, A. (1919). J. Physiol., 52, 409-415.
Krogh, A' (1979). J. Physiol., 52, 457-474.
McCracken, T.A., D.F. Bruley, D.D. Reneau, H.I. Bicher, and M.H. Knisely (1972). 1st Pacific Chemical Engineering Congress, Kyoto, Japan.
Opitz, E. and M. Schneider (1950). Erge. Physy. Chemie Exp. Pharm., 46, 126-260.
Reneau, D.D., D.F. Bruley, and M.H. Knisely (1969). AIChE, 15, 916-925.

ANALOG MODEL OF THE VERTEBRATE CONE RETINA

R. Siminoff

Institut für Physiologie der Freien Universität
Berlin, Arnimallee 22, 1000 Berlin, 33, F.R.G.

ABSTRACT

An electronic simulation of the generalized vertebrate cone retina is described using linear operational amplifiers for retinal elements, each of which consists of 2 components: a Linear Summator to algebraically add inputs and a Leaky Integrator to simulate synaptic delays for the outputs. Properties of retinal neurons are developed by the spatial organization of inputs and temporal interaction of antagonistic inputs, one of which has a longer delay. The model simulates the physiological properties of phasic and tonic neurons of the retina. The balance of center and surround field inputs to bipolar cells is shown to play a vital role in the development of the phasic and tonic channels. Negative feedback from horizontal cells to cones and electrical coupling between horizontal cells potentiate the effects of surround field input on center field responses.

KEYWORDS

Analog Simulation, Modelling, Retina, Vertebrate, Cone Mosaic, Phasic, Tonic, Horizontal Cell, Bipolar Cell, Amacrine Cell, Ganglion Cell, L-Type.

INTRODUCTION

This electronic simulation of the vertebrate cone retina is based on a theoretical, but empirical, model (Siminoff, 1980a,b, 1981a,b). A key feature of the model is that the spatial ordering of the cone retina is deterministic with rules for specifying how the entire cone mosaic is wired up to form the receptive fields of retinal neurons (Siminoff, 1980a). The bipolar cell (BC) is the basic building block for the visual system. The spatial organization of the BC is defined by its receptive field with the center field (CF) formed by direct inputs from a specified pattern of cones, while the antagonistic surround field (SF) is formed via horizontal cells (HC) acting as voltage inverters (Siminoff, 1980a). The properties of retinal neurons are developed by the spatial organization of inputs and the temporal interaction of antagonistic inputs, one of which has a longer delay. The complete model calls for both a color-coded C-channel and a non-color-coded L-channel. However, only the L-channel will be described and with a number of simplifications.

SPATIAL ORGANIZATION OF CONE MOSAIC

The cone retina consists of a 2-dimensional grid of red-, green- and blue- sensi-

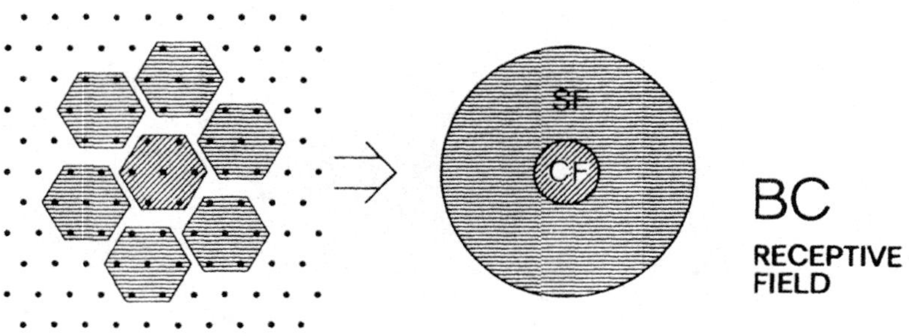

Fig. 1. Spatial Organization of Cones Into Unit Hexagons.

tive cones organized precisely by alternation of 2 primary rows (Siminoff, 1980a). However, the model is simplified by use of an approximately 11 X 11 grid of 115 photodiodes, each with the same spectral responses. As stressed many times (Siminoff, 1980a,b, 1981a,b), the most important feature of my model is that the cone mosaic can be ordered into unit hexagons (UH). The CF is represented by a UH of 7 cones (Fig. 1), while the SF is formed by 6 UHs, each of which forms the input to a surround-HC. The 6 HCs converge to form the SF of the BC. The entire retina can be organized in terms of these non-overlapping UHs and each UH forms the CF of a set of BCs and forms the SFs of 6 such sets of BCs. This leads to a topographic organization of the retina in terms of modules of UHs (Siminoff, 1980a).

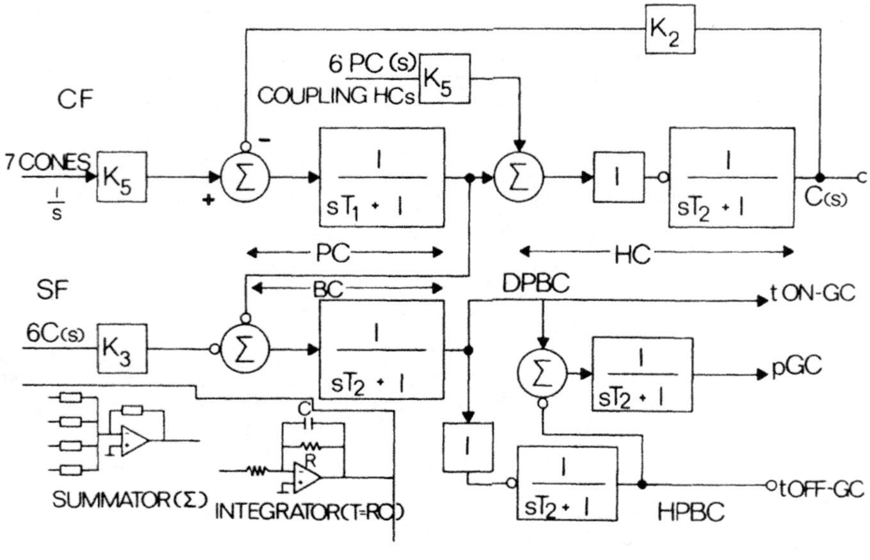

Fig. 2. Block Diagram of Analog Model.

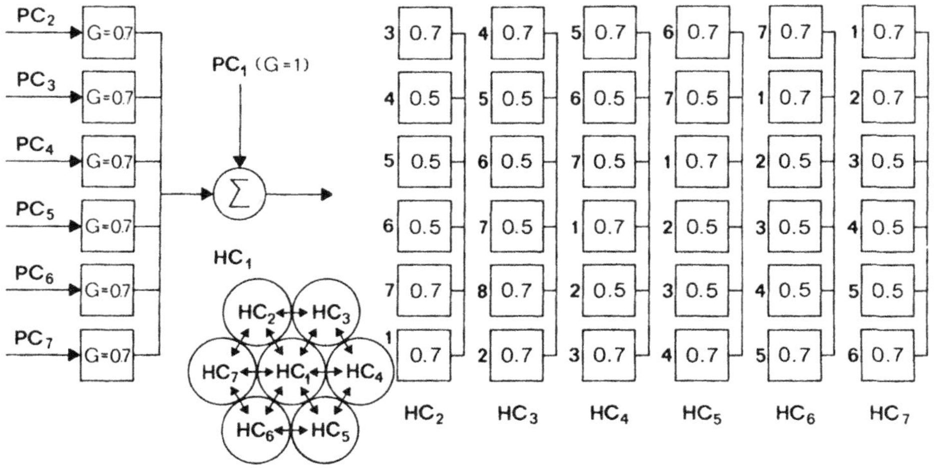

Fig. 3. Paradigm for Electrical Coupling of Horizontal Cells.

ANALOG MODEL

Linear operational amplifiers performing simple processes are used to make the retinal neurons, each of which consists of 2 components, a Summator ( ), which algebraically adds the inputs and a Leaky Integrator (Laplace Transform given by 1/(sT + 1)), which simulates the synaptic delay of the output. Operational amplifiers are used additionally for voltage inversion and gain control. The Summator of the BC (Fig. 2) receives its direct inputs from a single cone pedicle (PC), which is used in the analog model instead of 7 separate PCs from 7 separate cones to spare components. The SF of the BC is formed by convergence of 6 surround-HCs, each with a gain of K3. The HC, besides acting as the path for the SF of the BC, functions as a negative feedback to the PC with a gain of K2. HCs are coupled together electrotonically (Siminoff, 1980a) and this feature was simulated by networks of resistances at the output of each PC (Fig. 3). Two types of BCs are formed in parallel (Fig. 2); a depolarizing-center (DPBC) and a hyperpolarizing-center (HPBC) bipolar cell. For the HPBC, an Inverter (I) is used to reverse the voltage so that activation of the CF produces hyperpolarization (-) and activation of the SF produces depolarization (+). Tonic (t) ganglion cells (GC) are formed by direct input from the appropriate BC (Fig. 2). The GCs are simulated by voltage-to frequency converters so that the output of the GC is a train of standardized spikes with a frequency proportional to the positive analog input. For the OFF-GC, the negative input from the HPBC causes a decrease in the spontaneous activity. A phasic channel is produced by convergence of the HPBC and DPBC outputs to form the phasic (p) amacrine cell (AC), which in turn acts as the input for the pGC. A key feature to the formation of the pAC is that the HPBC input has an extra delay.

MODEL'S PERFORMANCE

The responses of the analog model were determined for both stationary and moving light spots of varying sizes and intensities. For stationary light spots, the tonic units produced non-adapting voltages (or changes in spike rates) proportional

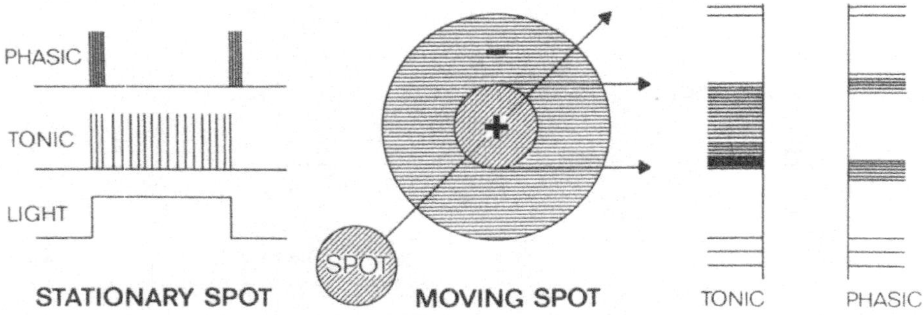

Fig. 4. Response Patterns of Phasic and Tonic Ganglion Cells to Stationary and Moving Light Spots.

to the light intensity, while for phasic units, brief transient voltages (or spike bursts) were seen only at the Onset and/or Offset (Fig. 4) of the light spot. Tonic units responded best to either stationary or slowly moving light spots when passing through the excitatory CF of the ON-GC (but inhibition of spontaneous activity for the OFF-GC), while phasic units responded best to faster moving light spots and only when either entering or leaving the excitatory CF (Fig. 4).

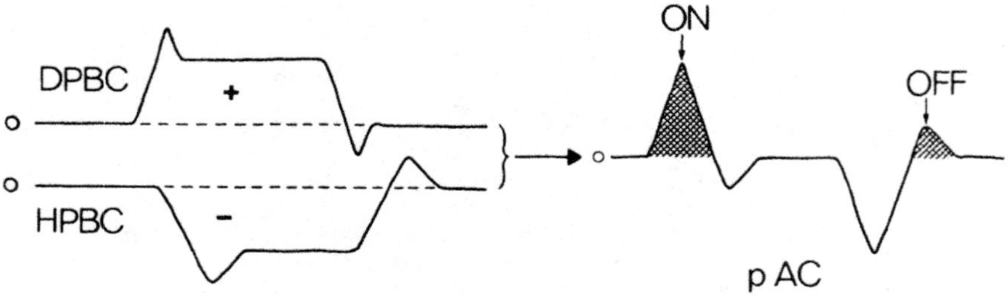

Fig. 5. Formation of Phasic Amacrine Cell by Convergence of DPBC and HPBC Inputs with Differing Delays.

The pAC is formed by convergence of the DPBC and HPBC with the HPBC input having an extra delay. Two conditions must be met for the phasic channel. (1). The 2 inputs must be equal and opposite to concel out any static levels (Fig. 5). The transient components are due to (a) the SF having an extra delay as compared to the CF, and (b) the HPBC having an extra delay compared to the DPBC. (2). The HPBC must have an extra delay which produces the positive OFF transient since the SF input of the HPBC (which is positive) has the longest delay. Thus, the pAC has a transient ON phase due to the CF input of the DPBC and a transient OFF phase due to the SF input of the HPBC. The negative components of the transients are lost since only positive voltages drive the GCs.

For tonic BCs, the response is the CF input decreased by the amount of SF input (Fig. 6). Thus, the tonic channel is seen best with a minimal SF input. The SF

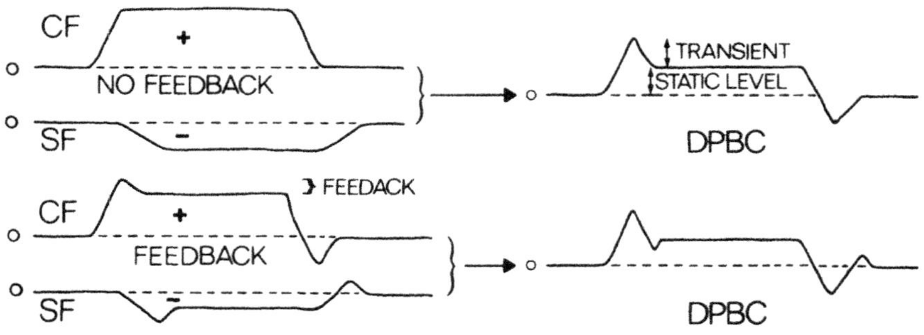

Fig. 6. Role of Negative Feedback and Surround Field Input on Receptive Field Organization of Retinal Neurons.

input has an extra delay as compared to the CF input, so that the SF input produces transient ON and OFF phases to the tonic responses of BCs (Fig. 6). Negative feedback from HCs to cones and electrical coupling of HCs act to potentiate the effects of the CF input, so unlike the tonic channel, the phasic channel is favored by a SF input. Thus, the formation of the phasic channel is done at the expense of the tonic channel and for each animal there must be a balance between the 2 channels dependent on the animal's visual specialization and needs.

When testing the model for linearity the non-linear photodiodes were by-passed and square wave voltages (Laplace transform given by 1/s) were introduced directly to the PCs. As to be expected, the model behaved as a stable linear system from the PCs to the GCs. The 2 criteria of linearity were observed; the law of superposition held and sinusoidal inputs passed through the system undistorted provided the working ranges of the operational amplifiers were not exceeded. The analog model of the L-channel of the vertebrate cone retina is described in detail in 3 papers submitted to Vision Research for publication.

ACKNOWLEDGEMENTS

This work was supported in part by funds from the Volkswagen Foundation while I was a Fellow of the Fritz Thyssen Foundation. The continued support and encouragements of Professor O.-J. Grüsser are acknowledged.

REFERENCES

Siminoff, R. (1980a). J. theor. Biol. 86, 673-708.
Siminoff, R. (1980b). J. theor. Biol. 87, 307-347.
Siminoff, R. (1981a). J. theor. Biol. 91, 437-476.
Siminoff, R. (1981b). Vision Res. 21, 1537-1539.

SIMULATION OF EPIDERMAL CELL MOVEMENT

Mahlon C. Smith
Mechanical Engineering Department
Michigan State University
East Lansing, Michigan 48824
U.S.A.

ABSTRACT

A two dimensional Monte Carlo computer simulation of epidermal cell movement is constructed using an hydraulic pressure transient postulated as arising from random mitosis of basal layer cells. Results show little lateral cell movement and essentially constant rate of progress of cells from the basal layer to the stratum corneum. Cell movements associated with "proliferative cell units" and wound healing cannot be accounted for by this postulation. Other forces contributing to cell dynamics, three dimensional simulations and stimuli and protocols necessary for more complex cell motions are discussed.

KEY WORDS

Cell motion; epidermal cells; simulation of cell motion

INTRODUCTION

In the steady state epidermis, basal cells divide to provide required new cells. This mitosis is not confined to the basal layer in the growing and abnormal skin (Sarsfield, 1978). In both steady state and growth, epidermal cells move from their point of generation to the surface of the epidermis or toward the "migrating edge" of the epidermis in a wound healing mode. The basis and details of this movement are not well known. Little information is available also concerning the forces producing cell movement, while cell self-movement in the laboratory has been described in terms of contratile and regulatory proteins that govern the generation of internal and external microfilament bundles.

The intent of the work reported herein is to simulate cell movement based on postulated force mechanisms and stimulation-response mechanisms.

REVIEW

Primary cell reproduction in the epidermis occurs in the basal cell layer, although in growth and wound healing, some suprabasal cells enter into mitosis. In the normal steady state the basal cells eventually (several days after leaving the basal layer) arrive at the lower layers of the stratum corneum in a flattened, keratinized state.

Electron micrographs of skin sections show that basal cells are bound to the basement membrane by attachments called hemi-desmosomes (Breathnoch, 1971). In addition, the cells are bound to one anouther at the cell membrane by similar attachments called desmosomes. It is inferred from various studies that the cells have the ability to form, dissolve, reform and break the desmosome attachments. The breaking of the attachment involves, apparently, the rupture of one of the cells' membranes (Mishima, Pinkus, 1968). In some instances, commonly observed, basal cells undergo mitosis such that the equatorial plane is perpendicular to the basement membrane (Bullough, Mitrani, 1978). It can be expected that this process, and, indeed, any basal cell movement requires also a breaking of the hemi-desmosomes. It may be noted, however, that in normal epidermis few detached desmosomes are observed.

There is evidence that, in some animals, cells do not participate in mitosis and journey to the surface of the epidermis in a random manner. In these instances the flattened keratinized cells of the stratum corneum have a hexagonal shape in the plan view whose edges match and interlock in a geometric pattern (Peck, Elias, Wetzel, 1977). Arguments have been given on experimental and phenomenological grounds concerning the inevitability or appropriateness of the ultimate hexagonal cell shapes (Menton, 1976). Some of these arguments contain the postulation that cells surrounding a core cell of the unit selectively engage in mitosis in preparation for keratinization and forming of the hexagonal shape (Potter, 1975).

Wound healing studies using a simple knife cut with wounds being microsectioned at short time intervals, have produced considerable detail on the healing process (Ordman, Gillman, 1966). However, the process is little understood in terms of stimulation and control of epidermal cell movement.

Results of experimental studies (Vaughn, Trinkhous, 1966; Bang, Daniels, Buck, 1978) have produced some results, but the movement of cells individually or in mass and the importance of hemidesmosomes in cell motion remain undetermined. Computer simulations have been used to calculate the formation of epidermal cell columns (Honda, Morita, Tanabe, 1979) and of cell migration (Goel, Rogers, 1978a) and engulfment (Goel, Rogers, 1978b).

Other models have been used to represent self-deformation of cells in various epithelia (Odell, Oster, Burnside, Alberch, 1980), and in determining the stability of cell membranes and cell colonies (Greenspan, 1976). All of these models and simulations have yielded some descriptions of cell motion not previously available.

PRESSURE TRANSIENT MODEL

The overall pressurized environment (due to atmospheric pressure) in which the skin exists insures that there must be a continuum of solids and liquids in the epidermis, i.e., no vacuums, open gas filled spaces or separations can exist or occur amongst the various parts of the epidermis or between it and the dermis. The importance of this fact is that no portion of the epidermis can move without causing movement of adjacent parts and indeed of every part of the epidermis, although vanishingly small at far distances.

It is postulated that cell movement in the normal epidermis is brought about by transient pressures occurring during cell expansion prior to mitosis. Thus, cellular movement in the epidermis is associated, in general, with mitosis, the generation of new cells. The stimulation for mitosis of a particular cell is not known. On an average, the concepts of mitotic control through such agents as the "mesenchymal factor" and "chalones," which stimulate and inhibit mitosis, respectively, can be called upon. For present purposes, it suffices to provide a probability of mitosis for any keratinocyte which is a function of age and position of the cell.

A Monte Carlo simulation in which random basal cell mitosis produces hydraulic pressure transients appropriate to the generation of a daughter cell shows that only very small horizontal cell motion occurs. The vertical motion is essentially at constant speed as shown by the results given in Fig. 1.

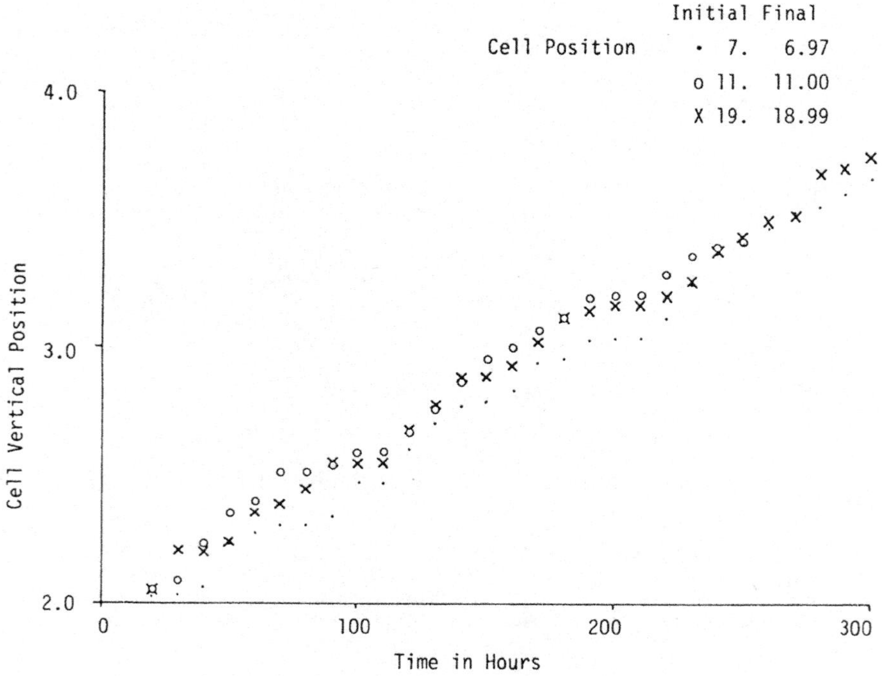

Fig. 1. Vertical position versus time.

INTER-CELLULAR DYNAMICS

Since continuity of the epidermal fluid mass requires any cell motion to be affected by the presence of the surrounding mass, a preliminary step would be to consider a single cell in an infinite homogeneous fluid and determine the forces needed to deform the cell membrane. In a simple instance, a single microfilament can be considered to be contracting. If the cell membrane is assumed to be massless and impervious to the fluid, then a classical fluid mechanical solution would yield the force in the tonofilament as a function of the acceleration and eventually steady state velocity of the desmosome attachment points. These would depend upon cell size, fluid density and viscosity, and desmosome velocity. Thus, an approximate method can be considered which consists primarily of assuming a quasi-steady, incompressible, viscous flow. The summation of all motions will be an approximation to the movement of the cell.

STIMULUS-RESPONSE

The concept of the proliferative cell unit and laboratory evidence regarding the healing of wounds requires epidermal cell motility which cannot be satisfied by simple pressure forces and movement with unbroken desmosome attachment. A realistic simulation of epidermal cell movement must include a description of cell self-movement. These movements are presumed to take place under the rules of a stimulation-response protocol. Various protocols may be postulated and used in simulation experiments. For example, in the proliferative unit concept, it may be that the unit cells contain an identifiable chemical species which stimulates the cell to leave the basal layer and assume its position as a roof for the unit.

REFERENCES

Bang, S. C., W. H. Daniels, and R. C. Buck (1978). Epidermal migration during the healing of suction blisters in rat skin: a scanning and transmission microscopic study. Am. J. Anat., 153, 177-192.
Breathnoch, J. and A. Churchill (1971). An atlas of the ultrastructure of human skin. London
Bullough, W. and E. Mitrani (1978). The significance of vertical mitosis in epidermis. Brit. J. Dermatology, 99, 603.
Goel, N.S. and G. Rogers (1978a). Computer simulation of engulfment and other movements of embryonic tissues. J. Theor. Biol., 71, 103-140.
Goel, N.S. and G. Rogers (1978b). Computer simulation of cellular movements: cell sorting, cellular migration through a mass of cells and contact inhibition. J. Theor. Biol., 71, 141-166.
Greenspan, H. P. (1976). On the growth and stability of cell cultures and solid tumors. J. Theor. Biol., 56, 229-242.
Honda, H., T. Morita, and A. Tanabe (1979). Establishment of epidermal cell columns in mammalian skin: computer simulation. J. Theor. Biol., 81, 745-749.
Menton, D. (1976). A liquid film model of tetrakkaidecahedral packing to account for the establishment of epidermal cell columns. J. of Invest. Dermatology, 66, No. 5.
Mishima, Y. and H. Pinkus (1968). Electron microscopy of keratin layer stripped human epidermis. J. of Invest. Dermatology, 50, No. 2, 89.
Odell, G., G. Oster, B. Burnside, and P. Alberch (1980). A mechanical model for epithelial morphogenesis. J. of Math. Biol., 9, No. 3.
Ordman, L. J. and T. Gillman (1966). Studies in the healing of cutaneous wounds. I. The healing of incisions through the skin of pigs. Archives of Surgery, 93, No. 6, 857.
Peck, G., P. Elias, and J. Wetzel (1977). Effects of retinoic acid on embryonic chick skin. J. of Invest. Dermatology, 69, No. 5.
Sarsfield, P. (1978). Epidermal expansion during the postnatal growth period in the mouse. Thesis, Dept. of Anatomy, U. College Galway, Galway, Ireland.
Vaughan, R. B. and J. P. Trinkhous (1966). Cell movements of epithelial cells in vitro. J. Cell Science, 1, 407-443.

Session 10

KNEE BIOMECHANICS

TRANSMISSIBILITY FUNCTION OF THE HEALTHY HUMAN KNEE (IN VIVO STUDY)

A. S. Voloshin and C. P. Burger

Department of Engineering Science and Mechanics
and Engineering Research Institute
Iowa State University, Ames, Iowa 50011, USA

ABSTRACT

The principles and results of a noninvasive in vivo procedure to evaluate the transmissibility function of the healthy knee are presented. The technique described here concentrates on the ways in which a shock wave is transmitted through the knee joint. Measurements, therefore, have been taken from two points close to the knee. The obtained signals were fed into the computer through a 12-bit resolution A/D board, stored, and later accessed for evaluation by the FFT procedure. These results allow us to calculate the knee's transmissibility function.

Thus, experimentally obtained transmissibility function of the human knee may serve as input information for constructing a dynamic model of the healthy human knee. It seems, therefore, that the use of the noninvasive in vivo technique in the assessment of the knee's transmissibility function holds promise as being both a challenging and rewarding endeavour. The potential for clinical use of this technique is worthy of further pursuit.

KEYWORDS

Human knee; transmissibility function; shock waves, diagnosis; accelerometry; digital data acquisition.

INTRODUCTION

The development of satisfactory methods for evaluating the conditions of the human knee in vivo may provide a useful tool for clinicians. To achieve a reliable technique for performing such measurements, the experimental procedure must be designed to obtain reproducible results by using a noninvasive in vivo approach. Since the knee is the largest joint in the human musculo-skeletal system, it has attracted the attention of researchers in an attempt to develop characterization procedures, as well as mathematical models, of this joint (Crowninshield and co-workers, 1976; Crowninshield, Pope, and Johnson, 1976; Pope and co-workers, 1976).

This paper describes a noninvasive in vivo technique to acquire data on bone vibration at two points just below and above the knee joint. After processing, these data yield an equivalent transmissibility function for the knee. Such a function reflects the knee's ability to transmit and dissipate dynamic forces and may be useful in diagnostic and prophylactic procedures.

METHODOLOGY

An accelerometric technique was used to measure bone vibrations at two points in close proximity to the knee joint: namely, at the tibial tuberosity (TT) and the medial femoral condyle (MFC). The accelerometers were held tightly to the points of interest by using commercially available knee braces which have a special construction that tends to keep constant, equal tension on all parts of the knee joint. The 15-meter-long cable from the power units to the minicomputer was soft and light, and, in addition, it was carried by an assistant following the subject. The subject walked naturally, with bare feet, along a hard-surface walkway eight-meters long. The step length was normal to the subject while average cadence was adjusted to 100 steps/min or 1.2 sec for a complete walking cycle (Fig. 1).

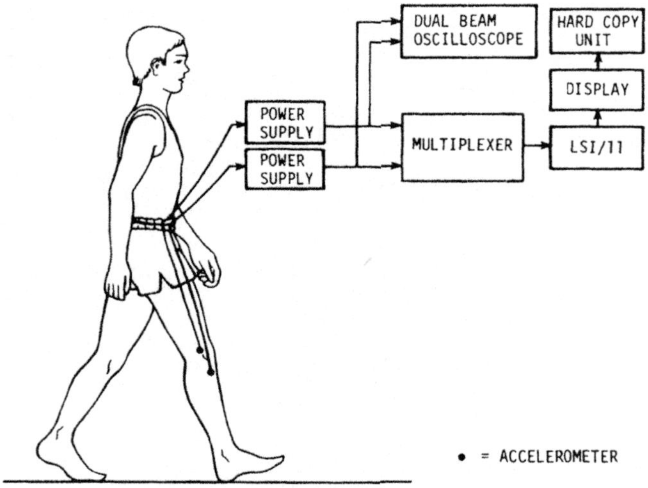

Fig. 1. Experimental set-up.

The sampling rate was 500 samples/sec/channel which allows for accurate evaluation of frequency components up to 250 Hz. The data were acquired over a period of 4 seconds, so more than three complete walking cycles were recorded and stored on a floppy disk for later evaluation. The experiments were performed on five male subjects, twenty years old. They were healthy and without any history of knee disorders.

RESULTS

Observation of the results obtained from the specified two points reveals that these are changes not only in amplitude of the measured signals (Voloshin and Wosk, 1980), but in frequency components as well. Figure 2 shows signals as they were recorded. It may be observed that the maximum amplitude of the signal from MFC is shifted in time relative to the corresponding maximum on TT. This shift is near 2-4 msec, which indicates that time is needed for the force impulse to propagate through the knee joint. No attempt has been made yet to accurately measure this time, since a much higher sample rate is needed, probably up to 5,000 samples/sec. The Fast Fourier Transfer (FFT) has been performed, and the flat weighting function was used. The first 32 harmonics are shown in Fig. 3. Notice that attenuation is increased for higher harmonics. Similar characteristics have been obtained for all subjects. The transmissibility function of the knee joint has been defined as

$$T(w) = \frac{MFC(w)}{TT(w)}$$

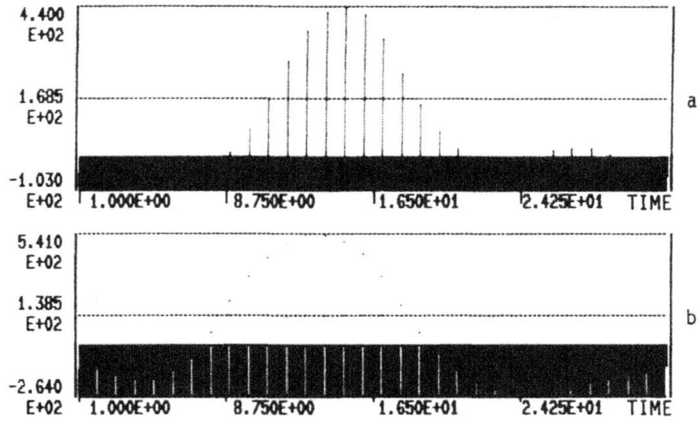

Fig. 2. Typical signal recorded on
a) medial femoral condyle b) tibial tuberosity.

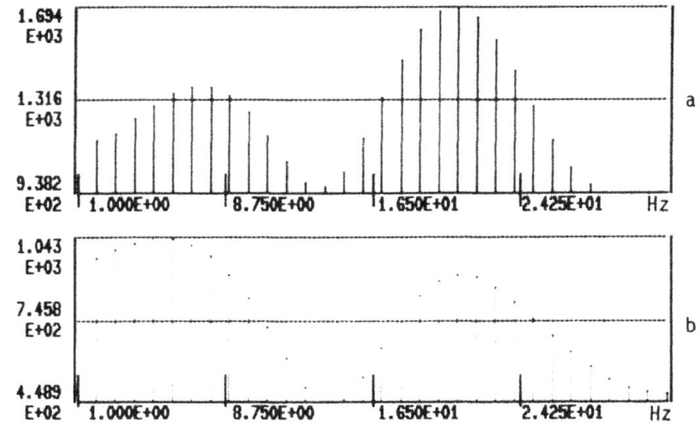

Fig. 3. Power spectrum for the recorded signal
a) medial femoral condyle b) tibial tuberosity.

and is shown in Fig. 4.

A simple two-degree-of-freedom rheological model of the knee has been considered for comparison purposes. The average parameters for K and C were obtained from the works of Crowninshield, Pope, and Johnson (1976) and Pope and co-workers (1976). The calculation shows that the resonant frequency is in the range of 5-20 Hz, depending on choice of parameters, which is consistent with the present finding.

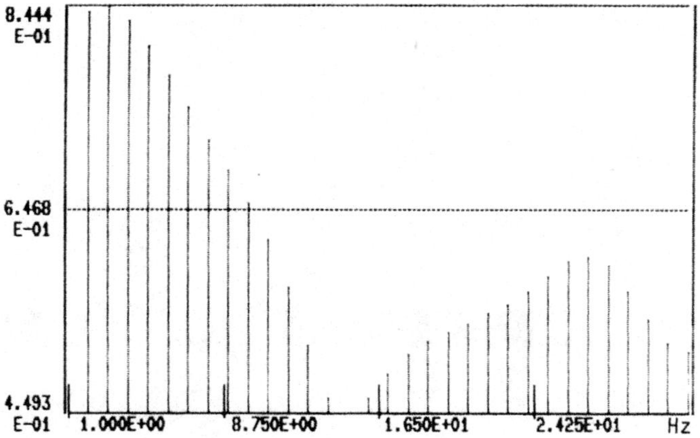

Fig. 4. Transmissibility function.

CONCLUSIONS

Experimentally, the results of the present and previous works (Voloshin, Wosk, and Brull, 1981; Voloshin and Wosk, 1980; Wosk and Voloshin, 1981) support the idea that the repetitive loading during gait generates intermittent waves which are propagated through the human musculoskeletal system. The technique presented here allows for noninvasive in vivo registration of those waves. Using a digital data acquisition approach gave the possibility of finding frequency components of the measured signals, as well as calculating transmissibility function. The results, in this case, were obtained for healthy subjects only. The next series of experiments will be conducted on subjects with different knee disorders. The obtained data will be correlated with knee disorders. This will allow us to develop a rheological model of the painful knee. The potential for clinical use of the described technique is worthy of further pursuit.

REFERENCES

Crowninshield, R. D., M. H. Pope, and R. J. Johnson (1976). An analytical model of the knee. J. Biomech., 9, 397-405.

Crowninshield, R. D., M. H. Pope, R. J. Johnson, and R. Milles (1976). The impedance of the human knee. J. Biomech., 9, 529-535.

Pope, M. H., R. D. Crowninshield, R. Milles, and R. J. Johnson (1976). The static and dynamic behavior of the human knee in vivo. J. Biomech., 9, 449-452.

Voloshin, A. S., and J. Wosk (1980). Shock absorbing capacity of the human knee (in vivo properties). Proceedings of the Special Conference of the Canadian Society of Biomechanics on "Human Locomotion I," London, Ontario, Canada, 104-105.

Voloshin, A. S., J. Wosk, and M. Brull (1981). Force wave transmission through the human locomotor system. ASME J. Biomech. Eng., 103, 48-50.

Wosk, J., and A. S. Voloshin (1981). Wave attenuation in skeletons of young healthy persons. J. Biomech., 14, 261-268.

A METHOD FOR QUANTIFICATION OF 3-DIMENSIONAL MOTION OF THE KNEE JOINT

Narender P. Reddy, Jeff Drew, Arnie Melby, III, and Robert E. Herron

Institute for Biomedical Engineering Research, University of Akron, Akron, Ohio 44325

ABSTRACT

A technique is developed for studying the 3-D, six degrees of freedom, motion of the tibia with respect to the femur in cadaver knee joints using a photo-encoded 3-D motion transducer. The method is being used in the author's laboratory for objective comparison of ligament injuries and repairs.

KEY WORDS

Knee joint, biomechanics, kinematics, displacements, stress test, 3-dimensional motion.

INTRODUCTION

The human knee joint represents a complex articulation with 3-D, 6 degrees of freedom, motion of the tibia with respect to the femur. Internal derangement of the knee, with damage to any of the knee structures, would alter the mechanics and 3-D motion of the joint. Consequently, successful assessment of ligament injuries and of various surgical repair/reconstruction procedures requires understanding of the biomechanics of the normal and ligament injured joint.

Biomechanics of the knee joint has been studied by a number of authors. In spite of the many excellent investigations, there exists a need for developing a method, with potential for clinical application, for assessing the ligament injury and for evaluating ligament reconstructions in terms of all the orthogonal components of the 3-D motion (flexion-extension, varus-valgus, internal-external rotations, medial-laterial, anterior-posterior and vertical displacements). The present paper describes a new technique developed by the authors for studying cadaver knee joints.

A 3-D motion sensor previously developed by Herron (1974) is adopted in the study to locate the position of various points in 3-D space (Fig. 1). The 3-D motion of the sensor arm is measured via photoencoders in the sensor unit. The digital output signals from the sensor provide 3-D information of a point, in space, in spherical co-ordinates (r,Θ,Φ). The instrument is sensitive to 0.002 cm in radial direction and 0.01 degrees in Θ, and Φ directions. The data is then recorded and sent to the University computer for later processing.

The femur is constrained by horizontally mounting it on a fixed table, and tibia is allowed to move with respect to the fixed femur. The flexion-extension motion of the tibia is achieved with a cable exerting a pulling force via the patellar tendon (Fig. 2). Special fixtures are placed on the tibia to locate four fixed points on the tibia. The line connecting points 2 and 3 is parallel to the long axis of the tibia. Point 1 is diametrically opposite to point 2 but proximal than point 2. The 3-dimensional motions of these four points are recorded, via the 3-D sensor with each position of the tibia.

The sensor output which is in spherical co-ordinates is first transformed to rectangular cartesion system (x,y,z). The origin of the co-ordinate system is then transformed to a location of the femur, close to the center of motion, with "z" axis directed along the long axis of the femur. The three angles of rotation (flexion-extension ($\Phi 1$) varus-valgus ($\Phi 2$), and axial rotation ($\Phi 3$) are then calculated using the principles of kinematics (table). Next the co-ordinate system is rotated such that one of the axes (w) is parallel to the longitudal axis of the tibia. The translational displacements are then calculated in the rotated axes by substracting the distances of point 2 (u,v,w) in the reference position of the tibia from the distance of point 2 in the displaced position (table). Thus, the three rotations and three translations are obtained. A computer program written in Fortran language is developed which calculates the displacements and rotations from the raw sensor data.

This procedure is currently being used in our laboratory to analyze the motion in normal and ligament damaged cadaver knees and for objective evaluation of various types of repairs by comparing the differences in displacements with and without applied loads (stress tests: A-P drawer, 30° Lachman's, 0° neutral pivot point, and varus-valgus tests) in normal, ligament injured, and repaired cadaver knee joints. As an example, figures (3,4, & 5) illustrate the differences in translational and rotational displacements with Lachman's, pivot point, and varus-valgus tests on normal cadaver knee, after anterior-cruciate removed, after combine, rotary instability, and after repairs. This way, an objective evaluation of knee injuries and ligament reconstructions can be achieved, and the full potential of the method can be realized only after these studies are complete.

REFERENCES

Herron, R.E. (1974). Experimental Determination of Mechanical Features of Children and Adults. Department of Transportation Report DOT-HS-231-2-397.

TABLE

dx, dy, and dz are the distances of line connecting points 2 and 3, and dx', dy', and dz' are distances of line connecting points 1 and 2 along the x, y, and z axes, Φ_1, Φ_2, and Φ_3 are rotations, respectively, about x, y and z.

Flexion Angle: $\Phi 1 = -\tan^{-1} (dy/dz)$

Varus-valgus Angle: $\Phi 2 = dx/(dz \cos \Phi_1 - dy \sin \Phi_1)$

Axial Rotation: $\Phi 3 - \tan^{-1} A/B$

where $A = C1\ dy' + S1\ dz'$

$B = C2\ dx' + S1\ S2\ dy' - S2\ C1\ dz'$

and $Si = \text{Sin}\ \Phi i$

$Ci = \text{Cos}\ \Phi i$

The distances of point 2 in the rotated axes are

$$u = (C_2C_c - S_1S_2S_3)dx + (S_1S_2C_3 + C_2S_3)dy - (S_2C_1)dz$$

$$v = (S_3C_1)dx + (C_1C_3)dy + S_1\ dz$$

$$W = (S_1C_3+C_1S_1S_3)\ dx + (S_2S_3-C_2S_1C_3)\ dy + (C_1C_2)\ dz$$

The translations are:
Medial-lateral displacement = U - U ref
Vertical displacement = V - V ref
A-P displacement = W - W ref

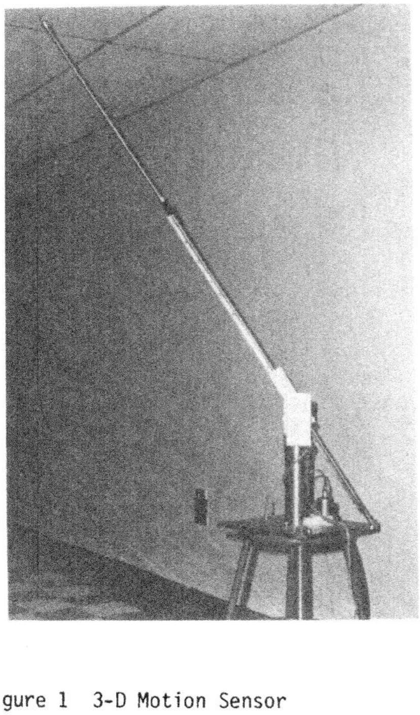

Figure 1 3-D Motion Sensor

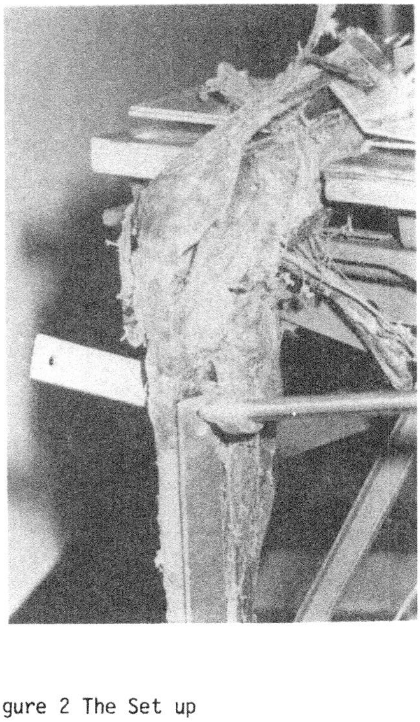

Figure 2 The Set up

Figure 3.

Axial Rotation in Degrees During 0° Neutral Pivot Point Test

A - Normal
B - Anterior Cruciate Removed
C - Combined Rotary Instability
D-G - Repairs 1,2,3 and 4

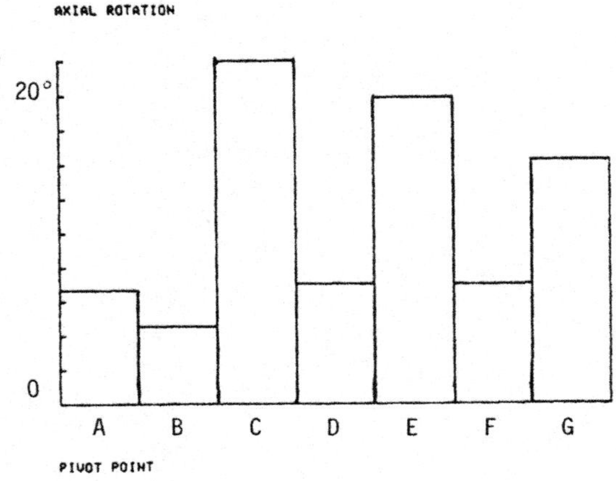

Figure 4.
Anterior Displacement in cm During 25° Lachman's Test

A - Normal
B - Anterior Cruciate Removed
C - Combined Rotary Instability
D-G - Repairs 1,2, 3 and 4

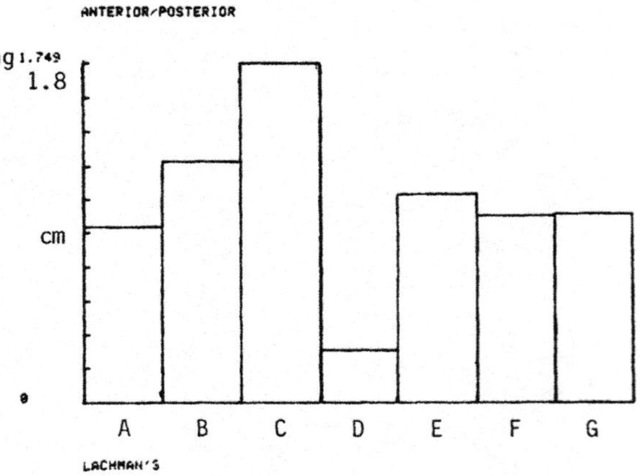

Figure 5.
Valgus Angle in Degrees During Valgus test at 0°.

A - Normal
B - Anterior Cruciate Removed
C - Combined Rotary Instability
D-G - Repairs 1,2,3, and 4

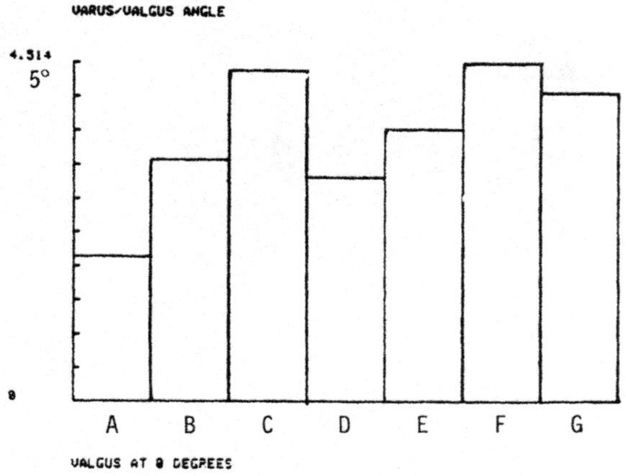

AN EXPERIMENTAL STUDY ON THE TIBIAL COMPONENT OF TOTAL KNEES

by

James D. Baril, M.D.*, Demetrios D. Raftopoulos, Ph.D.* & Youssef M. Salam*
*The University of Toledo
Toledo, Ohio 43606

ABSTRACT

Loosening of the tibial component of total knee arthroplasties has been a problem. Loosening rates from 10% to 76% have been reported. The purpose of this study was to investigate the fixation of tibial components of total knees. Testing was done with tibial components with either large central post or two smaller peripheral posts. Matched preserved cadaver tibias were utilized. Prostheses were implanted in the tibias and then loaded in a static fashion. Fracture loads were determined for each specimen. The tibias with the central posts withstood greater loads than those with peripheral posts. Eccentric loading of the prostheses decreased the amount of load withstood prior to fracture. The forces used to create these fractures (450-1300 Ks) exceeded physiologic loading.

KEYWORDS

Knee arthroplasties; loosening; fracture; central plug; tibias.

INTRODUCTION

Loosening of total knee arthroplasty components is an ever increasing problem. The estimated number of total knee arthroplasties performed in the United States has increased from 4,000 in 1972 to 40,000 in 1976. (Hori, 1978) Unfortunately, a high incidence of loosening has occurred with total knee arthroplasties. This problem has been particularly apparent in the tibial components of these implants.

Loosening of components has not been restricted to isolated designs but rather has been common in nearly all designs. The polycentric knee was one of the first total knee arthroplasties widely used. Gunston (1976) has reported an incidence of loosening of the tibial component in this implant in 10% of cases at two to seven years after implantation. Ilstrup (1976a) reported on 419 polycentric knees after two years of follow-up. This series revealed evidence of the formation of a radiolucent line in 41% of the knees implanted in patients with osteoarthritis and in 31% of the knees implanted in patients with rheumatoid arthritis. The development of radiolucent lines at the bone-cement interface is early evidence of loosening of the prosthesis. Ilstrup also reported an 8.5% incidence of frank loosening of the polycentric prosthesis when used in patients with osteoarthritis and in 1.9% of patients with rheumatoid arthritis. After geometric knee arthroplasties the incidence of radiolucent lines has been reported as high as 100% when utilized in patients with osteoarthritis. (Ilstrup 1976b) Actual loosening of the geometric prosthesis has occurred in 25% of the knees used in osteoarthritic knees. The incidence of the formation of radiolucent lines at the bone, cement interface after use of the geometric design in rheumatoid arthritis, has been reported at 25%. Although other designs have less loosening, the incidence is still high. An incidence of 7% loosening has been reported with the UCI knee. (Evanski, 1976) Ducondylar knee replacements have developed radiolucent lines about the bone-cement interface in 76% of implants when followed for at least three years. (Ranawat, 1976) In 26% of these knees, the size of the radiolucent lines has increased over a period of time.

Many attempts have been made to design total knees which are less likely to loosen. Most of the changes have centered around the tibial component as it loosens more frequently. Although all the new designs have differed in some details, the common feature of all has been the presence of a central post on the tibial component which is inserted into the cancellous bone of the proximal tibia. Designs have also varied with the materials used for this post. Some tibial components are made entirely of ultrahigh density polyethylene. Others are made of metal with only the articular surface consisting of the polyethylene. Long-term follow-up is not yet available on all these new designs. Insall (1979), however, has reported on the semiconstrained total condylar prosthesis. He reported that 22% of these prostheses developed radiolucent lines under the tibial components but not around the medullary post. Their follow-up was at least three years. Although these lines may be small, they suggest that this design, although more securely fixed to bone, is also likely to gradually loosen as time passes.

The aim of this study was to look at the problem of fixation of tibial components of total knees. Two types of components were tested, one with a central post and one with two peripheral posts. Both components were tested in compression.

MATERIALS AND TESTING

Paired preserved cadaver tibias were obtained from the anatomy laboratory of the Medical College of Ohio. All soft tissue was removed.

The tibial prostheses were manufactured in The University of Toledo machine shop. The prostheses were made entirely of steel. No attempt was made to incorporate any ultra high density polyethylene into the prosthesis as the articular surface was not important in this study. The central post prosthesis had a post which was about the size of a DePuy, Townley tibial component. The peripheral post prosthesis had smaller posts. Initially attempts were made to construct the peripheral posts to the same dimensions of the central post. However, the larger posts impinged on the medial and lateral cortices of the tibia and prevented seating of the prosthesis. Therefore the peripheral posts are considerably smaller. The superior surface of the tibial prosthesis was constructed flat.

The tibial prostheses were implanted in the bones by first making a 0.5 cm. horizontal cut through the tibial plateaus removing the articular surface. Cancellous bone roughly the shape of the plug or plugs was then curretted out. The prosthesis was then tested to see if it was well seated within the tibia. The polymethylmethacrylate cement was then mixed, inserted into the tibia and the prosthesis was held securely to the bone until the cement hardened. Since the bone was free of all soft tissue, the "operation" was technically perfect.

The distal end of the tibia was then potted into gypsum. A mechanical attachment was designed and built for the Instron in order to apply a combination of various compressive loads and moments on the tibial prosthesis.

The tibias were then placed into the Instron and loaded. The loads were applied at the rate of .0254 cm per minute. Essentially this rate of loading is a static load. All the bones were loaded at the same rate. A one contact point variable load was applied to the tibial prosthesis. This load was in compression only. The loading was carried out in two manners--(1) a central load through the middle of the prosthesis and the long axis of the tibia, and (2) an eccentric load 1.27 cm to either side of the center of the prosthesis (either medial or lateral). Loading was then applied until fracture occurred. The load required to produce a fracture was recorded. The mode of fracture was visually observed and recorded by means of photographs. Photographs were obtained in several dimensions. Corresponding radiographs were also obtained of the fractured specimens.

RESULTS AND CONCLUSIONS

Table 1 shows the results of the sample of the bones being tested. The fracture loads indicated are considered as static loads since the load was applied at a rate of 8.165 kg per second. The single post prosthesis held more load than the double post prosthesis when loaded centrally. The difference between the single post and double post prosthesis was 415 kg. The other pairs of bones were loaded eccentrically either in a lateral or medial position. In both of these tests also, the single post prosthesis held greater loads than the double post prosthesis prior to fracture. However, the amount of load was much less in the eccentrically loaded tibias then the centrally loaded tibias.

The manner of fracture was one of compression of the tibial plateau followed by longitudinal cracks in the shaft of the tibia. The plateau compression fractures occurred on the side of the bone to which the load was applied in the eccentrically loaded specimens.

We concluded from these tests that the implantation of the two post prosthesis significantly weakened the tibia. By decreasing the boney support with a two-post prosthesis, the tibial strength is decreased.

DISCUSSION

Although the single post prosthesis sustained more load, we should consider the fact that this force may exceed physiological requirements. From the geometry of the two designs one can determine that the two-post prosthesis should have better stability and probably better fixation. The load applied on the tibial condyles will be distributed over the two posts instead of one post and the surrounding cement will be subjected to less force.

Since the cement in a two-post prosthesis will be required to resist smaller forces, it may loosen less rapidly.

TABLE 1

	Pair 1	Pair 2	Pair 3
Tibia	Right Left	Right Left	Right Left
Load Position	Central	1.27 cm. medial	1.27 cm. lateral
Posts	single double	single double	single double
Fracture Load	1365 kg 950 ks	494 kg 453 Kg	780 kg 730 ks

ACKNOWLEDGEMENT

The authors would like to express their gratitude to the Department of Health of the State of Ohio which supported this work under Contract No. 729-A1. Also, we would like to thank Professor Frank Saul of the Medical College of Ohio at Toledo for his support in obtaining tibial specimens.

REFERENCES

Evanski, P.M., T.R. Waugh, C.F. Ortino, and S.H. Anzel (1976). UCI replacement. Cl. Ortho. & Rel. Res., 120, 33-38.

Gunston, F.H., and R.I. Mackenzie (1976). Complications of polycentric knee arthroplasty. Cl. Ortho. & Rel. Res., 120, 11-17.

Hori, R.Y., J.L. Lewis, J.R. Zimmerman, and C. Compue (1978). The number of total joint replacements in the United States. Cl. Ortho. & Rel. Res., 132, 46-52.

Ilstrup, D.M., J.J. Combs, R.S. Bryan, L.F.A. Peterson and M.D. Skolnik (1976a). A statistical evaluation of polycentric total knee arthroplasties. Cl. Ortho. & Rel. Res., 120, 18-26.

Ilstrup, D.M., M.B. Coventry, and M.D. Skolnik (1976b). A statistical evaluation of geometric total knee arthroplasties. Cl. Ortho. & Rel. Res., 120, 27-32.

Insall, J., N. Scott, and C.S. Ranawat (1979). The total condylar knee prosthesis. J. Bone Joint Surg., 61A, 173-180.

Ranawat, C.S., J. Insall, and J. Shine (1976). Duocondylar knee arthroplasty. Cl. Ortho & Rel. Res., 120, 76-82.

Session 11

ULTRASOUND IN MEDICINE

USE OF ULTRASOUND IN THE DIAGNOSIS OF SURGICAL DISEASES OF THE GALLBLADDER AND BILE DUCTS

EDWIN A. DEITCH, M.D.

LSU MEDICAL CENTER
ASSISTANT PROFESSOR DEPARTMENT OF SURGERY
1501 KINGS HIGHWAY
SHREVEPORT, LOUISIANA 71130

ABSTRACT

Three prospective studies were performed to clarify the clinical usefulness of gray-scale sonography in biliary tract (B.T.) disease. The first study compared the diagnostic accuracy of biliary tract sonography to the standard radiographic diagnostic test (OCG) in the diagnosis of elective disease of the biliary tract. Results of the study showed that biliary tract ultrasound was at least comparable to the OCG in the diagnosis of B.T. disease. The second study was performed to determine first, if the gallbladder wall could be accurately visualized by the sonographic techniques and if so, would a thickened G.B. wall reliably diagnose patients with acute disease of the B.T. Study of 106 patients and 100 controls documented the fact that the gallbladder wall was much thicker in patients with biliary tract disease than in the control population. Therefore, gallbladder wall thickening might be a potentially useful diagnostic criteria for the diagnosis of acute acalculous cholecystitis. Lastly, study of the bile duct size in 102 surgical patients established ultrasonic criteria by which to separate patients with medical from surgical jaundice.

KEY WORDS

Ultrasound, biliary tract, jaundice

INTRODUCTION

The diagnosis of disease of the biliary tract is fraught with errors. A major reason for the difficulty in diagnosing these diseases accurately is the fact that biliary tract disease is usually manifest by either abdominal pain or the development of jaundice (yellowing of the skin). However, these symptoms are not specific for biliary tract disease and can be caused by a variety of other diseases. Right upper quadrant pain which is the most frequent sign of biliary tract disease, can be caused by both medical and surgical diseases of the kidney, liver, pancreas, stomach, duodenum, small bowel, as well as by intrathoracic diseases such as heart attacks or pneumonias. To complicate matters, the stress of surgery in patients with nonsurgical diseases may result in death. However, on the other hand, if a patient has biliary tract problems and does not receive prompt operation the risk of complications and the subsequent mortality rate rises significantly. Therefore the key is accurate diagnosis, yet at best, decisions based on the history, physical examinations, and blood tests are associated with a 15% misdiagnosis rate. To reduce the rate of misdiagnosis, various radiologic diagnostic techniques have been developed. These radiographic techniques rely on the oral or parenteral intake of dyes. The most accurate radiologic contrast examination available today is the oral cholecystogram which relies on the oral ingestion of a dye which must then be absorbed by the gastrointestinal tract, be processed in the liver, excreted in the bile and then finally be concentrated in the gallbladder to be diagnostic (Deitch, 1980). However, this test is limited to patients with normal liver and gastrointestinal function. The presence of jaundice (a common biliary tract complaint) makes this test unreliable. Before the early 1970's these contrast radiographic techniques, based on the physiology of the biliary tract, were the only diagnostic techniques available. The development of medical ultrasonographic technology made available to the surgeon a new potential diagnostic tool. In contrast to the standard radiographic technique, ultrasonic visualization of the biliary tract is not dependent on normal physiology. Since ultrasound is a purely anatomic technique the risks of false positive tests due to physiologic abnormalities in other organs is avoided. The present studies were carried out to evaluate this imaging procedure in the clinical setting. Two basic clinical questions were asked. (1) Can patients with abdominal pain due to biliary tract disease be correctly identified from patients with nonbiliary tract complaints? (2) Can this technique separate patients with jaundice due to surgical diseases (biliary tract disease or cancer) from patients with jaundice due to medical diseases (hepatitis)?

PATIENT POPULATION AND METHODS

Over a 22 month period from August 1978 to June 1980 patients operated on for biliary tract disease were enrolled in one of three prospective studies. A total of 308 patients were studied.

Study one was performed to evaluate the reliability of Gray-scale sonography against the standard diagnostic radiographic technique (oral cholecystography). One hundred patients had both oral cholecystography and Gray-scale ultrasonographic evaluation of the biliary tract performed synchronously. The two tests were interpreted blindly and correlated with the results of surgery to determine the accuracy of ultrasound.

Study two was performed to evaluate the utility and accuracy of the ultrasonically measured gallbladder wall as a diagnostic criteria in biliary tract disease. To determine the sonographic size range of the gallbladder wall in the normal population 100 patients without biliary tract disease had the width of their gallbladder walls measured in the fasting state. Subsequently, a prospective blinded study of 106 patients undergoing cholecystectomy was performed to determine if the gallbladder wall could be accurately measured. Finally measurements of the gallbladder wall thickness were correlated with the patient's clinical presentation, hospital course and operative findings. The patients were divided into two groups, patients with acute versus chronic cholecystitis. The patients were further divided into two additional groups, those patients who had calculous (gallstones) cholecystitis versus patients with acalculous (no gallstones) cholecystitis.

Study three was performed to evaluate the reliability and clinical limitations of sonographic scanning of the biliary ducts in separating patients with surgical jaundice from patients with medical jaundice. A prospective study of bile duct size in 102 consecutive surgical patients with and without extrahepatic bile duct obstruction was performed. Preoperative ultrasonic measurements of the intrahepatic and extrahepatic biliary ducts were compared with surgical measurements of the bile ducts. Thus, the accuracy of the ultrasonic measurements of the bile ducts. Furthermore, the size range of these ducts in patients with surgical obstruction could be determined and compared with similar measurements in patients without obstruction.

In all these studies a unirad digital Gray-scale scanner using a 2.25MHz short or long focus transducer was used (Deitch, 1981a). Mineral oil was used as the coupling agent between the transducer and patient. Parasagital and transverse scans of the liver were performed at 1cm. intervals to visualize the interhepatic ducts. An individualized approach using transverse, longitudinal and intercostal scans in both the supine and left lateral decubitus positions was taken to evaluate the extrahepatic ductal structures. Once the gallbladder had been identified sonographically, the patients were placed in the left lateral decubitus position to facilitate identification of the gallbladder wall and to document movement of intraluminal contents. Gravity displacement of the intraluminal contents reduced distortion of the gallbladder wall thickness measurements by biliary sludge or gravel. Measurements of the gallbladder wall were taken at the free peritoneal surface in the region of the fundus and recorded to nearest half millimeters. Surgical measurements of the gallbladder wall and bile ducts were performed as previously described (Deitch, 1981b) without knowledge of the ultrasonic measurements.

RESULTS AND DISCUSSION

Study One: Accuracy of ultrasound in elective biliary tract disease.
Comparison of oral cholecystography (OCG) with ultrasound (US) of the gallbladder showed that in 90 of the 100 patients the two examinations agreed. In two of the ten patients in which the two tests were not concordant the ultrasonic examination was not technically adequate for evaluation. Two of the 8 evaluatable patients underwent surgery and in both the ultrasonic examinations were correct. The remaining 6 patients had a normal ultrasound and an abnormal OCG. In 4 of these patients the OCG's were false positive tested caused by occult liver disease. Table I shows a comparison of the accuracy of OCG versus US in this study.

Table I. Comparison OCG vs. US

	Sensivity +	Specificity *	Overall Accuracy
OCG	98.7%	77%	92%
US	96%	96.7%	96%

+ Sensivity = $\frac{\text{True test positives}}{\text{All positives in population}}$ * Specificity = $\frac{\text{True test negative}}{\text{All negatives in population}}$

Therefore, the results of this study indicates that ultrasound compares favorably with the standard contrast radiographic diagnostic criteria of gallbladder disease.

Study Two: Gallbladder wall thickness as an independent diagnostic criteria in biliary tract disease.

To determine what the normal ultrasonic thickness of the gallbladder wall is in patients without biliary tract disease one hundred volunteers were studied (Table 2).

Table 2. Normal Range of Gallbladder Wall Thickness

Gallbladder Wall in mm.		% of Population
Thin	≤2.5mm.	93%
Intermediate	2.5-3.0mm.	6%
Thick	>3.0mm.	1%

Having established the normal range of gallbladder wall thickness in the control population, the size range of the gallbladder wall was measured in 23 patients with acute cholecystitis and in 65 patients with chronic cholecystitis (Table 3).

Table 3. Range of Gallbladder Wall Thickness In Patients With Cholecystitis

Gallbladder Wall	Acute Cholecystitis (%)	Chronic Cholecystitis (%)
Thin	0	25%
Intermediate	5%	30%
Thick	95%	45%

The results indicated what in patients with acute cholecystitis essentially all the walls were thick and these results suggest that a thick gallbladder wall might identify the subgroups of patients with acute acalculous cholecystitis. However, before these ultrasonic results could be accepted it is important to verify that these indirect ultrasonic measurements accurately reflect the true anatomic thickness of the gallbladder wall. Therefore, in patients direct measurements of the gallbladder wall performed in a blinded fashion at surgery were compared with the ultrasonic measurements (Table 4).

Table 4. Comparison of Direct Surgical and Ultrasonic Measurements of Gallbladder Wall Thickness.

Difference in mm.	No Difference	0.5mm.	1.0mm.	$\geq$ 1.5mm.
% Patient Population	40%	21%	34%	5%

Therefore, it appears that in 95% of the cases the ultrasonic measurements agreed within one millimeter with direct measurements of the gallbladder wall. Therefore, based on this data US measurements of gallbladder wall thickness should be helpful in the diagnosis of acute acalculous cholecystitis.

Study three: Reliability and limitations of sonographic scanning of the bile ducts in the differential diagnosis of jaundice.

This study was initiated to determine the reliability and clinical limitations of sonographic scanning of the biliary ducts. One hundred two consecutive patients who underwent biliary sonography and surgery were studied. In 62% of the patients the sonography could visualize the extrahepatic ductal system while in 81% of the patients the intrahepatic system could be visualized. Sonographic and operative measurements of the common bile duct widths agreed with each other 84% of the time. Extrahepatic bile duct obstruction was present if extrahepatic bile ducts were 1cm. or wider, or if intrahepatic ducts were more than 0.5cm. ($p < 0.001$). Similarly, if the extrahepatic ducts were less than 0.8cm. and the intrahepatic ducts were less than 0.4cm. in width, then bile duct obstruction was not present ($p < 0.001$).

In summary, ultrasound has evolved into a reliable diagnostic technique for many of the diseases that effect the biliary tract. It's utility is based on the ability of ultrasound to accurately and directly visualize the biliary tree. Although it's clinical role had been constantly in evolution, it's potential as a research tool to understand biliary physiology, is only now beginning to be appreciated.

Deitch, E.A., and Engel, J.M. (1980). Ultrasound in elective biliary tract surgery. The American Journal of Surgery, 104, 277-283.
Deitch, E.A. (1981). Utility and accuracy of ultrasonically measured gallbladder wall as a diagnostic criteria in biliary tract disease. Digest Diseases and Sciences, 26, 686-693.
Deitch, E.A. (1981). In vivo measurements of the internal and external diameters of the common bile duct in man. Surgery, Gyn., and Ob., 152, 642-644.

ULTRASONIC QUANTIFICATION OF SKIN AND MUSCLE HEMODYNAMICS

P. K. Bhagat, W. T. Nickell, V. C. Wu
Wenner-Gren Biomedical Research Laboratory
Department of Mechanical Engineering
University of Kentucky, Lexington, Kentucky 40506

ABSTRACT

A multiple channel ultrasonic dimension gauge capable of noninvasive measurement of dimensional changes due to body fluid shifts is detailed.

KEYWORDS

Ultrasonic plethysmography; piezoelectric transducers; hemodynamics, blood volume redistribution.

INTRODUCTION

Plethysmographic measurements of limb hemodynamics have been used in hyper/hypo gravity research as well as in the diagnosis of circulatory disease. The measurement of small changes in tissue volume produced by venous occlusion or shifts in posture has been used for many years in cardiovascular research. It is also well established that time dependent whole body acceleration applied to human body produces significant alterations in the normal blood volume distribution. In order to ascertain the effectiveness of countermeasure devices and maneuvers, knowledge of blood volume redistribution in a non-invasive manner is essential. While mercury in rubber stain gauges are well established in research studies, operational difficulties have precluded their acceptance as a clinical diagnostic tool or in gravity related research.

Blood volume changes in the lower limb caused by venous occlusion and tilt table experiments have been reported by our group through use of a new technique, ultrasonic plethysmography (Bhagat and colleagues, 1980,1982). Recently blood redistribution in the superficial tissues during postural changes have been reported in the literature through ultrasonic measurement of tissue depth above bony structures (Kirsch and coworkers, 1980). In this paper, we report the development of a new instrument that combines both of the above approaches.

Ultrasonic plethysmography is based on the measurement of transit time required for an acoustic pulse to propagate from one point to another. If the pulse propagation takes place mainly through a homogeneous medium, the velocity of acoustic wave propagation can be assumed a constant. The measurement of transit time, then, provides an accurate measure of the traversed distance. Rushmer and coworkers (1956) were the first to use this principle for real time monitoring of left ventricular dimensions in laboratory animals. Measurement sites such as tissue depth at the thorax, forehead, tissue thickness across earlobes, midcalf region of the lower limb, digits provide a reasonable and practical baseline on blood volume shifts. While the size of calf muscle is in the range of 5-12 cms, when measured across the fleshy part, in the average adult population, tissue depth above the tibia is on the order of 5 mm to 1 cm. Thse variations in tissue thicknesses and their anatomical geometries impose restrictions on the instrument and transducers design.

Multiple Channel Ultrasonic Plethysmograph

A schematic diagram of the developed multiple channel ultrasonic plethysmograph is shown in Figure 1. It consists of control logic function block, transducer energizing circuit (transmitter pinger), receiver amplifier and comparator circuit (receiver network), time to voltage convertor and display select block.

Functional Block Diagram of the Multiple Channel Ultrasonic Dimension Gauge

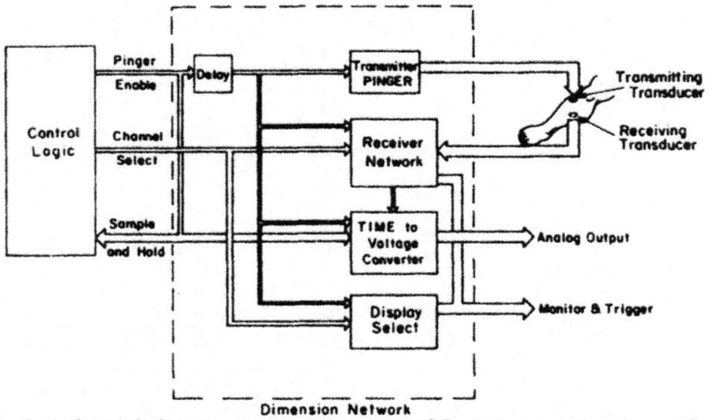

Operation of the developed instrument is controlled by a single chip, 8 bit, 8085 microprocessor system which includes a 2 kbit static RAM with I/O ports and timer (8155) and a 16 kbit EPROM with I/O ports (8755). A 4 MHz crystal provides the fundamental clock frequency to the system. Three signals, the pinger enable, the time to voltage (T/V) sample and hold, and receiving channel select are used in system operation. The transducer activation circuit is enabled by a time delayed positive going trigger signal from the control logic function block. Upon receipt of this clock pulse the transmitter pinger circuit provides a 200 V, 40 nsec negative going pulse which is applied to ultrasonic transducers (up to eight transducers) through rf transformer coupling. Due to this shock excitation transducers vibrate at their resonant frequency of 2-3 MHz resulting in transmission of an ultrasonic signal into the attached medium (tissue).

The purpose of time delay, 100 μsec, in the system is to provide for discharge of the charging capacitor in the time to voltage convertor circuit, thereby initializing the measurement process. Because the receiver amplifier has high gain, the possibility of false output from the comparator due to the low level interference may exist during the activation phase of the transmitted signal. This possibility is avoided through use of a disable (delay network) which inhibits the comparator output during the transmit cycle of the system. The control logic also provides for maintenance of the final output of the time to voltage convertor, through a sample and hold network so that several measurement channels can be displayed simultaneously. The time to voltage conversion is obtained through a capacitor connected across a transistor switch.

The ultrasonic pulse from a given transducer after propagation through the tissue layer is received by another ultrasonic transducer which is appropriately positioned on the skin surface. This piezo-electric transducer converts the ultrasonic energy into an electrical signal at its resonant frequency of 2-3 MHz. Two CD4052 analog switches, are used for selection of the appropriate received signal for input to the receiving amplifier. Each CD4052 switch is connected to a separate amplifier, thus, allowing for eight independent measurement sites. To achieve the highest resolution possible, it is necessary to detect the arrival of the received pulse accurately and reliably. To achieve this objective we adjust the gain of the received amplifier such that the amplifier is saturated on the second half cycle of the received signal. The polarity of the signal and the comparator level are set so that the comparator is triggered just above noise level on this part of the received signal. The instrument thus detects the first arrival and is not sensitive to changes in pulse shape of the received signal. The comparator output resets a flipflop and thus a time gate signal for the time to voltage convertor is obtained. The duration of time gate is equal to the transit time of the propagated ultrasonic pulse.

Once the output of time to voltage convertor has reached its final value for the given duration, this value is sampled and held until the next cycle. The subsequent clock pulse repeats the above process for measurement at another site. Once data is recorded for all independent measurement sites, the whole cycle is repeated. Our current system allows for future modification of the functions of the device without major alterations in the subsystem designs.

Fabrication of Appropriate Transducers

Due to specific anatomical geometries at the chosen sites, the shapes of the transducers and mountings have to be defined for specific measurements. Salient features of the transducers for (i) calf dimension, (ii) digit and (iii) skin depth application are described below.

The most critical problem with calf dimension transducers is to provide sufficient beam width to allow easy placement and to prevent small changes in orientation from altering the received pulse shape. Two types of transducers are used for the calf dimension measurement. Transmitting transducers consist of a 3 mm diameter flat disc of LTZ-2 ceramic. This element is embedded 1 mm behind a plastic convex lens, which serves to diverge the beam and to provide good contact with the tissue. Receiving transducers are constructed similarily from hemispherically shaped transducer elements, and are completely non-directional.

For small distance measurement across digits or other body parts, the directional characteristics of the transducer are less critical because of larger signal amplitude available; however, the transducer housings must be specially shaped for the location of interest. These transducers consist of 2 mm diameter discs of LTZ-2 transducer material embedded in plastic. The mounting is curved with a radius of 5 cm to give improved contact with the digit.

In the case of skin depth measurements both receiving and transmitting transducers are located on the same side of skin surface. Because of the small distance involved and the proximity of the receiving and transmitting transducers, the most difficult technical problem in design is to provide sufficient dampling so that the returning echo is not obscured by oscillations remaining from the transmitted signal. In our design two separate transducer elements are placed at an angle such that energy from the first, which is reflected from bone with a depth of 5 mm, is returned to the second transducer. The transducers are mounted in balsa wood, which provides light weight, and mechanical isolation. In tests with human subjects this design provides greater echo amplitude and low coupling between the two transducer elements.

Results and Discussion

Figure 2 shows the results of a venous occlusion experiment to simulatneously measure skin and muscle blood volume. Transducers were attached at midcalf so as to measure the depth of skin above the tibia and the distance across the soleus muscle. Limb expansion was produced by inflation of a thigh cuff to 50 mmHg. Because the skin is a small fraction of the thickness of the muscle, the upper trace records, for practical purposes, only changes in the volume of muscle mass.

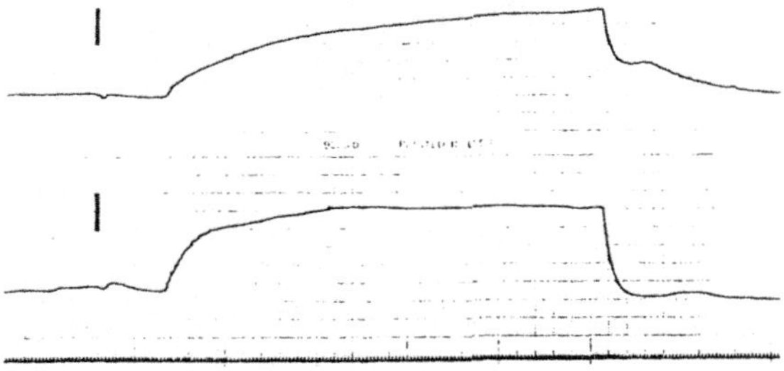

LEGEND: Simultaneous measurement of skin and muscle blood volume.
TOP: Muscle, Calibration Bar 1.11%
BOTTOM: Skin, Calibration Bar 0.72%
TIME: 1 Sec/mm

In this record pressure was applied to the occlusion cuff at the first arrow and released at the second arrow. In the muscle (top) record the initial slope of the volume increase is an accurate measurement of the blood flow into the soleus muscle. In the case of the skin volume record presented here, the flow cannot be determined with comparable confidence because an unknown quantity of blood may have flowed from skin to muscle through perforator channels. While the quantity of blood available is certainly insufficient to affect the muscle volume determinations, further study is required to determine whether this flow is sufficient to invalidate the determination of skin blood flow in the calf segment by this technique.

ACKNOWLEDGEMENT

The research reported here was supported in part by NASA contracts NAS9-15452 and ONR contract N00014-81-K-0325.

REFERENCES

Bhagat, P. K., J. F. Lafferty, D. Bowman, and M. P. Kadaba (1980). Aviat. Space Environ. Med., 51(2), 185-188.
Wu, V. C., W. T. Nickell, and P. K. Bhagat (1982). A practical ultrasonic plethysmograph. Aviat. Space Environ. Med. In Press.
Kirsch, K., J. Merke, and H. H. Szalkay. (1980). Pflugers Arch. 383, 195-201.
Rushmer, R. F., D. L. Franklin, and E. M. Ellis. (1956). Circ. Res. 684-688.

QUANTITATIVE MEASUREMENT OF FRACTURE HEALING BY ULTRASOUND

S. Saha, V.V. Rao, V. Malakanok, and J.A. Albright
Biomechanics Laboratory
Department of Orthopaedic Surgery
LSU Medical Center in Shreveport
P.O. Box 33932
Shreveport, Louisiana 71130

ABSTRACT

Ultrasonic techniques were used to monitor the rate of fracture healing in rabbits. Ultrasonic amplitude increased linearly with healing period. Bending strength of the healed ulnae also showed a significant positive correlation with the relative amplitude of the ultrasonic wave.

KEYWORDS

Ultrasound, fracture healing, mechanical strength.

INTRODUCTION

Radiographic examination is most commonly used to evaluate the progress of fracture healing. However the results are highly subjective and are of limited sensitivity. For instance, refracture of a bone which appeared completely healed in an x-ray examination, have been reported before (1). Therefore, there is a definite need for additional non-invasive means to measure the rate of fracture healing. This has prompted several investigators to attempt to use ultrasonic and vibration tests to measure the degree of union of healing fractures (2-4). Such attempts to use ultrasound were mainly based on the ultrasonic velocity change in the fractured bone compared to the control normals. Our objective was to evaluate if more sensitive tests could be developed to monitor fracture healing, based on the change in amplitude, velocity and frequency of an ultrasonic pulse.

MATERIALS AND METHODS

In a group of six adult New Zealand white rabbits, both ulnae of each animal was osteomized in the midshaft by using an oscillating saw. As a separate investigation, one fracture site in each animal was radiated to study the effect of radiation on bone healing (5). As attenuation of ultrasound in bone increases with increasing frequency, two low frequency (160 KHz) ultrasonic trasnducers were built and these were used to send and receive ultrasonic pulses across the fracture site. As shown in Figure 1 the experimental set-up contained a pulser (Metrotek MP215) and a receiver amplifier (Thornton). The signals were recorded on a digital oscilloscope (Nicolet, model Explorer III) and stored on floppy discs for subsequent analysis. The ultrasonic readings and radiographs of the healing bones were taken biweekly. After eight weeks of healing, the animals were sacrificed, their long bones were stripped of the soft tissue and ultrasonic readings were taken on each bone. These specimens were then tested mechanically in three point bending by using a servohydraulic Instron testing machine.

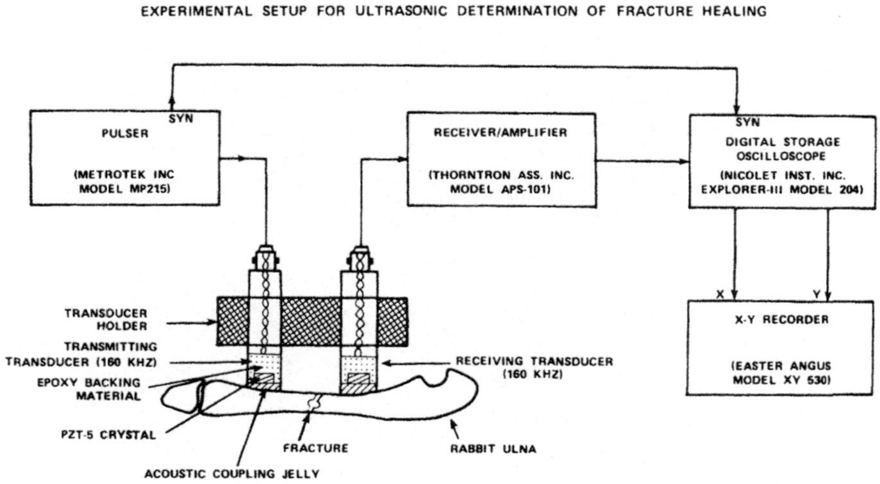

Figure 1. Experimental setup for ultrasonic determination of fracture healing

RESULTS AND DISCUSSION

To determine the relative amount of ultrasonic energy being transmitted by the soft tissue and the bone, ultrasonic readings were taken from a limb first, with the soft tissue intact and then with soft tissue removed from the area between the transducers. The results showed that most of the ultrasonic energy was being transmitted by the soft tissues surrounding a healing bone. A small but distinct portion of the wave was identified as the bone wave. As the wave velocity in bone (3000 to 3500 m/s) is much higher than that in the soft tissue (1500 m/s), the "bone wave" always preceded the large ultrasonic wave through the soft tissue (6).

Our results of in-vitro ultrasoinc tests showed a consistent linear increase in the normalized amplitude of ultrasonic pulse with increasing healing time.

Similarly results of our in-vitro tests in 12 healed (8 weeks) ulnae from 6 rabbits also showed a highly significant (p .01) positive correlation between the bending strength of the healed ulnae and the relative amplitude of the ultrasonic wave, (Figure 2). The linear regression equation between these two variables were (correlation coefficient r = 0.7607):

$$Y = -0.02 + 0.07x$$

where Y = Maximum bending moment in Nm

and X = amplitude of the ultrasonic wave

Statistically significant correlations were also obtained between the ratio of the average ultrasonic velocities across the fracture site and the ratio of mechanical strengths of the radiated side of the normally healing contralateral side.

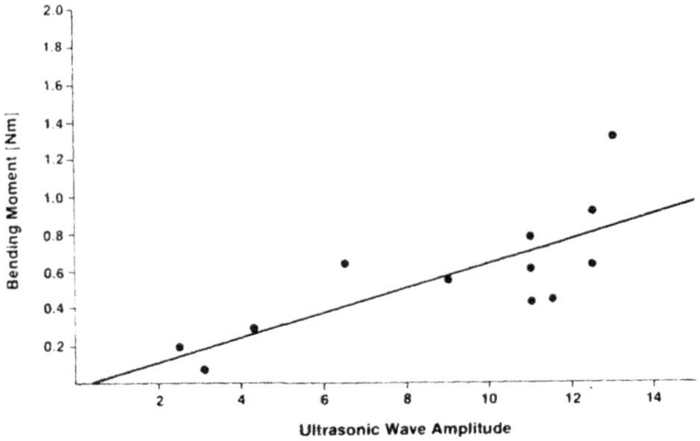

Figure 2. Relationship between bending strength (Nm) and ultrasonic wave amplitude.

CONCLUSION

These results indicate that ultrasonic monitoring of fracture healing could potentially be used as a non-invasive clinical tool for a quantitative determination of the rate of fracture healing.

ACKNOWLEDGEMENT

This research was partially supported by a USPHS research grant No. R01 AM 18360.

REFERENCES

1. Seimon, L.P. (1964) J. Bone Joint Surg., 46B:32
2. Brown, S.A. and Mayor, M.B. (1976) Biomed. Eng., 11:124.
3. Saha, S. and Pelker, R.R. (1976) Trans. Orthop. Res. Soc., 22nd Ann. Mtg., 1:219.
4. Sonstegard, D. A. and Matthews, L.S. (1976) J. Biomech., 9:689.
5. Saha, S., Datta, R. and Albright, J.A. (1982) Trans. Orthop. Res. Soc. 28th Ann. Mtg., 7:364.
6. Saha, S., Rao, V.V., Malakanok, V., Gross, B.D. and Albright, J.A. (1982) Trans. Orthop. Res. Soc. 28th Ann. Mtg., 7:376.

MEASUREMENT OF MANDIBULAR FRACTURE HEALING: A PRELIMINARY ULTRASONIC STUDY

B. D. Gross, S. Saha, V. V. Rao, and V. Malakanok

Departments of Orthopedic Surgery and Oral and Maxillofacial Surgery
LSU Medical Center in Shreveport
P.O. Box 33932
Shreveport, Louisiana 71130

ABSTRACT

Mandibular bone healing in fracture patients and bone graft patients was evaluated with ultrasound. This preliminary study indicates that this technique could potentially be a useful tool in monitoring the rate of mandibular bone healing.

KEYWORDS

Ultrasound; fracture healing; mandibular fracture

INTRODUCTION

Bony healing in the mandible is currently evaluated by manual and radiographic clinical examinations. The information gained is variable and difficult to quantitate. Several investigators have used vibration and wave propagation techniques in vivo to measure the rate of fracture healing (Abendschein and Hyatt, 1970; Sonstegard and Matthews, 1976; Brown and Mayor, 1976). Previously we have found a positive correlation between the amplitude of ultrasonic wave and the strength of healing rabbit ulna (Saha and co-workers, 1982a). Based on our positive findings of this animal study, we have started to use clinically a similar ultrasonic technique to evaluate the healing rate of mandibular fractures in patients. We have examined eleven patients in a planned series of 25, and some of these results are presented here.

MATERIALS AND METHODS

Two identical low frequency ultrasonic transducers were used to send and receive an ultrasonic pulse across a fracture site. An experimental instrumentation contained a pulsar (Metrotek MP 215) and a receiver amplifier (Thornton). The signals were recorded in a digital oscilloscope (Nicolet, Explorer III) and stored on floppy discs for subsequent analysis. Ultrasonic readings were taken from the area of fracture and similar recordings were made from the contralateral side, thus using the patient as his own control. Ultrasonic readings and radiographs were taken biweekly and two patients were followed for a total of twelve weeks. Figure 1 illustrates an example of the position of the ultrasonic transducers.

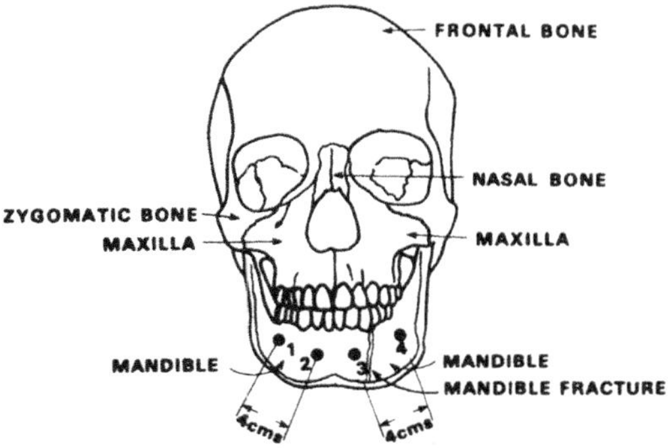

TRANSDUCER POSITIONS (1-2 AND 3-4)
FOR NON-INVASIVE DETERMINATION OF MANDIBLE
FRACTURE HEALING

FIGURE 1

Panaramic radiographs were made of each patient at the time of ultrasound evaluation. These radiographs were then graded by three separate surgeons as to their degree of healing. The grading, as shown in Table 1, ranged from a numerical value of one, meaning virtually no healing, to a numerical value of five, meaning complete consolidation of the fracture. The evaluators were not aware of the specific patient or the date of the x-ray.

TABLE 1

Fracture Healing	X-ray Healing Index
No evidence of healing sharp margins of fx. site.	1
Margins of fx. not sharply delineated - some radio-opacity.	2
Increased radio-opacity - some areas demonstrate healing.	3
Difficult to see fx. site - some parts of fx. line still radiolucent.	4
Complete healing - fx. site margins not visible.	5
Bone Graft Healing	
Obviously a fresh graft with particles easily delineated.	1
Particles seen but without clear and sharp edges.	2
Particles less delineated - consolidation becoming evident - density approaching that of stump edges.	3
Difficult to see the graft particles - graft appears consolidated.	4
Well consolidated graft - particles not seen.	5

RESULTS

After some experience, based on trial runs, we were able to isolate the ultrasonic wave transmitted by the bone from the much higher amplitude ultrasonic pulse transmitted by the soft tissue (Saha & co-workers, 1982b). In general, a linear increase was found in the normalized amplitude of the ultrasonic reading as well as in the x-ray healing index with increasing healing period (Figs. 2 & 3). Similarly the delay of the ultrasonic wave for the fractured side compared to the normal side also decreased as the healing progressed. Correlation between these ultrasonic findings and the radiographic evaluations taken on the same day was poor (Fig. 4). Ultrasonic readings were most dramatic for patients in which a large cancellous bone graft had been placed. For patients with hairline fractures it was difficult to distinguish ultrasonic readings from the fractured side compared to the contralateral control.

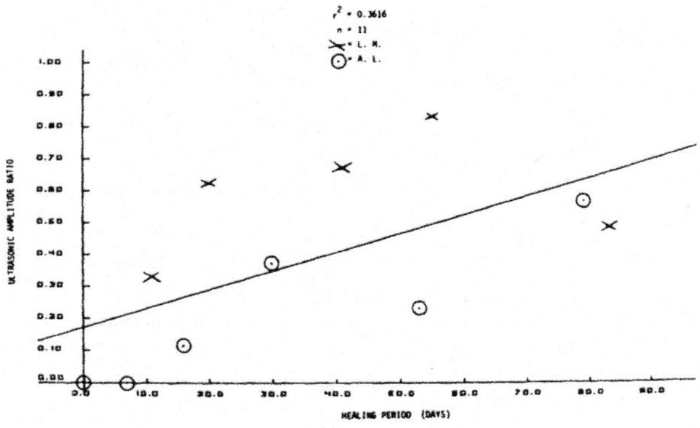

FIGURE 2 Healing Period (Days) vs. Ultrasonic Amplitude Ratio for a Fracture Patient (L.M., shown by cross) and a Bone Graft Patient (A.L., shown by circles).

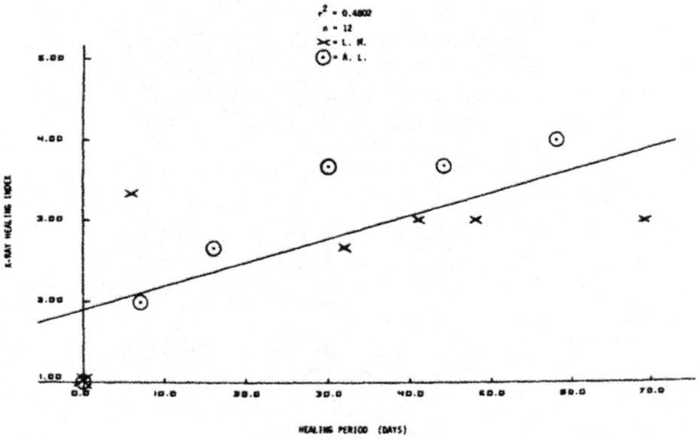

FIGURE 3 Actual Healing Period vs. X-ray Healing Scores for a Fracture Patient (L.M. shown by cross) and a Bone Graft Patient (A.L., shown by circles).

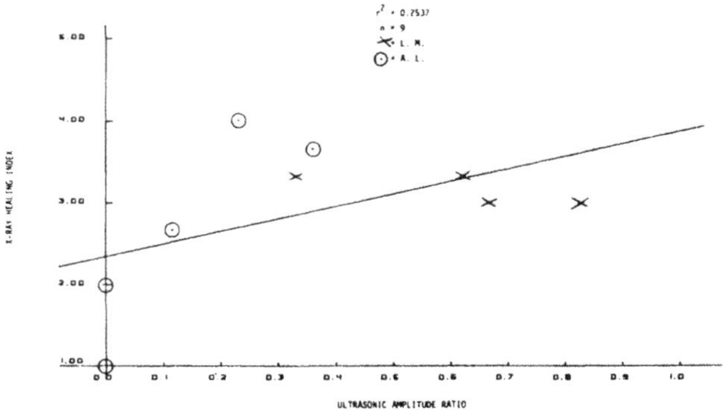

FIGURE 4 Ultrasonic Amplitude Ratio vs. X-ray Healing Scores for a Fracture Patient (L.M. shown by cross) and a Bone Graft Patient (A.L. shown by circles)

CONCLUSION

This preliminary study indicates that ultrasonic technique could potentially be used as a valuable tool for monitoring the rate of mandibular fracture healing in selected patients. It also appears that this may be a valuable tool in monitoring healing of bone grafts. We are presently continuing the study to increase the number of patients included in our series.

REFERENCES

1. Abendschein, W. and Hyatt, G.W. (1970) Clin. Orthop. 69:294-301.
2. Brown, S.A. and Mayor, M.B. (1976) Biomed. Eng., 11:124-136.
3. Saha, S., Rao, V.V., Malakanok, V. and Albright, J.A. (1982a) Proc. 1st. Southern Biomed. Eng. Conf. (ed by S. Saha), Pergamon Press, N.Y.
4. Saha, S., Rao, V.V., Malakanok, V., Gross, B.D. and Albright, J.A. (1982b) Trans. Orthop. Res. Soc. 28th. Ann. Mtg., 7:376.
5. Sonstegard, D.A. and Matthews, L.S. (1976) J. Biomech., 9:689-694.

Session 12

BIOCHEMICAL ENGINEERING

MODELING A MEMBRANE BLOOD POUCH FOR DIALYSIS DEGLYCEROLIZATION OF RBC

R.J. Ponzio*, A. Zelman*, G.M. Odell*, D. Gisser*, C. Kleinstreuer*, T. Eastlund** and A. Britten**

*Rensselaer Polytechnic Institute, Troy, New York
**American National Red Cross, Blood Resource Services, Albany, New York

ABSTRACT

We have developed a dialysis process for removing the high concentration of glycerol necessary for cryopreservation of RBC. Glycerolized blood is placed in a transport cell which permits a dialysate solution to flow over the external surface of the pouch. Glycerol migrates from the blood, across the dialysis membrane into the dialysate and is carried out of the transport cell with the dialysate. In order to prevent osmotic hemolysis, NaCℓ is programmed to enter the pouch, then the glycerol and NaCℓ concentration are gradually reduced to an acceptable concentration within 30 minutes.

A mathematical model based upon mass balance and the time dependent Fick diffusion equations was formulated and an analytic solution was attained. The model was used with experimental data to estimate values for the mass transport coefficients of glycerol and NaCℓ. The model is shown to be valid over a wide range of experimental operating conditions and is valuable in assessing potential modifications of the system.

KEYWORDS

Cryopreservation; dialysis; deglycerolization; modeling; red blood cells.

INTRODUCTION

Deglycerolization of packed red blood cells has very recently been accomplished by a method which utilizes a "membrane blood pouch" and which shows considerable promise in reducing the present high cost of cryopreserved blood cells (Zelman, 1980, 1981a, 1981b).

The membrane blood pouch is filled with glycerolized blood and is placed into a transport cell, which is constructed so that dialysate flows over the surface of the pouch. Glycerol migrates rapidly from the blood, across the dialysis membrane, into the dialysate flow, and is carried out of the transport cell. NaCℓ is dialyzed into the membrane pouch to prevent hemolysis, then is removed by further dialysis.

EXPERIMENTAL METHODS

The membrane blood pouch, composed of polycarbonate membrane material (American

Membrane, Covina, CA), has a total surface area, A_m, of 0.2840 m^2. It is filled with a volume, V_B, of 0.350 L of glycerolized red blood cells (3.5 M glycerol, C^o_{Bg}) at 25% Hct. The composition of the dialysate solution is given by Zelman (1981b). Dialysate flow rates ranged from 0.4 L/min to 2.4 L/min at 37 ± 0.5°C.

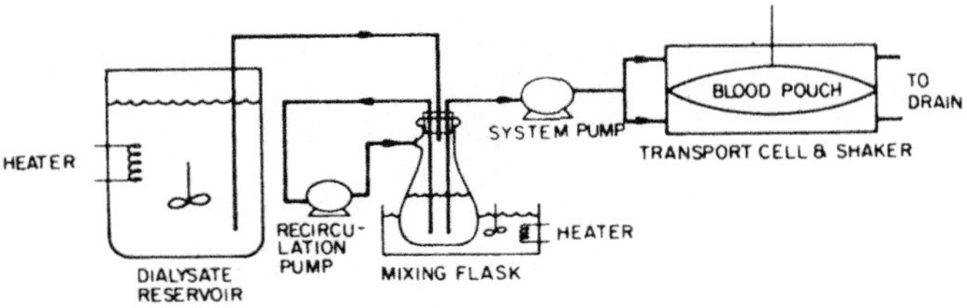

Fig. 1. Schematic diagram of the dialysate and transport cell systems.

The dialysate, C^o_D, flows into the mixing flask which is initially filled with a high concentration of NaCℓ, C^o_f. The concentration of electrolyte in the dialysate entering the transport cell, $C_D(t)$, follows the equation:

$$C_D(t) = C^o_D + (C^o_f - C^o_D) \exp(-Qt/V_f) \tag{1}$$

where V_f = volume of the mixing flask, t = time and Q = dialysate flow rate. The transport cell is mounted on a mechanical shaker (Eberbach Corp., Ann Arbor, MI), which contributes to mixing the membrane pouch and the dialysate channel. The shaker agitation rates ranged from 75 to 230 rpm. Variations of solute concentration in the dialysate channel are further reduced by inserting plastic screens in the dialysate channel.

ANALYTICAL METHODS

The assumptions used in the development of the model are summarized:

1. The contents of the membrane pouch and dialysate channel are well mixed.
2. Volume transport across the membrane pouch is zero.
3. Glycerol and electrolyte transport are uncoupled.
4. Concentration equilibrium is assumed across the red blood cell's membrane.

The initial conditions are: $\overline{C}_{Dg}(t = 0) = 0$, $C_{Bg}(t = 0) = C^o_{Bg}$, $C_{Be}(t = 0) = C^o_{Be}$ and $\overline{C}_{De}(t = 0) = C^o_{De}$. The governing differential equations are:

$$\frac{d}{dt}\begin{Bmatrix} C_{Bg} \\ \overline{C}_{Dg} \end{Bmatrix} = \begin{bmatrix} -K_gA_mV_B^{-1} & +K_gA_mV_B^{-1} \\ +K_gA_mV_D^{-1} & -(K_gA_m + Q)V_D^{-1} \end{bmatrix} \begin{Bmatrix} C_{Bg} \\ C_{Dg} \end{Bmatrix} \tag{2}$$

$$\frac{d}{dt}\begin{Bmatrix} C_{Be} \\ \overline{C}_{De} \end{Bmatrix} = \begin{bmatrix} -K_eA_mV_B^{-1} & K_eA_mV_B^{-1} \\ K_eA_mV_D^{-1} & -(K_eA_m + Q)V_D^{-1} \end{bmatrix} \begin{Bmatrix} C_{Be} \\ C_{De} \end{Bmatrix} + \begin{bmatrix} 0 \\ Q\overline{C}_D(t)V_D^{-1} \end{bmatrix} \tag{3}$$

Where V_D = volume of dialysate; C_{Be}, C_{Bg} = concentration of glycerol and electrolytes in the blood; $\bar{C}_{Dg}$, $\bar{C}_{De}$ = average concentration of glycerol and electrolytes in the dialysate. The function in equation (3), $C_D(t)$, is prescribed in equation (1). Analytic solutions to equations (2) and (3) were attained (Boyce, 1977). The model was fit to experimental data by estimating values for the mass transport coefficients (Dixon,1979; Ralston, 1978).

RESULTS

Experiments showed that the mass transport coefficients, K_i, varied with the dialysate flow rate and the shaker agitation rate. The mass transport coefficients were expressed analytically as:

$$K_j \times 10^6 = \alpha_0 + (\alpha_1/rpm) + (\alpha_2/Q) \quad (4)$$

Table 1 list the parametric values fitting equation (4) to data points and values for the correlation coefficient, R^2, and reliability coefficient R_K, i.e., standard error of curve fit, as well as values for the percentage difference, Δ%, between the asymtotic value of the mass transport coefficient and the value at maximum shaker agitation rate and maximum dialysate flow rate.

Table 1 Mass Transport Coefficients for Glycerol and NaCℓ.

Solute	Membrane Blood Pouch Contents	R^2 (dimensionless)	$R_k \times 10^6$ ($m\ sec^{-1}$)	Δ% (%)	α_0 ($m\ sec^{-1}$)	α_1 ($m\ rpm\ sec^{-1}$)	α_2 ($m\ ml\ sec^{-1}\ min^{-1}$)
$K_g \times 10^6$	blood	0.94	± 0.27	15.8	4.55	-140.7	-323.4
$K_e \times 10^6$	blood	0.95	± 0.34	9.6	5.30	-97.3	-250.91

DISCUSSION

Increasing the NaCℓ concentration during deglycerolization prevented osmotic hemolysis. Also, since NaCℓ is impermeable to the RBC, it was used to reduce the volume of the RBC, and thus shorten dialysis time.

The assumption of equilibrium of glycerol across the RBC membrane was impossible to justify experimentally. However NaCℓ had to be dialyzed into the membrane pouch to prevent osmotic hemolysis which indicates a fault in the design of the model. Yet the success of the characterization indicates this fault is not a significant factor.

The mass transport coefficients were observed to increase with increasing dialysate flow rate and shaker agitation rate. Although the flow rate increases the rate of removal of glycerol from the membrane pouch by increasing the concentration gradient across the membrane, this effect is taken into account in the modeling of the system. However the dialysate flow rate contributes to the mixing of the contents of the membrane pouch and the solution in the dialysate channel. Since the degree of mixing is not considered in the modeling of the system, its effect is observed in the predicted values of the mass transport coefficients.

A sensitivity analysis of the model for the total membrane area revealed that the time constants for the glycerol concentration were significantly affected by the membrane area. From the glycerol versus time curve a mono-exponential was fit and a single time constant for the process determined. Figure 2 shows the inverse of the mono-exponential time constant versus the total surface area of the membrane pouch. The length of the deglycerolization process is considered to be five times the inverse time constant. Figure 2 provides a rational basis for pouch size selection.

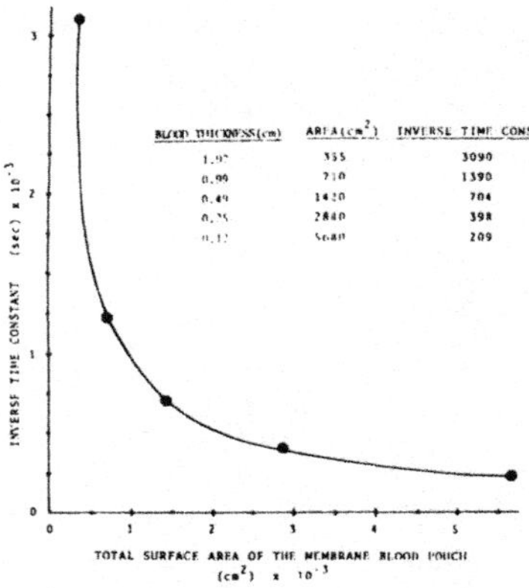

Fig. 2. Membrane pouch size.

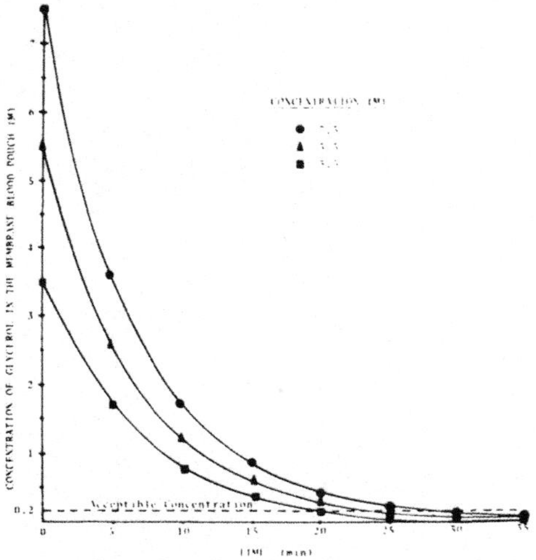

Fig. 3. Predicted concentration.

Our experiments indicate that hemolysis due to freezing and thawing can be completely avoided by elevating the glycerol concentration to 7.5 M. In Fig. 3, the model predicts (using the parameters Q = 1.2 L/min, rpm = 160, $A_m = 0.2840\ m^2$, V_B = 0.350 L and V_D = 0.240 L) that the deglycerolization time is not appreciably extended for initial glycerol concentrations of 7.5 M and 5.5 M relative to 3.5 M, which is the value presently used by the American National Red Cross. The model provides a means to evaluate potential modifications of the system and justifies the effort made by clearly defining regions of the system's operation for maximum efficiency.

ACKNOWLEDGEMENT

This research was supported by Grant (NIH) HL 24466. The technical assistance of Dawn Barthel is acknowledged.

REFERENCES

1. Boyce, W.E. and R.C. DiPrima (1977). In Elementary Differential Equations and Boundary Value Problems, Third Edition. John Wiley & Sons, New York. pp. 329-332.
2. Dixon, W.U. and M.B. Brown (1979). In BMDP-79, Biomedical Computer Programs, P-Series. Univ of California Press, CA. pp. 484-498.
3. Ralston, M. and R.I. Jennrich (1978). In Derivative-Free Non-linear Regression. Proceedings of Computer Science and Statistics: Tenth Annual Symposium on the Interface. U.S. Government Printing Office. pp. 312-322.
4. Zelman, A., D. Smith, J. Ryan, D. Gisser, and R. Stephen (1980). Electro-osmosis and dialysis processing of cryopreserved red blood cells using a membrane blood bag. Trans Am Soc Artif Intern Organs, 26:518-522.
5. Zelman, A., D. Gisser, D. Smith, and R. Stephen (1981). Dialysis processing of cryopreserved red blood cells, In F. Albin (ed.), Synthetic Membranes, Vol. 2. pp. 109-118.
6. Zelman, A., D .Gisser, D. Mathison, and R. Stephen (1981). Dialysis processing of cryopreserved red blood cells using a sterile membrane blood bag. Proc 2nd World Congress of Chemical Engineering, 5:239-242.

EFFECT OF DISTANCE ON THE MIGRATION OF PMN LEUKOCYTES UNDER AGAROSE TOWARDS CASEIN

M. Gould*, R. B. Beard**, S.P. Edkins***, and M. Hoag***

*Senior Research Scientist, EM Science, Gibbstown, N.J. and Graduate Student in Biomedical Science, Drexel Univ., Philadelphia, PA

**Professor of Electrical and Computer Engineering and Biomedical Engineering & Science Institute, Drexel Univ. Philadelphia, PA

***Graduate Students in Biomedical Science, Drexel University Philadelphia, PA

ABSTRACT

The chemotactic index A/B from migration patterns checks well with the chemotactic index determined from the ratio of the relative percent absorbance due to the stained migrating polymorphonuclear leukocytes (PMN) in the front of a well to the relative percent absorbance due to the strained random migrating PMN's on the rear of the well. An automatic computing densitometer used for determining the relative percent absorbance should become a valuable clinical instrument for determining chemotactic indices (C.I.). The C.I.'s for PMN's from a healthy male produced a linear regression of C.I. on the distance from the chemoattractant casein. For wells close to the chemoattractant a beta distribution appears to fit the population distribution migrating towards the chemoattractant and an exponential distribution, the random migration.

KEYWORDS Leukocytes, PMN's, Chemotactic Index, Densitometer, Distance Variation

INTRODUCTION

Over the past eight years a study has been conducted at Drexel University on the chemotactic response of polymorphonuclear leukocytes (PMN) in the presence of metal particle suspensions (1,2,3,4,5). Early migration and chemotacis of PMN leukocytes to metal particles have been measured using a Boyden chamber and an agarose technique (1,3,5). More recently a technique based on the measurement of the deflection angle of migrating leukocytes under the influence of a calibrated electric field and a positive chemotactic material has permitted the quantification of the chemotactic force (7). This study has been undertaken to give a greater insight into the forces acting on PMN's in agarose gel. The characteristic tear drop shape formed by the migrating PMN's is studied after staining as a function of its distance from the channel of chemoactive casein. An automatic computing densitometer has been used to monitor the population distribution of the leukocytes in the tear-drop-shaped migrating pattern. The chemotactic indices, C.I., determined from the leading edge are compared for various well distances from the chemoattractant material.

The density plot for the front and the rear of the well was divided up into intervals and the relative population of the PMN's determined for each interval. The beta distribution was first considered since it is a probability distribution appropriate for a random variable whose values are bounded between finite limits (8,9). The leading edge of the tear drop and the surface of the well form the two finite boundaries. The exponential distribution was also considered since it is a useful general purpose probability density function. A Chi-square test was used for testing the validity of the assumed distributions.

THEORY

The density function of the beta distribution is given in the following relationship (8,9):

$$f_x(x) = \frac{1}{B(q,r)} \frac{(x-a)^{q-1}(b-x)^{2-1}}{(b-a)^{q+2-1}} \qquad a \leq x \leq b \tag{1}$$

where a and b are the bounded finite limits. a is taken as starting point of the PMN migration at the well and is assumed zero. b is taken as the leading edge of the migrating PMN's.

The beta function, B (q,r), is related to the gamma function as follows:

$$B(q,r) = \frac{\Gamma(q)\ \Gamma(r)}{\Gamma(q+r)} \tag{2}$$

The parameters q and r of the beta distribution can be calculated from the mean, $\overline{x}$, and variance, s^2, using the expressions:

$$\overline{x} = a + \frac{q}{q+r}\ (b-a) \tag{3}$$

$$s^2 = \frac{qr}{(q+r)^2(q+r+1)}\ (b-a)^2 \tag{4}$$

The probability between $x=x_1$ and $x=x_2$, $P(x_1<x\leq x_2)$ can be shown to be equal to:

$$P(x_1<x\leq x_2) = \frac{1}{B(q,r)}\ [B_u(q,r)-B_v(q,r)]$$

$$P(x_1<x\leq x_2) = I_u(q,r) - I_v(q,r) \tag{5}$$

where $I_u(q,r)$ and $I_v(q,r)$ are incomplete beta function ratios with $u = \frac{x_2-a}{b-a}$ and $v = \frac{x_1-a}{b-a}$,

the incomplete beta function ratio tables are given in Pearson, et al. (9). The exponential density distribution function has teh relationship

$$f_x(x) = \lambda e^{-\lambda x}$$

where λ is a constant which is experimentally determined at x=0. In the Chi-square test for the validity of the assumed distribution, a 5% significance level was considered.

EXPERIMENTAL PROCEDURES

1) Separation of PMN Leukocytes (11)

The separation of PMN leukocytes is accomplished using 3% dextran in 0.9% saline (3,4). This method is employed using approximately thirty mls of heparized whole blood (19 USP units of sodium heparin per ml). The red blood cells are hypotonically lysed until the hemoglobin measured is zero. The white blood cell buttons are washed and resuspended to contain 500,000 cells/3µl in normal Hank's balanced salt solution (HBSS) without Ca^{++} and Mg^{++} (4,10).

2) Agarose Tissue Medium and Plates (11)

The methodology used in the preparation of the agarose plates is based on the technique of Nelson (6) and Wittels (3). 25 ml of a combined mixture of salt and agarose is pipetted into 100 x 100 x 15mm square tissue culture plates. The cast plates are stored, before cutting wells and longitudinal channels, in a 4°C refrigerator covered with wet paper towels to keep the plates moist. Figure 1 shows the agarose plate with the wells and longitudinal channels cut out.

The channels (alleys) are filled with 200 µl of test solutions of HBSS and casein. The wells are filled with 3 µl of the harvested cells. The plates are placed in an incubator with the temperature set at 37°C and the CO_2 concentration set at 5% for 24 hours. After 24 hours the plates are removed from the incubator, fixed, and stained with Wright stain or Harleco Hemocolor.

3) Chemotactic Index from Migration Patterns

The chemotactic index was determined by two methods. One method consisted of determining the chemotactic ratio of A/B from the migration patterns illustrated in Figure 2. Another method of determining the C.I. is the ratio of the relative percent absorbance due to the stained PMN's in the front of the well to the relative percent absorbance due to the stained PMN's on the rear of the well. A Gelman automatic computing densitometer, model ACD 18, at a 640 nm wavelength was used to determine the C.I.'s.

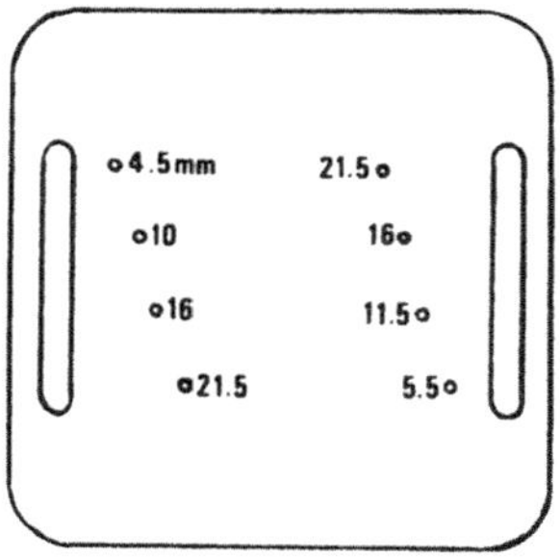

Figure 1: Agarose Plate with Wells and Longitudinal Channels

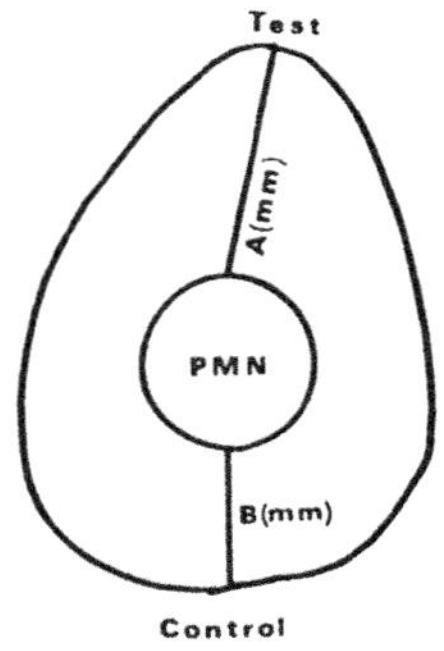

Figure 2: Migration Patterns in Agarose Gel

4) The Gelman automatic computing densitometer was used to study the population distribution around the wells in the agarose gel. In order to obtain the greatest sensitivity the scan length of the densitometer was set at 24 mm or 1.71 mm/square. The automatic computing densitometer was programmed to determine the

relative percent absorbance at 0.86 intervals in the front and rear of the well. Using this data the mean and variance were determined and used to calculate the theoretical frequency of occurrence of the PMN population based on beta and exponential distributions. The Chi-square validity test for a 5% significance level was used to determine the goodness of the assumed distribution.

RESULTS AND DISCUSSION

Figures 3 and 4 illustrate the migration patterns of PMN's toward chemoattractant casein. Figure 3 is an overall view showing how the wells are placed at various distances from the channels of casein. The increased migration of the PMN's for the wells closest to the channels is shown in Figure 4.

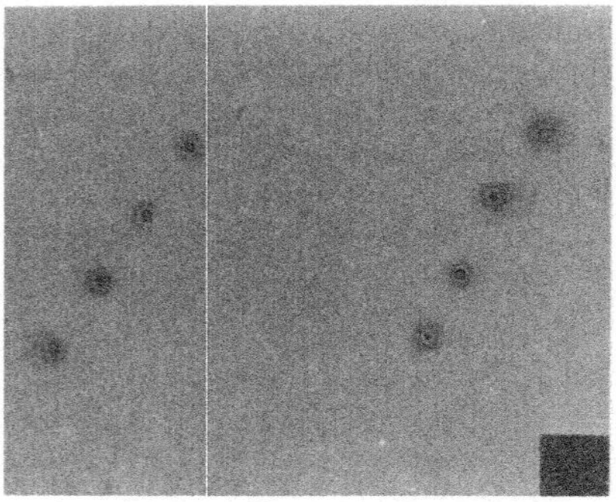

Figure 3: Migration Patterns of PMN Leukocytes from a Young Healthy Male Toward Casein

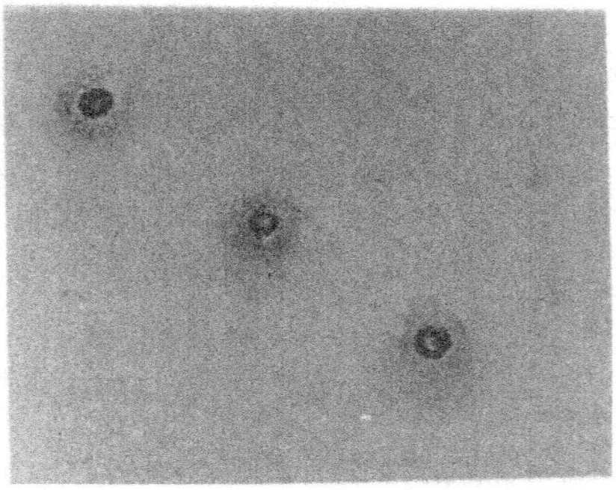

Figure 4: Close-up View of Migrating Patterns

The chemotactic indices, C.I., based on the leading edge and densitometer readings are compared in the linear regression curves in Figures 5 and 6, based on 87 data points. The dotted lines represent 95% confidence intervals.

It is noted that the regression lines of the chemotactic index on the distance from the positive control casein are C.I. = - 0.090x + 3.1. The close agreement between the two methods of determining the chemotactic index demonstrates that the use of a densitometer for automatically obtaining C.I.'s devoid of the potential for inaccuracies due to human error holds great promise for clinical applications.

The following Figures 7, 8, 9, and 10 are illustrative graphs of the automatic densitometer readings for the stained plates. Figure 7 is for well #1 with a distance of 5.5 mm from the chemoattractant casein. The front of the well fits rather well to a beta distribution with $q = 1.49$ and $r = 2.45$ according to the Chi-square validity test. The rear of the well where there is random migration appears to follow an exponential distribution with $\lambda = 0.62$.

The beta distribution, however, does not follow the PMN distribution in front of wells 2, 3, and 4. An exponential distribution appears to better represent the PMN distribution in the front and rear of these wells. The λ's of the exponential distribution in the rear of the wells for the random migration are quite close in value, as expected. The front and rear λ's for the assumed exponential distribution of the PMN are not the same for the other wells due to the increased population of PMN's towards the chemoattractant casein.

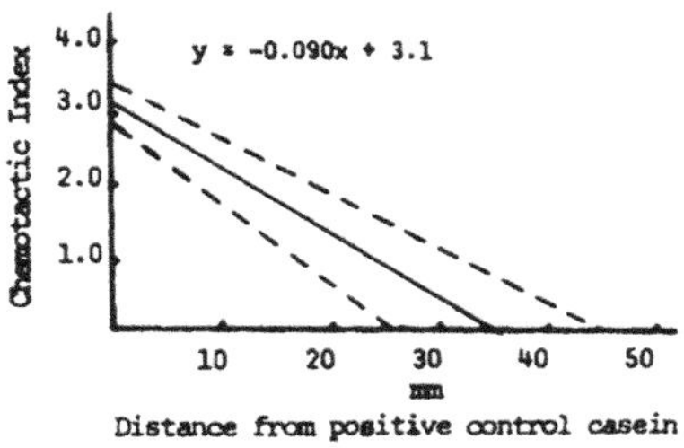

Figure 5: Linear Regression of Densitometer C.I.'s on Distance from Chemoattractant. Dotted Lines are 95% Confidence Limit.

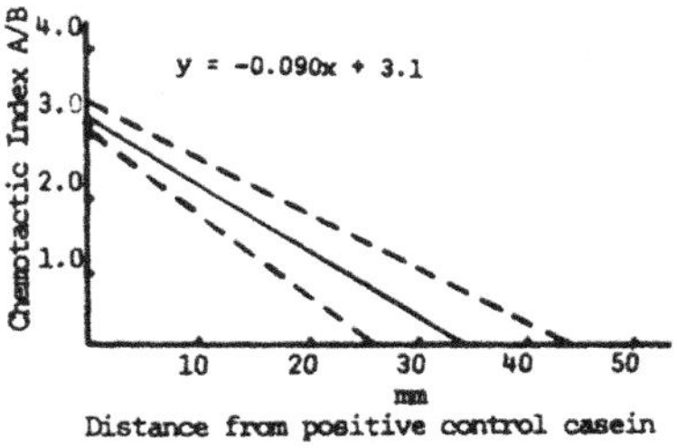

Figure 6: Linear Regression of Leading Edge C.I.'s on Distance from Chemoattractant. Dotted Lines are 95% Confidence Limit.

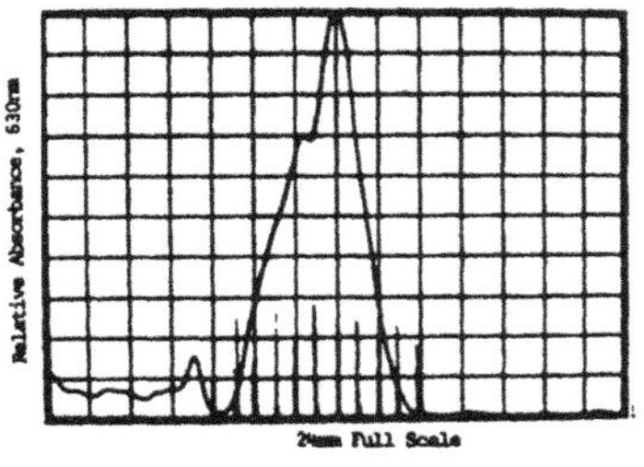

Figure 7: Relative Absorbance of PMN's around Well #1 5.5 mm from Channel of Casein.

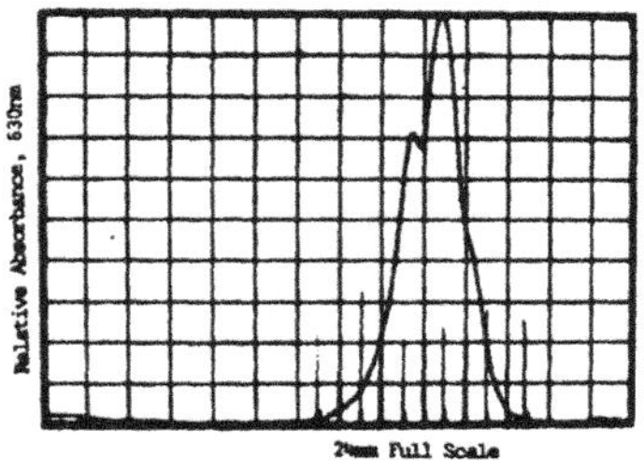

Figure 8: Relative Absorbance of PMN's around Well #2 10.0mm from Channel of Casein.

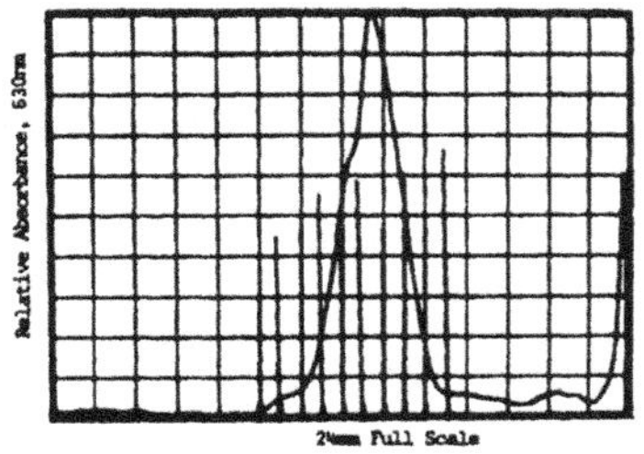

Figure 9: Relative Absorbance of PMN's around Well #3 16.0 mm from Channel of Casein

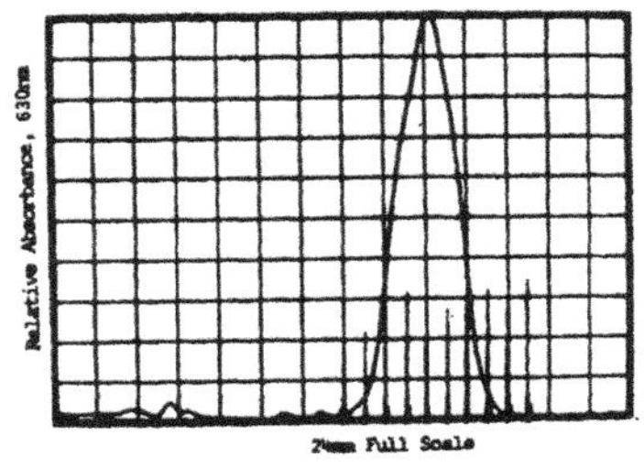

Figure 10: Relative Absorbance of PMN's around Well #4 21.5 mm from Channel of Casein.

SUMMARY

For determining chemotactic indices based on population of PMN's, the use of an automatic computing densitometer checks well with presently used techniques based on the leading edge. The densitometer approach offers a technique which can be made independent of human variations in reading the leading edge. Thus, the densitometer technique has a potential for becoming a valuable clinical diagnostic instrument for measuring chemotactic indices. In addition, the automatic computing densitometer has the capability of being a research tool for studying the population distribution of PMN's.

ACKNOWLEDGEMENTS

The authors acknowledge the assistance of M. Beard, M. Brownstein and S. Vulimiri. Acknowledgement is also given for partial support from a Biomedical Research Support Grant and The Gelman Instrument Company.

REFERENCES

(1) Lynn, S., Master's Thesis, Drexel University, 1974.

(2) Levine, S., Master's Thesis, Drexel University, 1976.

(3) Wittles, H.L., Master's Thesis, Drexel University, 1977.

(4) Strable, D.L., Master's Thesis, Drexel University, 1978.

(5) Strable, D.L., and Beard, R.B., 7th Northeast Bioengineering Conference, March 1979.

(6) Nelson, R.D., J. Immunology; 115, 1650, 1975.

(7) Gould, M., Beard, R.B., and DellaVecchia, 10th Northeast Bioengineering Conference, Dartmouth, Hanover, N.H., March 1982.

(8) Ang , A.H.S., and Tang, W.H.: Probability Concepts in Engineering, Vol. I, John Wiley, 1975.

(9) Pearson, K., Pearson, E.S., and Johnson, N.L.: Tables of Incomplete Beta - Function, Cambridge University Press, Cambridge, England, 1968.

(10)Abramson, H.A., and Moyer, L.S., J. Gen. Physiol. 19, 601-7 (1936).

(11)Gould, M., Master's Thesis, Drexel University, expected 1982.

THE INFLUENCE OF BACTERIAL ENDOTOXIN ON THE DIELECTRIC RELAXATION OF BOVINE SERUM ALBUMIN

S.L.R.Vulimiri[*], R.B.Beard[**], J.Bigdeli[*], J.H.Adler[***], and J.J.Aleo[+]

[*]Biomedical Engineering & Science Institute, Drexel Univ., Philadelphia, Pennsylvania
[**]Electrical & Computer Engineering Dept., Drexel Univ., Philadelphia, Pennsylvania
[***]Biological Sciences Department, Drexel University, Philadelphia, Pennsylvania
[+]Pathology Department, Dental School, Temple University, Philadelphia, Pennsylvania

ABSTRACT

Dielectric studies on bovine serum albumin, BSA, with the addition of endotoxin produced skewed Cole-Cole plots and a decrease in the dielectric increment. Addition of Zn^{+2} enhanced the interaction of the endotoxin with BSA to decrease markedly the dielectric increment. Dielectric measurements on lipid A and polysaccharides isolated from endotoxin in the presence and absence of the divalent zinc ion illustrated that it is the polysaccharides which are greatly influenced by the zinc ion. The addition of vitamin C to BSA-endotoxin solutions produced changes in the dispersion curves as expected from the resulting pH change.

KEYWORDS Endotoxin, Dielectric Relaxation, Bovine Serum Albumin (BSA)

INTRODUCTION

During acute inflammation, serum zinc concentrations are reduced due to an increase in the absorption and retention of zinc by tissues. Zinc protects against bacterial endotoxin (lipopolysaccharide) by stabilizing the membranes of target cells and suppressing the release of lysosomal hydrolases which cause tissue necrosis (Pekarek, 1975). Endotoxin-challenged cultures have shown that endotoxin limits the proliferation of fibroblasts. When added to the endotoxin-challenged cultures, zinc chloride reverses the endotoxin-induced effects. Endotoxin has been shown to depress the binding of zinc to serum proteins. Variations in zinc uptake have been measured by the use of radioactive zinc, ^{65}Zn (Aleo, 1976). This study has been undertaken to determine the influence of endotoxin on bovine serum albumin, BSA, using dielectric and viscosity measurements. These measurements have been used to study the interaction between endotoxin and albumin in the absence and presence of the divalent cation Zn^{+2} and of vitamin C at different pH's. The polysaccharide and lipid A portions of the endotoxin molecule were isolated in order to study their influence on BSA.

THEORY

Serum albumins are small molecules having a molecular weight of between 65,000 and 69,000 with a high solubility, net negative charge and diversity of ligand binding affinities (Svedberg, 1930). The dipole moment of protein is due to permanent dipole moments and ionic processes such as fluctuations of protons and counter-ions. Thus, protein solutions have four dispersions, one due to counter-ion fluctuations in the frequency range of 10^2 to 10^4 Hz, a dispersion centered around 10^6 Hz caused by the orientation of the macromolecule, the bound water relaxation in the 10^7 to 10^9 Hz frequency range, and the water dispersion in the range of 10^{10} Hz (Schwan, 1963). In this article the dispersion studied for BSA is the orientation or Debye type of relaxation. The BSA monomer has been described as a prolate ellipsoid with dielectric relaxations for rotation about the two axes. The dipole moment of BSA is due to amino acid residues and charge groups about the hydrodynamic center of the macromolecule. The highly symmetrical BSA molecule has a dipole moment of 348D for the monomer.

The behavior of the above biological molecules in an alternating field can be represented by the Debye equations as a complex permittivity, ε^* (Schwan, 1963; Daniel, 1967).

$$\varepsilon^*(\omega) = \varepsilon'(\omega) - j\varepsilon''(\omega) = \varepsilon_\infty + \frac{\varepsilon_o - \varepsilon_\infty}{1+j\omega\tau} \tag{1}$$

where $\varepsilon'(\omega)$ and $\varepsilon''(\omega)$ are the real and imaginary components respectively of $\varepsilon^*(\omega)$. ε_o and ε_∞ are the asymptotic values of the permittivity at low and high frequencies, respectively. The relaxation time, τ, is related to the critical frequency, $f_c = \frac{\varepsilon_o + \varepsilon_\infty}{2}$, as follows:

$$\tau = \frac{1}{2\pi fc} \tag{2}$$

Figures 1 and 2 illustrate the Cole-Cole plot and a dielectric dispersion curve.

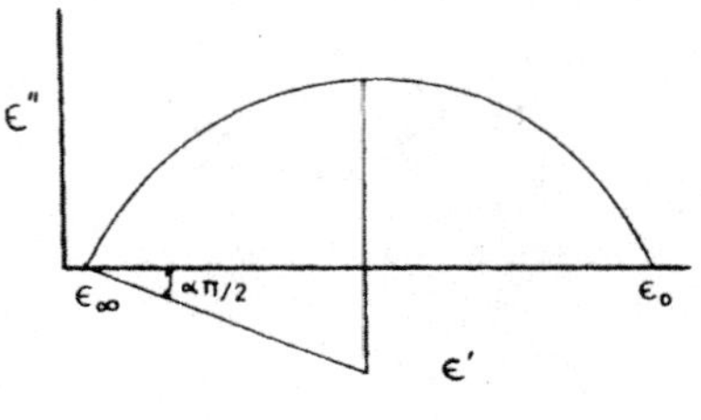

Fig.1. Cole-Cole plot for a system with multiple relaxation times.

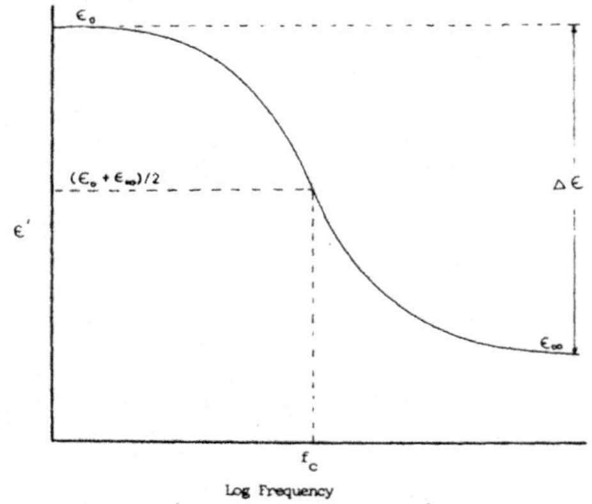

Fig.2. Dielectric dispersion curve.

When there are multiple relaxations due to molecular interactions and nonspherical molecules, the center of the semi-circle of the Cole-Cole plot is displaced from the real axis, ε'. Cole and Cole (1941) modified the Debye equation to account for a distribution of relaxation times as follows:

$$\varepsilon^* = \frac{\varepsilon_o - \varepsilon_\infty}{1+(j\omega\tau)^{1-\alpha}} \tag{3}$$

where α is an experimentally determined constant.

When there are a series of relaxation mechanisms which become of decreasing importance at the higher frequencies, an asymmetrical or skewed Cole-Cole plot develops which can be represented by the Davidson and Cole (1951) modified equation:

$$\varepsilon^* = \varepsilon_\infty + \frac{\varepsilon_o - \varepsilon_\infty}{(1+j\omega\tau)^\beta} \tag{4}$$

where β, which defines the shape of the arc in the plot, is an empirical constant.

The relaxation time, τ, of Debye for a macroscopic sphere is:

$$\tau = \frac{4\pi}{kT} \eta r^3 \tag{5}$$

where k is Boltzmann's coefficient; T, the absolute temperature; η, the shear viscosity; r, the radius of the sphere.

The modification by Perrin and Fisher for an ellipsoid, (Boettcher, 1952), is:

$$\tau_i = \frac{4\pi}{kT} \eta \text{ abc } f_i \tag{6}$$

where a, b, and c are the axes of the ellipsoid and f_i is a function of the axial ratios. Therefore, viscosity measurements of the protein and endotoxin solutions were made to correlate with the relaxation times.

Endotoxins are the surface antigens of gram-negative bacteria which play an important role in the hose-bacteria interaction (Westphal, 1975). The two components making up endotoxins are a heteropolysaccharide and a covalently linked lipid, lipid A. The polysaccharide portion is hydrophilic as are the BSA molecules, which are effective carriers of zinc. The endotoxically active lipid A is primarily made up of glucosamine, phosphate and four fatty acids. Removal of the fatty acid residues reduces the biological activity of the molecule (Luderitz, 1973).

The pH is monitored since BSA undergoes pH-dependent transitions. Dialysis was carried out between 3°C and 5°C and the impedance measurements were taken within two hours to minimize denaturation of the protein.

MATERIALS

The BSA (Fraction V, pH 5.2), obtained from Miles Laboratories, Inc., has a 98% purity by electrophoresis. The 35 g/l BSA solutions were dialyzed against precooled double distilled water for 72 hours of 3-5°C using membrane tubing (molecular weight cutoff at 12,000). Lipopolysaccharides from E. Coli were obtained from Sigma Chemical Co. and J. Bigdeli (1980). The polysaccharide portion of the endotoxin molecule was obtained by isolating it using hydroxylaminolysis, with an 80% yield. The lipid A was isolated from the lipopolysaccharide by acetic acid hydrolysis according to Galanos, et al. (Galanos, 1969).

EXPERIMENTAL PROCEDURES

Admittance measurements over the frequency range of 0.5 MHz to 150 MHz were made using a Boonton RX Meter. Measurements over the 0.15 MHz to 3 MHz range were made using a Wayne Kerr B201 electrical impedance bridge. The impedance cells were jacketed to keep the solutions at 37.5°C. The platinum electrodes were platinized to reduce electrode polarization. The cell and electrodes are soaked for a minimum of 24 hours in distilled water before adjusting the background solution conductivity with 0.01N KCl. Impedance measurements were also taken on background solutions to compensate for the variation in solution conductivity due to the adsorption of ions on the platinized electrodes. The (background imepdance) measurements are taken within the same time frame as those for the sample measurements. Viscosity measurements were made at 37.5°C using Cannon Fenske Modified Ostwald kinematic viscometers in a TEM V-70 water bath.

RESULTS

Upon the addition of endotoxin (LPS), the dielectric dispersion of BSA produced skewed Cole-Cole plots and a decrease in dielectric increment, as shown in Fig.3. Endotoxin with or without zinc did not display a dispersion over the frequency range of 0.15 MHz to 150 MHz, as illustrated in Fig.4.

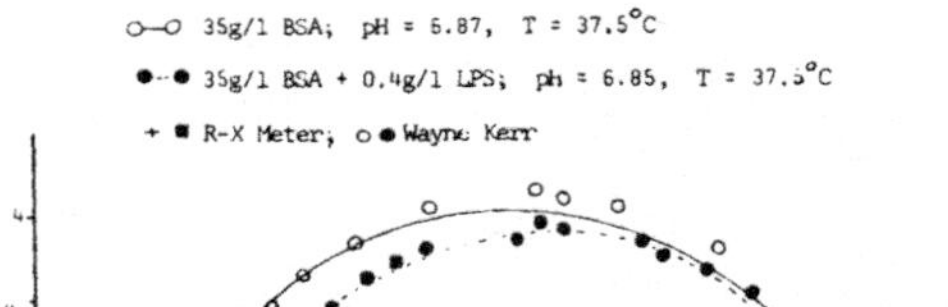

Fig.3 Cole-Cole plot of BSA at pH=6.87 and BSA-LPS at pH=6.85.

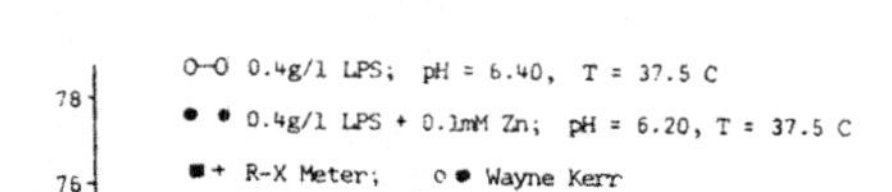

Fig.4 Dispersion curve of lipopolysacharides (LPS) at pH=6.40 and LPS with Zn at pH=6.20.

Figure 5 demonstrates that the addition of endotoxin to BSA in the presence of the divalent zinc ion produced an enhanced asymmetrical Cole-Cole plot with a marked decrease in the dielectric increment. However, the Cole-Cole plots of the polysaccharides isolated from endotoxin demonstrate that they do have a relaxation which is enhanced by the addition of zinc, as shown in Fig.6. The dispersions of BSA-lipid A complexes were similar to those obtained for BSA in the presence or in the absence of zinc. The addition of vitamin C to the BSA solutions lowered the pH and the dielectric increment. Neither endotoxin nor vitamin C noticeably affected the viscosity of the BSA solutions.

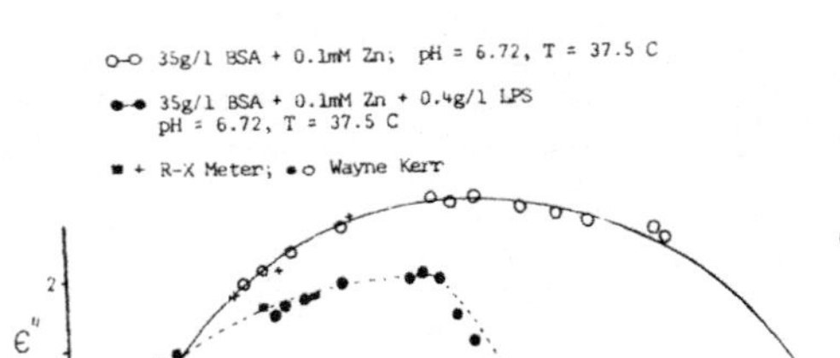

Fig.5 Cole-Cole plot of BSA at pH=6.72 and BSA-LPS at pH=6.72 in the presence of Zn.

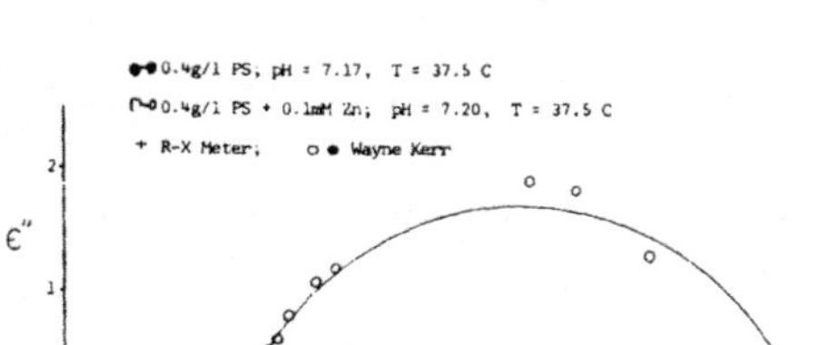

Fig.6 Cole-Cole plots of polysacharides (PS) at pH=7.17 and of PS in the presence of Zn at pH=7.20.

DISCUSSION

The measured dispersions of BSA checked well with previously published data (Takashima, 1969). Upon the addition of endotoxin to BSA, there is a corresponding decrease in the dielectric increment. The skewed Cole-Cole plots describing this phenomenon indicate an interaction involving the shielding of some of the BSA charge groups. The dielectric curve, which displays a narrow dispersion at low frequencies and a broad dispersion at high frequencies, implies a single relaxation at low frequencies with multiple relaxations at the higher frequencies. The enhancement of the interaction of endotoxin with BSA upon the addition of the divalent zinc ion is not understood at this time. However, this study indicates that it is the interaction of zinc with the polysaccharide of the lipopolysaccharide (endotoxin) which enhances the dielectric dispersion. Previous dielectric measurements at Drexel on muccopolysaccharides have shown a dispersion in the same frequency range as that of the dispersions for the polysaccharides of the endotoxin (Dellavecchia, 1972). The addition of vitamin C to BSA and endotoxin solutions with the resulting decrease in pH produced a shift in the dispersion curves of BSA as expected due to the change in pH (Takashima, 1969). The negligible variation in the viscosity of BSA upon the addition of vitamin C and endotoxin was unexpected even for the low concentrations used. Further dielectric studies are being undertaken on BSA using other divalent ions.

ACKNOWLEDGEMENTS

This research was partially supported by a B.R.S.G. grant. The authors wish to thank Dr. Shiro Takashima for his many contributions.

REFERENCES

Aleo, J.J. (1976). Role of Zinc^{+2} in Endotoxin-Challenged Fibroblast Cultures. Microbis. Let., 3,191-197.

Bigdeli, J. (1980). M.S. Thesis, Drexel University.

Boettcher, T.C.F. (1952). Theory of Dielectric Polarization. Elsevier, Amsterdam.

Cole, K.S.; and Cole, R.H. (1941). Dispersion and Absorption in Dielectrics. 1. Alternating Current Characteristics. J. Chem. Phys., 9, 342-351.

Daniel, V.V. (1967). Dielectric Relaxation, Acad. Press.

Davidson, D.W.; and Cole, R.H. (1951). Dielectric Relaxation in Glycerol, Propylene Glycol and n-Propanol. J. Chem. Phys., 19, 1484-1490.

Dellavecchia, M. (1972). M.S. Thesis, Drexel University.

Galanos, C.; Luderitz, O.; and Westphal, O. (1969). Eur. J. Biochem., 19, 245-249.

Luderitz, O.; Galanos, C.; Lehmann, V. (1973). Lipid A: Chemical Structure and Biological Activity. J. Infec. Dis., 128, S17-S29.

Pekarek, R.S.; and Evans, G.W. (1975). Effect of Acute Infection and Endotoxemia on Zinc Absorption in the Rat. Soc. Exp. Biol. Med., 150, 755-758.

Schwan, H.P. (1963). Determination of Biological Impedances. Physical Techniques in Biological Research, New York: Academic Press, Vol. VI, Chapter 6.

Svedberg, T.; and Sjogren, B. (1930). The pH Stability Regions of Serum Albumin and of Serum Globulin. J. Amer. Chem. Soc., 52, 2855-2863.

Takashima, S. (1969). Dielectric Properties of Proteins - 1. Dielectric Relaxation. Physical Principles and Techniques of Protein Chemistry. Chapter 6, Part A. ed. S.J. Leach, Academic Press.

Westphal, O. (1975). Bacterial Endotoxins. Int. Archs. Allergy Appl. Immun.,49, 1.

MASS TRANSFER CHARACTERIZATION IN HOLLOW FIBER KIDNEYS

Thomas A. Armer* and Thomas R. Hanley**

*Department of Chemical Engineering, Tulane University, New Orleans, Louisiana

**Department of Chemical Engineering, Rose-Hulman Institute of Technology, Terre Haute, Indiana

ABSTRACT

This paper describes the in vitro characterization of the mass transfer behavior in a Cordis-Dow hollow fiber artificial kidney. The purpose of this investigation was to establish the behavior of the dialyzer device, apart from the patient-artificial kidney system, so as to serve as a reference by which in vivo performance may be correlated to the particular design and operating parameters of the capillary artificial kidney. This characterization allows specific therapy based upon the variation of surface area, blood and dialysate flow rates, transmembrane pressure and the solute spectrum to be removed. The characterization also allows the optimization of therapy using present designs and direct improvement of future designs to more accurately duplicate natural kidney function.

The mass transfer behavior of the Cordis-Dow hollow fiber artificial kidney was characterized by an in vitro experimental analysis. A total of 4,860 base data points describing the operational and performance parameters of the 1.3, 1.8 and 2.5 square meter dialysis units were amassed to establish sodium chloride (NaCl) and disodium N-octadecylsulfosuccinate (A-18) removal rates. Data reduction by computer yielded clearances, clearance efficiencies, phenomenological coefficients, and rate constants based upon mass balances, nonequilibrium thermodynamics, and kinetic models of the dialysis system.

KEYWORDS

Hollow Fiber Artificial Kidney; Mass Transfer; Dialysis; Ultrafiltration; Middle Molecules; In Vitro Characterization

INTRODUCTION

Hollow fiber, or capillary type, artificial kidneys (HFAK) were developed in the early 1970's on the basis of technology from industrial reverse osmosis processes. In HFAK units, blood is pumped through bundles of capillary tubes approximately 200 microns in diameter with wall thicknesses of 30 microns. The tube bundles are contained in a compact, cylindrical, acrylic cartridge which serves as an envelope for counter-current dialysate flow. Surface areas in the range of 0.6 to 2.5 square meters are commercially available with the membrane material usually cellulose acetate or cuprophan. The hollow fiber configuration offers the advantages of (1) high structural strength without the cumbersome static membrane support;

(2) high surface areas with low priming volumes (less than 200 millimeters); (3) compactness; and (4) disposability. The latter feature assures pyrogen and infection free operation. Ultrafiltration in HFAK's is induced by application of a subambient pressure to the dialysate side and has proven effective in removal of excess body fluid. HFAK's are widely used in therapeutic applications due to their efficiency and ease of preparation and disposal.

Maximum operational performance in fulfillment of individual dialysis requirements is of major concern in artificial kidney design and clinical implementation. Malchesky (1974) has cited functional requirements, operational features, and patient oriented considerations as areas in need of improvement and suggests systems which minimize both time and cost requirements yet ensure efficient, controlled functional behavior are highly desirable. To date, however, the lack of reproducibility of performance and the inability to define a uniformly applicable performance index still deter design and subsequent operation of a dialyzer. In vitro comparisons of commercial dialyzers using the standard urea clearance performance index as a function of dialysate flowrate show significant differences between varying dialyzer designs and even between those of the same type (Babb, et al, 1974). An exact characterization of mass transfer of specific dialyzer design is necessary in formulating a realistic evaluation of performance. Equally important to the accessment and development of optimal performance is the establishment of a comprehensive performance index that can be related to the whole range of patient needs.

Concurrent to the elucidation of the biochemical nature of uremia, dialyzer systems which can selectively remove known toxins must be developed. Only a rigorous characterization of artificial kidney behavior will enable development of such systems. A definition of the extent and conditions of removal of a spectrum of solutes is prerequisite to an ordered dialysis procedure. In vivo characterizations of dialyzer are difficult to perform with reproduceable results. Fluctuations in blood composition, the difficulty of real time quantitative and qualitative measurement of blood constituents, membrane occlusion, and physiological complications hamper interpretation of dialyzer behavior. Colton, et al, (1971) have noted the complexity of a rigorous treatment of solute transport in blood due to the effects of diffusion in plasma and into erythrocytes, solute interactions, and the associated multicomponent diffusion with shear dependent diffusion phenomena. Kozinski and Lightfoot (1972) have illustrated the problems associated with protein sedimentation and solute-protein binding in the modeling of ultrafiltration. In seeking a reliable data base, studies are often conducted in vitro, thereby allowing evaluation of artificial kidney behavior under conditions where critical parameters are subject to greater control. This approach is justified in that in vitro removal rates approximate those of in vivo situations, especially for smaller solutes. Klein, Colton, Farrell, et al, (1972) have recommended an in vitro data base as "a firm basis for evaluation of the suitability of the dialyzer for further development and in vivo evaluation."

EXPERIMENTAL

The apparatus shown in Figure 1 was assembled for the purpose of measuring clearance and overall mass transfer coefficients of sodium chloride, and disodium N-ocadecylsulfosuccinate (A-18). Hollow fiber artificial kidney units manufactured by Cordis-Dow with surface areas of 1.3, 1.8 and 2.5 square meters were tested. Step input and impulse forcing functions were used to evaluate the system in steady state and unsteady state modes of operation. In steady state mode, a solution of constant concentration of either NaCl of A-18 was supplied to the blood inlet of an artificial kidney unit under conditions of fixed Q_{Bin}, Q_{Din} and ΔP. Blood outlet concentration was monitored on the chart recorder until a steady

Figure 1. DIALYSANCE MODE OPERATIONAL DIAGRAM

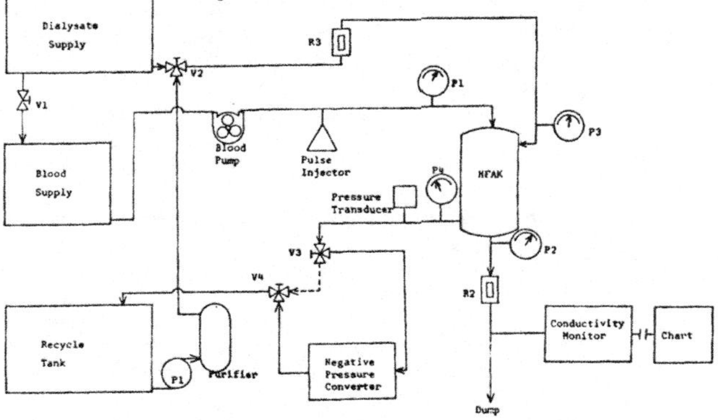

state, i.e., $dC/dt = 0$ plateau was reached, indicative of a typical response to a step forcing function. The outlet concentration was then recorded, Q_{Bin}, Q_{Din} or ΔP varied and the entire procedure repeated to evaluate a new steady state outlet concentration at the new conditions.

RESULTS

Cordis-Dow Hollow Fiber "D" series artificial kidneys - 1.3, 1.8 and 2.5 square meters surface area - were experimentally characterized in a steady-state analysis of mass transfer of sodium chloride and disodium N-octadecylsulfosuccinate, at all possible permutations over the specified operational ranges of variables Q_{Bin}, Q_{Din}, and ΔP. Convective contribution to overall transport behavior was gauged by evaluations at $\Delta P \approx 0$ and $\Delta P > 0$. The results for the ultrafiltration testing for the 1.8 square meter unit are presented in Figure 2.

A correlation for the general hollow fiber configuration was developed utilizing the Stanton number, $N_{ST} = K_o/V_B$, where V_B if the bulk velocity of the blood flow through an individual tube, and K_o is the overall mass transfer coefficient. The flow and geometry parameter was represented by the Reynolds number $\rho V_B D_F/\mu$ where D_F is the fiber diameter, ρ and μ are the solution density and viscosity. Figure 3 shows the NaCl and A-18 characterizations for dialysis and ultrafiltration after fitting the data to second order polynomials.

DISCUSSION AND CONCLUSIONS

Sodium chloride and A-18 solute removal in the hollow fiber dialyzer is primarily dependent upon the diffusive mechanism. As expected, the larger A-18 molecule (radius = 10.5Å) has clearances which are less than half those of the small (2.4 Å) NaCl molecule. Transfer of the A-18 molecule was also observed to be approximately ten times more dependent upon convective transport (Peclet number) than the NaCl molecule, which is quite capable of being convected, but is just as readily diffused back across the membrane along its concentration gradient.

Sodium chloride clearance values concur with literature citations (within 12 percent) for urea which is close to NaCl in molecular dimensions. Thus, an in vitro data base for small molecules has been established which validates the testing procedure and data analysis employed in this study.

Clearances for A-18 in the isobaric (dialysance) mode were within 15 percent of literature citation for bromosulfophthalein (BSP) values. The molecular radii of

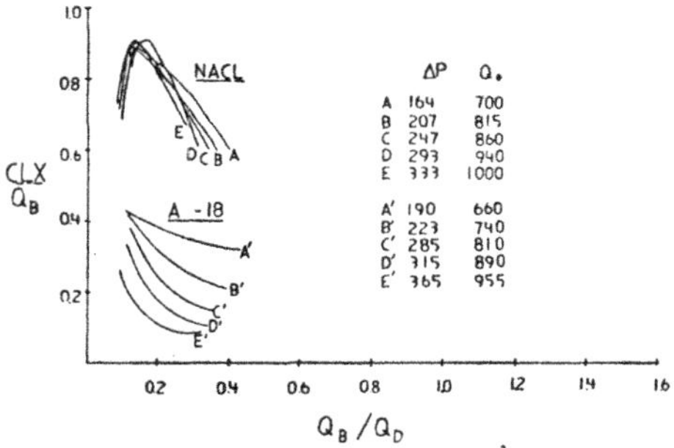

Figure 2. CLEARANCE EFFICIENCIES FOR 1.5M^2 HFAK

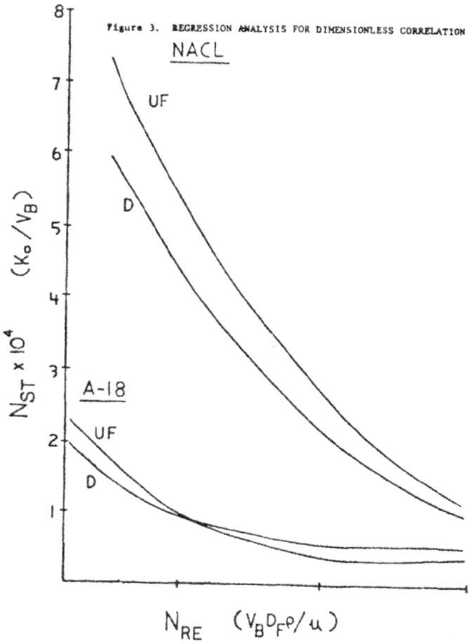

Figure 3. REGRESSION ANALYSIS FOR DIMENSIONLESS CORRELATION

these solutes was also close at 10.5 Å (A-B) and 10.1 Å (BSP). The marked deviation of A-18 efficiency from dialysance and NaCl behavior under conditions of transmembrane pressure gradients is attributable to convective fluxes acting against the normal ultrafiltrative flux. Pore occlusion or boundary layer sedimentation as observed in proteins could also contribute to this effect.

Ultrafiltration contributes less than five percent to the removal of NaCl or A-18. At high transmembrane pressures the A-18 transfer is actually reduced by pore occlusion, and boundary layer sedimentation under the influence of the initially high ultrafiltration rate and the affinity of the A-18 molecule for the cellulose acetate membrane surface. Such pore occlusion and other concentrative effects at the blood side membrane interface would generate osmotic pressure gradients of sufficient magnitude to induce the large convective (dialysate to blood) fluxes observed with this solute. Sodium chloride, though readily convected to the dialysate side of the membrane is just as readily diffused back into blood side fluid under the influence of a concentration gradient or counter convective currents as might be found towards the blood side outlet. Such an effect has been reported and would serve to cancel the majority of convective contribution.

ACKNOWLEDGEMENT

This work was supported by an N.I.H. Biomedical Research Support Grant at Tulane.

REFERENCES

Babb, A.L., M.L. Strand, D.A. Urelli, J. Milutinovic and B.H. Schribner (1972). Quantitative Description of Dialysis Treatment: A Dialysis Index, presented to the Conference on Adequacy of Dialysis, Monterey, CA.
Colton, C.K., K.A. Smith, E.W. Merrill, and S. Friedman (1971). Diffusion of Urea in Flowing Blood, AIChE J., 17(4), 800-808.
DHEW Pub. No. (NIH) 72-103 (1972). Evaluation of Hemodialyzers, 72-103.
Kozinski, A.A., and E.N. Lightfoot (1972). A General Example of Boundary Layer Filtration, AIChE J., 18(5), 1030-1040.
Malchesky, P.S. (1974). Patient Oriented Dialyzer Selection and Design, Biomat. Med. Dev. Art. Org., 2(3), 249-266.

EFFECT OF ADENOSINE ON VASCULAR WALL TISSUE PO_2

Donald G. Buerk

Department of Chemical Engineering,
Northwestern University,
Evanston, IL 60201

ABSTRACT

Measurements of PO_2 levels on the outer surface and in vascular wall tissue were made from the thoracic aorta and femoral artery (N = 5 dogs) with recessed oxygen electrodes. Adenosine (.25 mg/kg) caused a transient increase in the vascular wall PO_2 of the thoracic aorta by a maximum of + 23.2 ± 6.0 (SE) torr from the control level of 37.3 ± 8.3 (SE) torr ($p < .025$). No significant change was measured in the femoral artery. The PO_2 responses indicate different vasodilatory capabilities for the vasa vasorum in the outer walls of the two vessels and may explain why some blood vessels are more susceptible to atherosclerosis.

KEYWORDS

Adenosine, atherosclerosis, femoral artery, thoracic aorta, oxygen electrode, vasa vasorum, vascular wall PO_2.

INTRODUCTION

The medial layer of the vascular wall of larger blood vessels in dogs and humans is supplied with oxygen both from the vessel lumen and from the microcirculation (vasa vasorum) in the outer wall. For normal conditions *in vivo*, the oxygen supply from the vasa vasorum is necessary to maintain oxygenation of vascular wall tissue above hypoxic levels (Buerk and Goldstick, 1981). This supply is essential since our measurements of PO_2 profiles (Buerk et al., 1981) suggest that the inner wall has a higher oxygen consumption rate compared to the outer wall, limiting the amount of wall tissue that can be supplied by diffusion from the lumen. Disturbances of the vasa vasorum blood supply can lead to medial wall necrosis in the dog thoracic aorta (Wilens et al., 1965; Heistad et al., 1981a), presumably due to tissue hypoxia or anoxia. Computer modeling of the oxygen supply to the vascular wall also confirms the importance of vasa vasorum blood flow to maintain adequate oxygenation of the wall (Schneiderman and Goldstick, 1978). In the present study, we have examined the vasodilatory capacity of the vasa vasorum by measuring the PO_2 changes on the vessel surface and in the outer vascular wall tissue supplied by vasa vasorum in the dog thoracic aorta and femoral artery. We attempted to increase oxygen supply to the wall by increasing vasa vasorum blood flow using the vasodilator adenosine. This information should be useful for assessing the ability of the vasa vasorum to regulate oxygen supply under other conditions.

METHODS

A sturdy oxygen electrode was constructed for tissue PO_2 measurements by embedding a 10 μm diameter, glass-coated gold wire cathode in a 32 gauge hypodermic needle with epoxy. The electrode is similar to the design described by Whalen and Spande (1980), except that only a single cathode was embedded. The hypodermic needle was silver plated, serving as the anode for the polarographic reduction of oxygen at -0.7 volt. The current sensitivity to oxygen was reduced to less than 5 picoamperes per torr after electrochemically recessing the gold cathode and applying a membrane. Stir sensitivity was less than 2%. A surface electrode was constructed from the same type of glass-coated wire, with a silver wire anode and a thermistor embedded into a lucite assembly. The cathode was recessed and a membrane applied. The face of the probe was curved to accommodate the cylindrical surface of the blood vessels being studied. Both electrodes were calibrated at 37°C in physiological saline solutions equilibrated at known oxygen concentrations prior to and at frequent intervals during each experiment.

Five dogs were anesthetized with sodium pentobarbitol (30 mg/kg) and artificially ventilated to maintain normal blood gases. Blood pressure was monitored from a catheter in the carotid artery. Surface and tissue PO_2 measurements were first obtained from the femoral artery and later from the middle third of the descending thoracic aorta. The aorta was exposed by entering the thoracic cavity through an incision between the ribs of the left chest wall. Positive pressure was applied to the respirator after opening the chest. The surface PO_2 probe was positioned against the blood vessles, serving to damp pulsatile motions at the tissue PO_2 measurement site. The hypodermic needle electrode was advanced by a micromanipulator to depths between 0.5 to 1 mm from the outer surface of the thoracic aorta, or between 0.2 to 0.4 mm for the femoral artery. Blood pressure, tissue PO_2, surface PO_2 and the vessel surface temperature were recorded on FM magnetic tape. Responses were obtained following a bolus injection of adenosine (.25 mg/kg) via a catheter in the jugular vein.

RESULTS

Tape recorded PO_2 responses were digitized by a minicomputer at 1 Hz, averaged every 10 seconds and the resulting PO_2 changes are plotted in Fig. 1 for the two blood vessels. The average surface and vascular wall PO_2 levels during the two minute control period preceding the injection of adenosine are listed in Table 1, along with the maximum increase (ΔPO_2). The maximum increases in surface and vascular wall PO_2 for the femoral artery were not significantly different from control levels. The maximum increase in surface PO_2 for the thoracic aorta was also not significant. However, a significant ($p < .025$) increase in vascular wall PO_2 was measured in the thoracic aorta, rising to a maximum around 90 seconds. In most cases, the vascular wall PO_2 returned to control in less than 3 minutes, but remained elevated at one site for over 10 minutes. At sites in the femoral artery in 2 dogs, a similar but smaller increase in vascular wall PO_2 was measured. In both blood vessels an initial decline in surface and vascular wall PO_2 was often observed. The decline was more pronounced in the surface PO_2 mesurements, with average decreases of -3.4 and -1.0 torr at 20 seconds for the thoracic aorta and femoral artery respectively. The PO_2 decrease tended to follow the transient fall in systemic blood pressure following adenosine. The blood pressure decreased appoximately 20% before returning to control within 30 to 45 seconds. No changes in surface temperature were observed.

TABLE 1. Surface and Vascular Wall PO_2 Responses to Adenosine

BLOOD VESSEL	SURFACE PO_2 (torr) Control PO_2 ± SE	SURFACE PO_2 (torr) Maximum ΔPO_2 ± SE	VASCULAR WALL PO_2 (torr) Control PO_2 ± SE	VASCULAR WALL PO_2 (torr) Maximum ΔPO ± SE
Thoracic aorta	51.4 ± 7.5	+ 1.4 ± 5.2	37.3 ± 8.3	+23.2 ± 6.0*
Femoral artery	60.8 ± 4.8	+ 1.4 ± 0.7	53.7 ± 5.8	+0.3 ± 0.6

(N = 5 dogs)

*(p < .025)

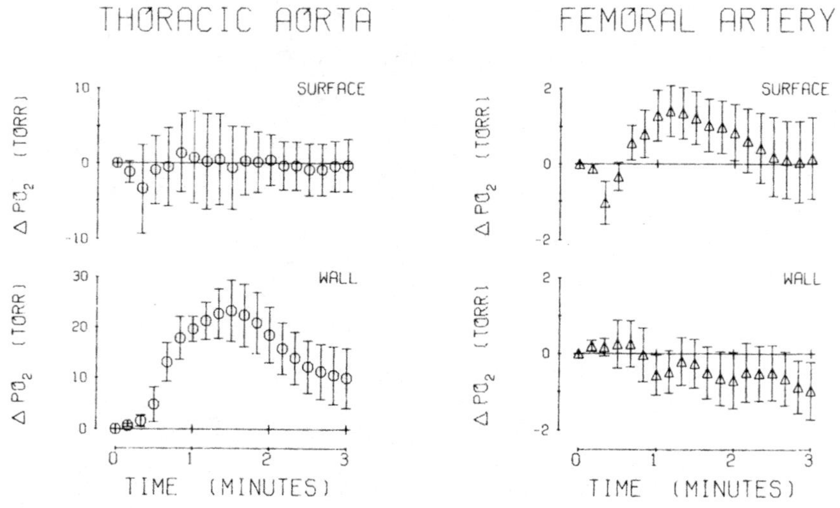

Fig. 1. Changes in surface and vascular wall PO_2 of thoracic aorta (circles ± SE) and femoral artery (triangles ± SE) following adenosine (N = 5 dogs).

DISCUSSION

Heistad et al. (1981a) have reported large increases in vasa vasorum blood flow in both middle and outer layers of the dog thoracic aorta during infusion of adenosine. Only the vascular wall PO_2 responses measured in the middle layer of the thoracic aorta were consistent with dilation of the vasa vasorum in the present study. The thoracic aorta surface PO_2 did not increase, although this response may have been masked by the transient fall in blood pressure. It is also possible that the vasa vasorum in the outer wall respond more slowly to adenosine. Adenosine had little effect on the femoral artery. These results suggest that there is a difference between the vasodilatory capabilities of the vasa vasorum for these normal blood vessels. The thoracic aorta appears to be better protected from medial wall hypoxia than the femoral artery. It is well established that diffusion distances are longer and oxygen requirements increased for atherosclerotic vessels. Heistad et al. (1981b) have found proliferation of new vessels in the thoracic and abdominal aortas of atherosclerotic monkeys, presumably to increase the oxygen supply. It is reasonable to expect that the oxygen supply is also important during the progression of atherosclerosis. There are several conditions where the supply could be reduced. Sacks (1975) has suggested that hypertension could decrease the supply, possibly by compressing the vasa vasorum. Buerk and Goldstick (1982) have recently reported that carbon monoxide in the range experienced by heavy cigarette smokers can reduce vascular wall PO_2 in the dog thoracic aorta and femoral artery. The femoral artery surface PO_2 decreased more than the thoracic aorta surface PO_2 ten minutes after a brief exposure to carbon monoxide. The difference between the two vessels is consistent with the results of the present study. If peripheral vessels are more susceptible to vascular wall tissue anoxia or hypoxia, the greater incidence of atherosclerotic disease in these vessels may be explained.

ACKNOWLEDGEMENTS

This study has been supported by U.S.P.H.S. N.I.A.M.D.D. Award 06292 and by a grant from Evanston Hospital.

REFERENCES

Buerk, D.G. and T.K. Goldstick (1981). An evaluation of the importance of the vasa vasorum for oxygenation of vascular wall tissue. Microvas. Res., 21, 238. Abstract.

Buerk, D.G., T.K. Goldstick, J.T. Ernest and P.B. Dobrin (1981). Oxygen tension profiles and oxygen consumption inhomogeneities in the arterial wall: Implications for atherogenesis. Adv. Physiol. Sci., 25, 23-24.

Buerk, D.G. and T.K. Goldstick (1982). Vascular wall PO_2 following carbon monoxide exposure. Adv. in Exp. Med. and Biol. (in press).

Heistad, D.D., M.L. Marcus, G.E. Larsen and M.L. Armstrong (1981a). Role of vasa vasorum in nourishment of the aorta. Am. J. Physiol., 240, H781-H787.

Heistad, D.D., M.L. Armstrong and M.L. Marcus (1981b). Hyperemia of the aortic wall in atherosclerotic monkeys. Circ. Res., 48, 669-675.

Sacks, A.H. (1975). The vasa vasorum as a link between hypertension and arteriosclerosis. Angiology, 26, 385-390.

Schneiderman, G., and T.K. Goldstick (1978). Significance of luminal plasma layer resistance in arterial wall oxygen supply. Atherosclerosis, 31 11-20.

Whalen, W.J. and J.I. Spande (1980). A hypodermic needle PO_2 electrode. J. Appl. Physiol., 48, 186-187.

Wilens, S.L., J.A. Malcolm and J.M. Vasquez (1965). Experimental infarction (medial necrosis) of the dog's aorta. Am. J. Pathol., 47, 695-711.

Session 13
IMPLANT

Comparative Study of Carbon and Graphite Cloth
With Marlex and Dacron Mesh in the Repair of
Abdominal Wound Defects

*S. Saha, M.D.,** T. B. Pietsch, M.D., ***S. Mack, *S. Saha, Ph.D.
*S. Pal, Ph.D., *J. A. Albright, M.D.
*Dept. of Orthopaedic Surgery, LSU Medical Center, Shreveport, LA. 71130
**Dept. of Surgery, LSU Medical Center, Shreveport, LA 71130
***Medical Student

ABSTRACT

In this study, carbon and graphite cloth were compared with marlex and dacron mesh in the repair of artifically created ventral hernia in rats. Both carbon and graphite cloth were found to be of comparable mechanical strength with marlex and dacron mesh.

KEYWORDS

Abdominal wound defect, carbon cloth, graphite cloth, marlex mesh, dacron mesh, hernia repair, mechnical strength.

INTRODUCTION

Abdominal wall defects, whether congenital or acquired, are common surgical problems which affect all age groups in either sex. Surgeons have long been trying to use various materials for repair of such wound defects. Numerous materials including fascia lata, steel wire, and various synthetic materials such as nylon mesh, have been used with varying results. Presently marlex mesh and dacron mesh enjoy the most popularity as an implant material for such repair. These are strong materials which are not biodegradable. They stimulate very little collagen formation except initially, (Forster et al, 1978). On prolonged use, they provide protection against the recurrence. Being a permanent foreign body, they act as a potential source of infection. In fact the two major complications following such hernia repair are infection and recurrence of hernia. This often requires removal of the implant by a second surgical procedure.

Recently filamentous carbon fibers have been used by various investigators for ligament or tendon repair (Jenkins et al, 1977). Carbon fibers were shown to enhance collagen tissue formation at the site, and they helped to form a neotendon to act as secondary ligament or tendon. These fibers are not biodegradable, but they fragment after a certain period and may be found in regional lymph nodes (Ochiso, Matsnoka, 1979). There was no evidence of an increased rate of infection, nor have other adverse effects such as tumor formation etc. been documented.

In this study we have compared the effectiveness and suitability of woven cloth of braided filamentous carbon and graphite fibers in-vivo, with conventional marlex and dacron mesh in the repair of artificially created wound defects in rats. After several weeks of implantation, the materials were also tested mechanically to evaluate their strength and stiffness properties.

MATERIALS AND METHODS

Seventeen white wister rats of either sex, with average body weight of 257 gm. (range, 140 - 382 gms), were selected. The animals were divided into four different groups according to the type of implants used, e.g. group I: carbon; group 2: graphite; group 3: marlex; and group 4; dacron.

The animals were anesthesized by intraperitonal injection of Nembutal 3.5 mg/100 gm of body weight. The abdomen was opened with a midline skin incision. A rectangular piece of anterior abdominal wall (3 cm x 2.5 cm) was removed from each animal. In five animals, the defect was repaired with a piece of carbon (cloth made of braided filamentous carbon) fibers, (3 cm x 2.5 cm). The carbon cloth was sutured to the margins of the defect with 4 '0' marsilene, as shown in Fig. 1.

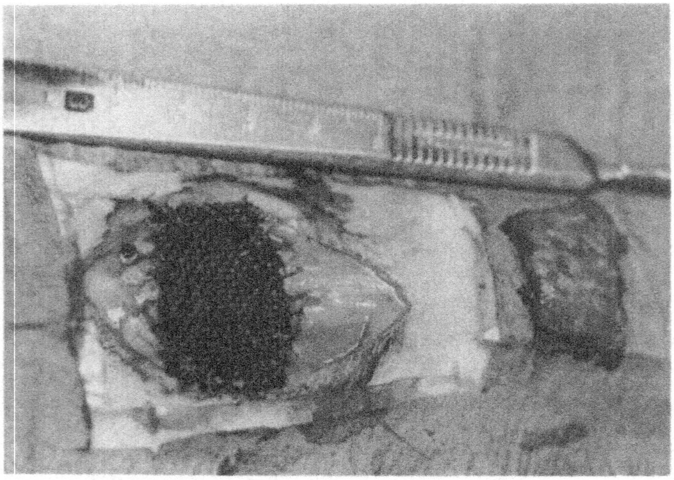

Figure 1. Abdominal wound defect covered with carbon cloth. The segment of the abdominal wall removed is also shown.

The skin was closed over the implant. In groups 2,3, and 4, similar repairs were made using graphite cloth, marlex mesh, and dacron mesh respectively. No dressing was applied. The animals were allowed to move freely in the cage with easy access to food and water.

Five weeks postoperatively, each amimal was sacrificed by intracardiac potassium chloride injection. Skin was opened through the previous incision and it was dissected from the underlying tissue. The implant materials along with 3 cm. margin of anterior abdominal wall were removed after separating omental or intestinal adhesions from the inner surface of the implant. From each specimen, long samples (2 cm. long and 0.5 cm. wide) were prepared. In each sample, the thickness was measured at the level of muscle, suture lines and implant material, using a micrometer.

RESULTS

All the animals moved freely in their cages and behaved normally. There were no postoperative deaths or wound infections. On clinical observation there was no evidence of hernia.

All specimens of carbon and graphite cloth were covered by a shiny membrane on either surface, along with tissue growing through the mesh. In case of marlex and dacron mesh, such tissue growth was minimal or absent. On the peritoneal surface, in groups 1 and 2, omental adhesions at the center of the implants were seen. In two animals, a small portion of the intestine was also adhered. All these adhesions were easily dissectible. Fewer adhesions were found in groups 3 and 4. There was no evidence of suture line disruption in any animal.

Mechanical test results are shown in Table 1. The ultimate load carrying capacity was highest in marlex and lowest in Dacron samples. Ultimate strengths of both carbon and graphite cloth samples were 90% of that of marlex and approximately 110% of that of Dacron mesh.

Table 1

Materials	Ultimate Strength (MN/m^2)	Ultimate Strain	Stiffness N/cm.
Carbon (N-18)	5.83 ± 2.5	0.45 ± 0.15	8.81 ± 4.23
Graphite (N-18)	5.81 ± 3.09	0.36 ± 0.14	12.27 ± 7.62
Marlex (N-13)	6.48 ± 3.21	0.407 ± 0.076	7.76 ± 2.64
Dacron (N-4)	5.22 ± 0.92	0.747 ± 0.286	6.58 ± 2.41

The deformation (ultimate strain) was highest in Dacron but lowest in graphite. In carbon and marlex samples, the deformation was comparable. Similarly stiffness was highest in graphite samples but lowest in Dacron, and in marlex it was lower than carbon and graphite but higher than dacron.

Table 2 shows the percentage of the distribution of the site of breakage in different groups. It shows that most of the carbon specimens failed from the adjacent muscles, whereas almost half of the marlex specimens failed through the implant materials. Dacron samples failed at the suture line in 50% of the cases, whereas graphite samples failed only in 25% of the cases.

Table 2

Materials	Broke at soft tissue	Broke at implant materials	Broke at suture line
Carbon (N-18)	16 (89%)	1 (5.5%)	1 (5.5%)
Graphite (N-18)	9 (50%)	4 (22%)	5 (28%)
Marlex (N-13)	7 (54%)	6 (47%)	0
Dacron (N-4)	2 (50%)	0	2 (50%)

DISCUSSION

No major adverse tissue reaction due to carbon or graphite cloth was found in this study. The adhesions in the carbon and graphite specimens could be easily avoided by placing the implant extraperitoneally. The mechanical properties of the implanted carbon and graphite samples are comparable to those of marlex and dacron mesh. Our previous experience with carbon fibers in ligament reconstruction demonstrated its capacity for collagen tissue formation (Saha et al, 1982), similar to that found by others (Alexander et al, 1981). Such tissue when fully grown should act as a normal biological tissue, and therefore would decrease the danger of associated infection or foreign body reaction. At the same time fibrous tissue thus grown at the site, is likely to be able to support the abdominal contents and prevent recurrence of the hernia.

Results of our study indicates that carbon and graphite cloth could be a suitable material for implantation in the repair of abdominal wound defects. We are continuing this study to evaluate long term behavior of such implants.

REFERENCES

Alexander, H., Parsons, J.R., Strauchler, I.D., Corcoran, S.F., Gona, O., Mayott, C., and Weiss, A.B. (1981) Canine patellar tendon replacement. Ortho. Review, Vol. 10, No. 11, pp. 41-51.

Forster, I.W. Ralis, Z.A., McKibbin, B., Jenkins, D.H.R. (1978) Biological reaction to carbon fiber implants. Clinical Ortho. and Related Research, 131, pp. 299-307.

Jenkins, D.R.H., Forster, I.W., McKibbin, B., Ralis, Z.A. (1977) Induction of tendon and ligament formation by carbon implants. J. Bone and Joint Surg., 59B: pp. 53-57.

Ochiso, Y., Matsnoka, O. (1979) Distribution of colloidal carbon in lymphnodes of mice injected by different routes. J. Exp. Med., Vol. 49, pp. 223-231.

Saha, S., Keating, E.M., Saha, S., Pal, S., Albright, J.A. (1982) Biomechanical evaluation of filamentous carbon fiber implant in patellar tendon replacement. Trans. of 8th Annual meeting of Soc. for Biomaterials, Vol. 5.

BIOMECHANICS OF THE KUNTSCHER INTRAMEDULLARY ROD

A.F. Tencer*#, K. Johnson#, T.W. Carmichael+

* Biomedical Engineering, Univ of Texas at Arlington, Texas
\# Department of Orthopedics, U. Texas Health Science Center Dallas, TX
\+ Rehabilitation Engineering, VA Medical Center, Dallas, Texas

ABSTRACT

A kuntscher intramedullary rod has been instrumented with load and deflection cells to allow measurement of anterior-posterior bending, as well as compressive forces on the rod during insertion into cadaver femur specimens. Calibration of the devices has shown that high sensitivity (1.6 N/mv), linearity (± 2 %) and low crosstalk (> 8 %) occur in all measurement modes. Trial results of insertion of the device in five fresh cadaver femurs show a possible correlation between shape and length of the isthmus of the medullary canal and normal force and deflection of the rod.

KEYWORDS

Intramedullary Rod, mechanical loading, load and deflection measurement, isthmus length and shape.

INTRODUCTION

The technique of femur fracture fixation using intramedullary rodding is attractive surgically because the procedure does not require a direct incision at the fracture site. This reduces the potential for morbidity and enhances fracture healing. However the Kuntscher rod is a load sharing device of relatively low stiffness, compared with the intact femur, as shown in recent biomechanical studies (Allen,1968, Laurence,1969 , Martens 1973). In order to understand the biomechanics of the device, Pilster (1980) measured the anterior-posterior and slot deflections at one point ten cm from the distal end of a kuntscher tibial rod during insertion. The reamed diameter of the isthmus was found to significantly affect slot compression in the rod, but not its overall deflection. The present study was initiated to further define the mechanical and geometric factors affecting the working stiffness of the implanted intramedullary rod.

OBJECTIVES

The objectives of the study are to develop and calibrate a deflection cell instrumented Kuntscher rod and to determine the magnitude and location of the support loads at the interface between the rod and the wall of the medullary canal. These can be determined from knowledge of the rod deflections, contact areas, and magnitudes of the compressive loads at particular locations. From the support loads at the rod-bone interface the actual working stiffness of the implanted device can be determined and this can be correlated with the results of mechanical loading tests. In this report, the accuracy of the instrumented rod is demonstrated. Rod deflections during implantation, contact areas, and mechanical stiffness in anterior-posterior bending, with and without axial loading are presented.

METHODS

A 14mm diameter standard Kuntscher intramedullary rod (OEC, 36 cm long) was instrumented with compliant spring steel load cells to which wire strain gauges were mounted. These were positioned in the rod as shown in Fig. 1. The green, white, red, and orange load cells measure the compressive force required to deform the slot of the rod, while the blue load cell measures anterior-posterior (YZ plane) deflection. Fig 2, and 3 show the results of calibration of the instrumented rod. All load cells showed a high degree of linearity and repeatability in their outputs.

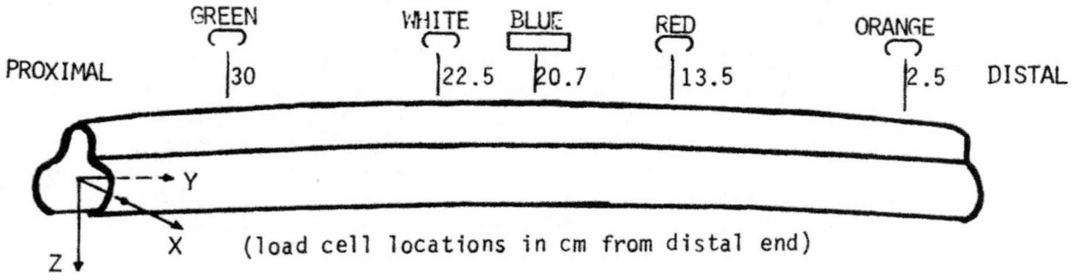

Fig 1. Load and deflection cell locations in IM rod

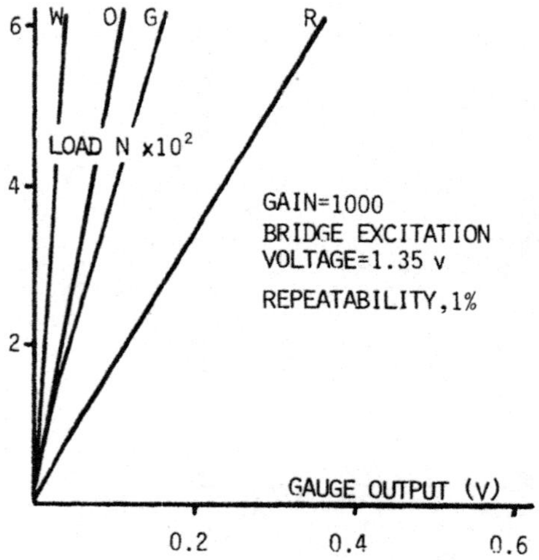

Fig 2. Slot load cell calibration

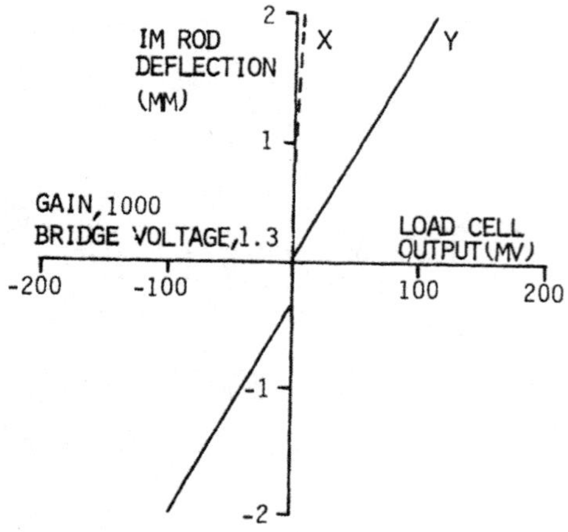

Fig 3. AP deflection cell calibration

Fresh cadaver femur specimens, obtained from autopsy specimens were first reamed to 14.5 mm diameter using standard surgical procedures. The location at which reaming commenced and ceased were noted so that the center of the isthmus, the narrowest region of the medullary canal, could be identified. The specimens were then mounted into aluminum cups using a rigid epoxy potting compound (West System 501) and supported pinned-free. The unsupported span of each specimen was kept constant and a four point bending load applied by the plunger of a universal materials tester. Additionally a jig, constructed to allow application of an axial compressive load to simulate body weight, was mounted to the specimen. The specimen was then retested in bending with axial loads of 450N and 1000 N. The instrumented rod was then inserted into the intact specimen and the load cell outputs recorded at various insertion distances. The specimen was sawed transversely at the center of the isthmus, to simulate a simple fracture, and the femur-rod system retested in bending with and without axial loading. After mechanical testing, the transverse saw cut was sealed and a low viscosity epoxy was injected into the medullary canal, to allow determination of the rod-canal contact areas.

RESULTS

As shown in Fig 4, there was some data spread in the load deflection curves of in-

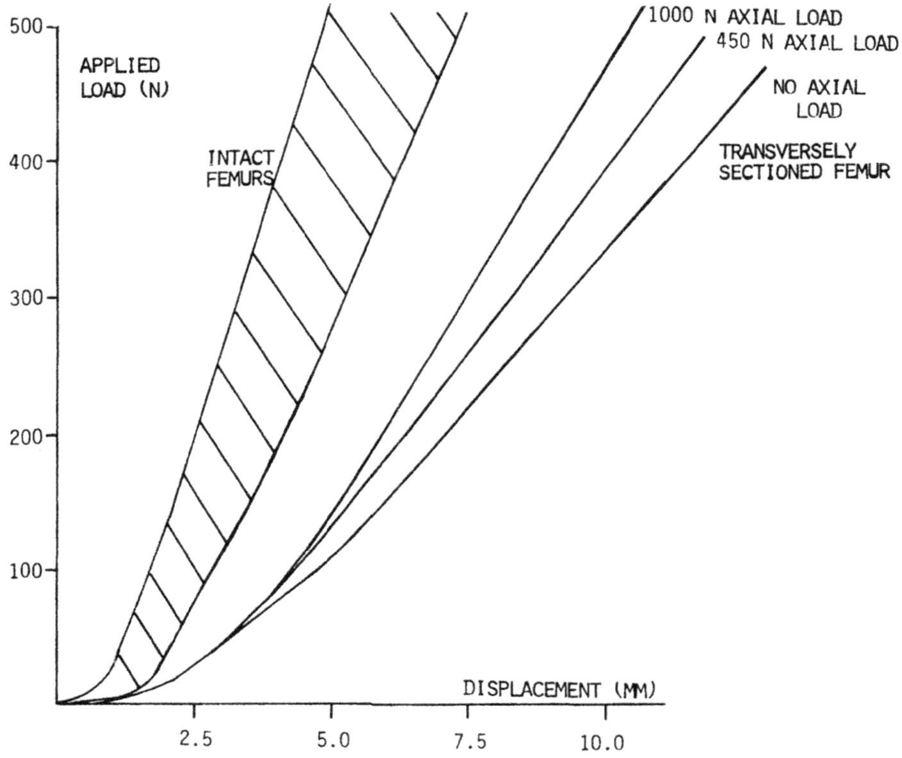

Fig 4. Load-deflection results of intact and IM rod stabilized femurs

tact specimens. These results combine the load-deflection data of both AP and ML bending, as well as the results with and without axial loading, there being no distinguishable difference in a specimen. Axial loading, as would be expected, resulted in no significant change in the load-deflection properties of the specimen. After implantation of the rod and sawing of the femur, the stiffness of system was significantly reduced. However, application of an axial load did increase system stiffness, most likely because the axial load drives the fracture components together, thus allowing the transfer of compressive stresses in the bone in bending.

During rod insertion, variations occur in the rod slot compression loads and its AP profile. Considering that the medullary canal has been reamed oversize with respect to the rod, it is only the mismatch in rod medullary canal curvature which retains the rod in its position. Hence, as shown in Fig 5, the rod length and its curvature should be carefully matched to that of the femur to produce maximum support at the rod-medullary canal interface. Fig 6, shows a typical contact area diagram of an implanted rod. In this case the distal component of the fractured femur was physically able to slide on the rod, as is evidenced by the single contact area. The rod was wedged into the proximal component and retained there by three point contact. The proximal contact region is governed by the location of the starting hole, hence proximal support of the rod is highly dependent on location of the starting hole.

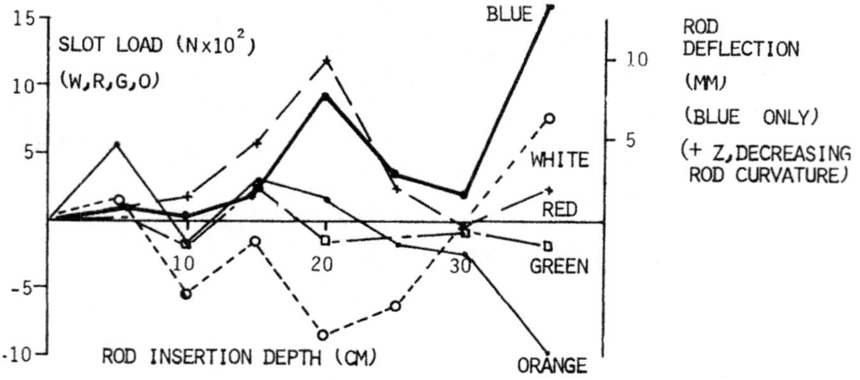

Fig 5. Load and deflection in IM rod during insertion

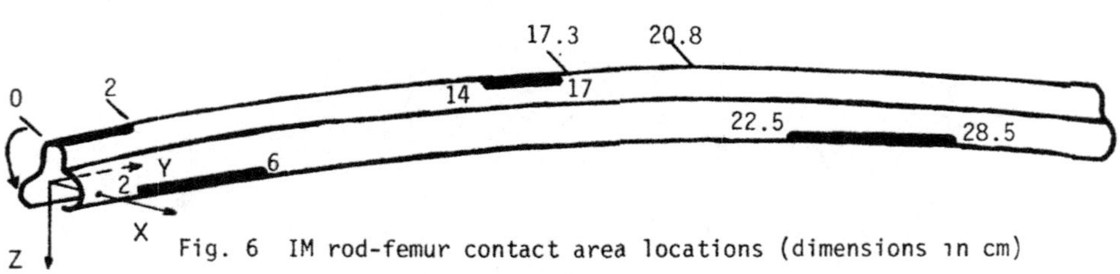

Fig. 6 IM rod-femur contact area locations (dimensions in cm)

Lawrence M. Freeman, M.A.R. Swanson S, A.V., JBJS, Vol.51B, Pp 754-768, 1969
Allen, W.C., Piotrowski, M.S., Burstein, A.H. Frankel, V.H. Clin Orthop, Vol.60, Pp 13-20, 1968
Martens, M., Frankel, V.H., Burstein, A.H. Injury, Vol.4 Pp 18-24, 1973

PERFORMANCE CHARACTERISTICE OF TOTAL HIP FEMORAL COMPONENTS AS A FUNCTION OF PROSTHESIS MODULUS

J. Gurtowski; L. Stern; A. Musella; M.T. Manley

Orthopaedic Research Laboratories, Health Sciences Center, State University of New York at Stony Brook, Stony Brook, N.Y. 11794

ABSTRACT

A study of total hip femoral components has been performed which demonstrates the effects of changes in prosthesis modulus upon strain levels and subsidence data measured in vitro. The distinguishing feature of this project is that femoral specimens of identical geometry but fabricated from dissimilar alloys were sequentially implanted in the same femoral specimens so that the relative performance of the components could be directly compared. It was found that femoral components fabricated from titanium alloy produce higher levels of calcar strain than cobalt-chromium components when collar/calcar contact is preserved. However, after collar/calcar contact was removed it was shown that cobalt-chromium stems have a greater propensity for subsidence into the medullary cavity and that the increased rate of subsidence of these stiffer stems produces concomitantly greater circumferential strain levels in the proximal cortical shell.

KEYWORDS

Femur, total hip, in vitro study, femoral component modulus.

INTRODUCTION

Loosening of the femoral component following total hip arthroplasty still remains a problem of clinical significance. The mechanisms causing implant loosening have been blamed on a combination of factors which include stress shielding of the calcar femorale and the proximal femoral cortex (Oh, 1978; Markolf, 1980; Charnley, 1973) fracture of the cement mantle (Charnley 1965) and micromotion at the implant/cement interface or the cement/bone interface (Beckenbaugh, 1978; Lanyon, 1981). The inferior mechanical characteristics of PMMA bone cement leaves little doubt that motion of the femoral component due to resorption of the calcar femorale or remodelling of the proximal femur will cause destruction of the integrity of the cement mantle and recent trends in prosthesis design have been towards stems of a lower material modulus in an attempt to minimize stress shielding and reduce proximal femoral resorption. Various authors have shown that stems fabricated from titanium alloy produce strain levels in the proximal femur which approximate those measured in the intact bone (Lewis 1981; Tarr, 1979; Crowningshield, 1980), but most investigators have ignored the effects of stem geometry and of femoral characteristics upon measured cortical strain and the results presented are probably highly influenced by these uncontrolled variables. A recent study (Manley 1982) has shown that significant differences in calcar strain can be demonstrated when changes are made in the projected medial area of the femoral component stem or when identical femoral components are implanted into unmatched cadaveric femora.

The study described here was thus designed to quantitate the effects of stem modulus upon implant performance when other variables such as component geometry and femoral characteristics are held constant.

MATERIALS

In order to ensure that the results obtained in this study were independent of stem geometry two separate series of prostheses were tested. In the first series (Osteonics Geometry) the performance of a cobalt/chromium Osteonics HS1 medium stem was directly compared to the performance of an Osteonics titanium alloy stem of identical geometry, neck length, head size and collar angle. In the second series (Harris Geometry) the performance of a cobalt-chromium Howmedica Harris HD2 was directly compared with the performance of a titanium alloy Hexcel Anitomic stem. Slight modifications to the original stem geometry were made in the Harris series to ensure constant geometry testing. The A/P geometry of the Osteonics and the Harris series is shown schematically in Fig. 1.

Recent studies on the interface between PMMA and prosthesis stems have shown a variable adhesive interface between stem and cement which if enhanced improves the load carrying capacity of the implant. (Barb, 1981; Raab, 1980) In order to remove this variable interface from our study, each prosthesis was coated with a thin layer of latex rubber before implantation to prevent uncontrolled adhesion of the PMMA to the stem. Howmedica Simplex bone cement was used throughout.

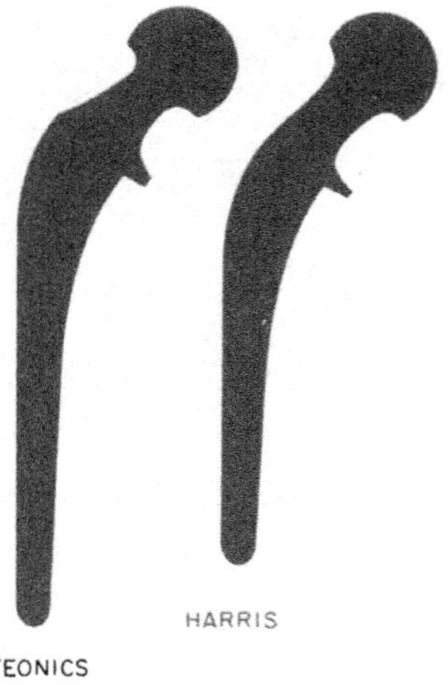

Fig. 1

All testing was performed in fixed cadaveric femora which were defatted and prepared with standard instrumentation for the stem geometry of the series under investigation. In each specimen the calcar femorale was prepared by filing to ensure intimate contact with the prosthesis collar. Uniaxial strain gauges aligned circumferentially with the axis of the femoral neck or shaft were bonded to each femoral specimen 5mm. below the ressection cut on both the medial aspect of the calcar and the posterior femoral neck, and to the medial shaft at locations corresponding to stem tip and one-half stem length. Longitudinal gauges were bonded adjacent to the circumferential gauge at one-half stem length and adjacent to the gauge on the medial aspect of the calcar. Prosthesis subsidence was measured with an extensometer mounted between the prosthesis neck and the femoral shaft.

METHODS

Loading of each specimen was between the head of the implant and a ball joint inserted into the medullary cavity 300mm. distal to the center of the femoral head. The loading vector was applied at an angle of 20° to the longitudinal axis of the femoral shaft. Load was applied in 25lb. increments every five seconds up to a maximum load of 1,000lbs, (4.48kN) in an MTS load frame. Strain gauge and subsidence data was recorded automatically at each loading increment. Once the specimen had been subjected to this loading regime, a sawcut was made through the cement layer trapped between the prosthesis collar and the calcar and the test was repeated. Upon completion of the two tests the prosthesis was carefully removed from the femur and the next stem in the series implanted in the same femur after careful removal of the previous cement mantle. Care was taken to ensure that subsequent implantation of each prosthesis was in the same orientation within the medulla and as a control the orientation of each stem was checked by X-ray and corrected if necessary. Once both the titanium alloy and the cobalt-chromium stems of the same geometry had been tested in one femoral specimen, the first prosthesis in the series was reimplanted and retested to ensure repeatability of the data and to ascertain whether the testing procedures had modified the mechanical characteristics of the cadaveric specimen.

Testing was repeated in five cadaveric femora for each series so that repeatability of the experimental trends could be confirmed.

RESULTS

For both the Osteonics and the Harris series strain and subsidence data recorded during retesting of the first prosthesis were always within 5% of the values achieved during initial testing of the stem in the same bone. We thus concluded that our experimental methods did not significantly affect the material characteristics of the cadaveric specimens and that our comparative data was not influenced by uncontrolled femoral variables.

Strain gauge data recorded from both series of stems by the circumferential gauges and the longitudinal gauge mounted on the femoral neck are summarized in Fig. 2. It can be seen that similar trends were observed in the comparative performance of the cobalt-chromium vs. the titanium alloy stems for both the Osteonics and the Harris geometries. Comparison of both circumferential and longitudinal strain data in the calcar shows that when collar/calcar contact is preserved, the titanium alloy stem is more effective at loading the calcar than the cobalt-chromium stem of the same geometry. When collar/calcar contact is removed longitudinal strain levels with both stems is reduced by about 50% to that seen when collar/calcar contact is preserved. By comparison removal of collar/calcar contact causes an increase in circumferential strain levels measured in the calcar; the increase in circumferential tensile strain being greatest when cobalt-chromium stems were implanted.

Subsidence data for both the Harris and the Osteonics geometries is summarized in Fig. 3. In both geometries it can be demonstrated that removal of collar/calcar contact allows prosthesis subsidence to occur. However, the results consistently showed that the stiffer cobalt-chromium stems begin to subside at a lower loading magnitude than the less stiff titanium alloy stems and that the final subsidence level measured when 1,000 lbs. load was applied was always greater for the cobalt-chromium stems than for the titanium alloy stems of identical geometry.

CONCLUSIONS

The results of this study have shown the effects of changes in prosthesis stem modulus upon strain levels and strain subsidence in the proximal femur. The experimental methods developed have allowed the performance of prostheses to be directly compared with one another without the results being modified or masked by changes in cadaveric femoral characteristics or by changes in femoral component geometry. We have shown that if collar/calcar congruency is not achieved during surgery, or if such congruency is destroyed thereafter due to resorption of the calcar femorale, a cobalt-chromium stem will have a greater tendancy to subside into the medullary space than a less stiff titanium alloy stem of identical geometry. The greater subsidence of the cobalt-chromium stem explains the larger increase in circumferential strain seen in testing of these stiffer stems after collar/calcar contact is removed. If collar/calcar congruency is achieved at surgery and is maintained thereafter the titanium alloy stems were found to produce greater circumferential and longitudinal strains in the calcar femorale than the stiffer cobalt-chromium stems. The increases in circumferential and longitudinal strains were found to be19% and 30% respectively. These figures are lower than the enhanced calcar strain predicted by other authors, although the smaller increases found here can possibly be explained by the more rigorous controls on stem geometry, stem position and femoral characteristics.

Although the findings from this study support the concept that a less stiff stem is desirable to enhance loading of the calcar femorale and to prevent stress shielding, the subsidence and circumferential strain data provides additional evidence of the advantages of titanium alloy femoral component stems. As the tensile properties of PMMA are particularly poor, the lower rate and lower magnitude of subsidence exhibited by the titanium stems which produces a lower level of tensile stresses in the cement mantle, will help to protect the mantle's integrity. Thus, the use of titanium alloy stems can be supported on the grounds of a lesser tendency to subside and this along with more effective calcar loading when collar/calcar contact is preserved should help to prolong the functional integrity of the implant.

REFERENCES

Barb W.; von Recum A.F.; Kenner G.H.; Park J.B. Trans. 7th Meeting Society for Biomaterials, Troy, 1981

Beckenbaugh R.D.; Ilstrup D.M. J. Bone & Joint Surg. 60A 306-312, 1978

Charnley J; Cupic Z. Clin. Orthop. & Rel. Res. 95 9-25, 1973

Charnley J; Kettlewell J. J. Bone & Joint Surg. 47B 57-60, 1965

Crowningshield R.D.; Pedersen D.R.; Brand R.A. Transactions of ASME 102 230-233, 1980

Lanyon L.E.; Paul I.L.; Rubin C.T.; Thrasher E.L.; DeLaura A.B.; Rose R.M.; Radin E.L. J. Bone & Joint Surg. 63A 989-1001, 1981

Lewis J.K.; Kramer G.M.; Wixson R.L.; Askew M.J. Trans. 27th Ann ORS 6 75, 1981

Manley, M.T.; Gurtowski J.; Musella A.; Spiegel N.; Stern L. Effect of Titanium Hip Stem Geometry on Circumferential Strains in the Proximal Femur. Proc. Northeast Bioeng. Conf., Dartmouth, 1982

Markolf K.K.; Amstutz H.C.; Hirschowitz D.L. J. Bone & Joint Surg. 62A:1315-1323, 1980

Oh I; Harris W.H. J. Bone & Joint Surg. 60A 75-85, 1978

Tarr R.R.; Lewis J.L.; Jaycox D.; Sarmiento A.; Schmidt J; Latta L.L. Trans 25th Ann. ORS 4 34, 1979

Raab S. Ph.D. Thesis, McGill University, Montreal, 1980

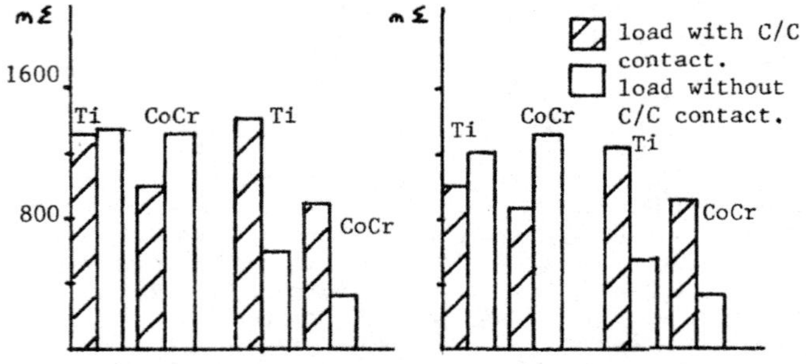

Fig. 2

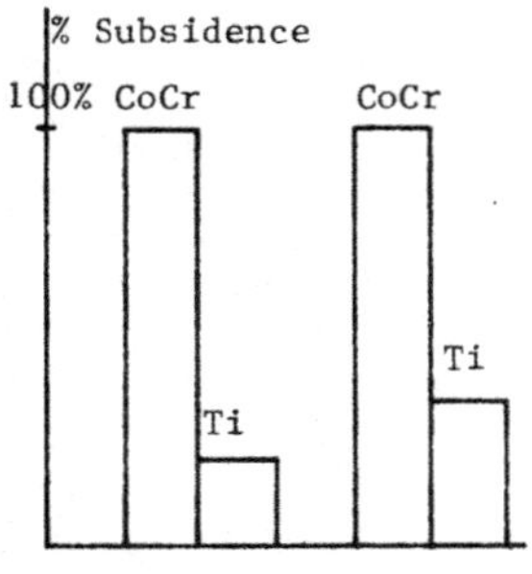

Fig. 3

LONG TERM EVALUATION OF PRECOATED CANINE FEMORAL PROSTHESIS

J.B. Park, W. Barb*, and J. P. Davies

Department of Biomedical Engineering, Tulane University
New Orleans, La 70118
*Intermedics, Intraocular, Inc.,
Pasadena, Ca 91106

ABSTRACT

Precoating of a canine femoral hip prosthesis with PMMA cement has been investigated in order to minimize incidence of implant loosening. A bilateral implantation, one for the uncoated, the other for the coated prosthesis was made using the same bone cement used to precoat. After 1, 3, 6, and 13 months both femora were excised and sectioned for mechanical push-out test to assess the interfacial strengths among bone, cement and prosthesis.

The precoated prostheses had twice higher interfacial shear strength (12.3 MPa) between prostheses/cement interface than the uncoated implants at all implant period. The interfacial shear strengths of the bone/cement of the precoated implant had much higher value up to 6 months (2 MPa) but became close to the value of the uncoated implants after 13 months (0.7 MPa) due to the bone remodelling process. The present results indicate that the loosening of implant/cement interface can be minimized by precoating but the initial better fixation at the bone/cement interface will diminish to that of uncoated impalnts after 13 months.

INTRODUCTION

The current method of hip replacement, a method first used by Charnley in 1958 (Charnley,1960 & 1970; Eftekhar, 1978), in which the stem of a femoral prosthesis is secured in the intramedullary cavity with polymethyl methacrylate (PMMA), has proven to be a highly successful surgical procedure. The fixation depends upon the mechanical interlock between the PMMA and the irregularities of the medullary bone surface and the stem of the prosthesis. The stability of the implanted prosthesis depends on the quantity and quality of the mechanical interlocking. Loosening of the prosthesis is being encountered in increasing numbers, years after implantation, in cases which were described as having a good initial fixation (Amstutz and co-workers, 1976). The loosening has been blamed on the inherent problems of using PMMA for fixation (Willert and Semlitsch, 1976). The problems are:

1. Thermal necrosis at the bone/cement interface due to high heat of polymerization.
2. Chemical necrosis caused by monomer release during and after insertion.
3. Shrinkage of the cement and lack of good cement distribution around the implant.

By precoating the prosthesis, the amount of unpolymerized PMMA used at the time of surgery can be reduced (Barb and co-workers, 1982; Park and co-workers, 1982). This in turn reduces the heat of polymerization, the amount of monomer release and shrinkage of the cement. Precoating also ensures good cement distribution and thus better stress distribution results. However, the extra added cost of precoating, handling and sterilization are some of the drawbacks.

MATERIALS AND METHODS

Prosthesis Precoating

In this study canine hip prosthesis (cat. no. 8590-105, Zimmer USA, Warsaw, IN) with sand-blasted circular cross-sectional shaft was used. A two millimeter layer of cement was coated on the stem by injecting the cement in its lower viscosity state into a nylon split mold as shown in Figure 1. The implants were then stored in a partial vacuum desicator for one day, rinsed with distilled water the next day, dried, wrapped in sterilization bags (ABCO Inc., Milwaukee, WI) and then sterilized with ethylene oxide (CR Bard Inc., Murray Hill, NY). The implants were degassed and stored in a partial vacuum desiccator until use.

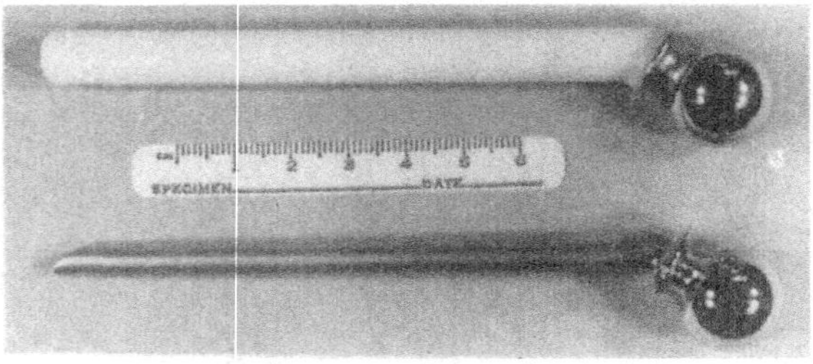

Figure 1. Femoral prosthesis before and after (top) precoating with PMMA bone cement by using a pair of molds.

Surgical Procedure

Random source dogs were used to implant the precoated prosthesis and one uncoated prosthesis by performing a bilateral hemiarthroplasty using Gorman's approach, osteotomy of the trochanter major (Piermattei and Greeley, 1966). Radiographs were taken of each leg from a bilateral view.

Post-Surgical Procedure

Prophylactic distrycillin was administered twice daily for three days. By the third day most dogs stood and walked unassisted, at which time they were returned to the outdoor kennel.

The dogs were sacrificed at a predetermined time using a commercial barbituate solution. Both femora were excised, radiographed and placed in an ice-saline bath. The femora were then mounted on a Bronwill VWR scientific sectioning saw. The individual femur was then cut with a diamond blade into 2 mm thick sections for mechanical testing as shown in Figure 2 and 7 mm sections were made for histo-morphological assessment which will be published elsewhere (Park and co-workers, 1982).

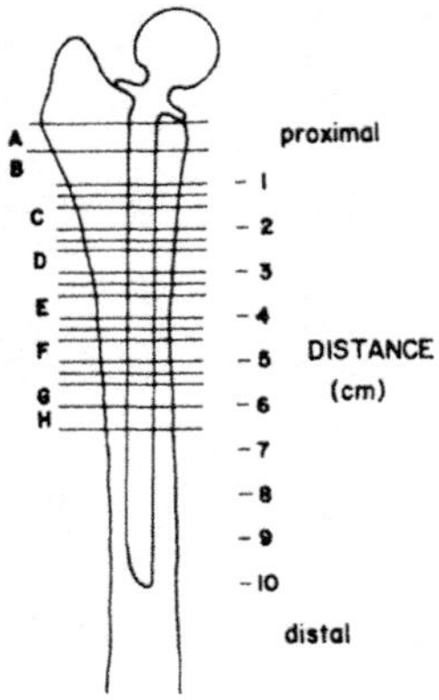

Figure 2. Sectioning scheme for mechanical (numbers) and histo-morphological test specimens (letters) along the longitudinal axis of the femur.

Mechanical Testing

The 2 mm sections were allowed to reach room temperature before testing. The sample dimensions were measured and the specimen was placed on a push-out jig and loaded via an Instron mechanical testing machine in a compression mode at a cross head rate of 2.54 mm/min. to failure. The load versus displacement was recorded. This data and the nominal area between the interfaces were used to calculate the shear stress.

RESULTS AND DISCUSSION

The summary of results of the maximum interfacial shear strengths of the prosthesis/cement and bone/cement interfaces is given in Table 1. Figures 3 and 4 show the plot of maximum interfacial shear versus implant period for the precoated and uncoated prostheses respectively.

Table 1. Summary of Results

Implant Period (mon.)	Maximum Interfacial Shear Strength (MPa)			
	Precoated	Uncoated	Precoated	Uncoated
	Cement/Bone		Cement/Prosthesis	
1	2.3±1.7(24)	1.6±1.4(38)	12.4±2.0(36)	5.7±1.0(33)
3	2.1±1.3(16)	1.2±1.0(40)	12.1±1.0(36)	4.8±1.7(46)
6	1.6±1.9(22)	0.8±0.8(41)	12.6±2.0(45)	6.0±1.8(34)
13	0.7±0.3(34)	0.5±0.3(39)	12.3±1.9(34)	5.5±0.6(35)

Mean ± One standard dev. (No. of samples).

Figure 3 shows that both the uncoated and precoated prostheses had a decrease in interfacial shear strength at the prosthesis/cement interfaces from one to three months followed by a slight increase from three to six months. The decrease may be due to the stress relaxation of the locked in hoop stresses caused by the initial shrinkage of the cement followed by an increase in the strength of the cement because of molecular rearrangement and late polymerization. Overall, however, the interfacial shear strength of the prosthesis/cement interface did not change significantly for the 13 month implantation period for both the precoated and uncoated implants. This indicated that the stress relaxation at this interface was not significant for the 13 month implantation period for both the precoated and uncoated prostheses. What is significant is that the interfacial shear stress *per se* for the precoated prosthesis is over two times higher than the uncoated prosthesis at the prosthesis/cement interface.

Figure 4 shows the steady decrease with time of the interfacial shear strength at the cement/bone interface for both uncoated and precoated prostheses. The initial high value for the interfacial shear stress for the precoated prosthesis may be due to the increased pressure in the precoated prosthesis than the uncoated since the former is larger than the latter. However, the interfacial shear stress for the precoated prosthesis does reach the level of the uncoated prosthesis by 13 months. This indicated that the bone remodelling at the cement/bone interface was considerable and resulted in weakening of the mechanical interlock between the bone and cement. Considerable bone remodeling is not unexpected when a load bearing prosthesis such as the hip prosthesis is implanted.

SUMMARY

The results of the mechanical interfacial shear stress tests show that precoating the prosthesis before insertion into the medullary canal results in higher interfacial shear strengths between the prosthesis and cement, up to 13 months and up to 6 months at the cement/bone interface. The decline of interfacial shear stress of the precoated implant to the value of the uncoated implant from 6 to 13 months at the cement/bone interface is the result of bone remodeling.

Overall the precoated implant showed improved initial and long term in vivo stability over the uncoated prosthesis at the prosthesis/cement interface and to a lesser degree at the cement/bone interface.

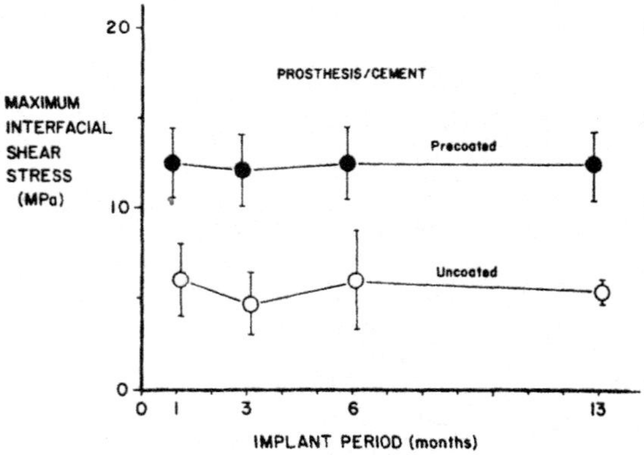

Figure 3. Maximum Interfacial shear stress vs. implant period for prosthesis/cement interface.

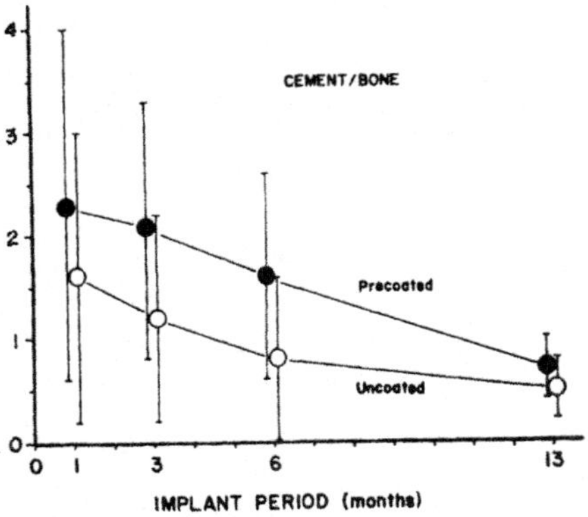

Figure 4. Maximum Interfacial shear stress vs. implant period for bone/cement interface.

ACKNOWLEDGEMENTS

The help of Dr. G.H. Kenner (Clemson University) is gratefully acknowledged. This work is supported by National Institute of Arthritis, Metabolism and Digestive Diseases Grant No. AM 30591.

REFERENCES

Amstutz, H.C., K.L. Markolf, G.M. McNiece, and T.A. Gruen (1976), in The Hip. C.V. Mosby Co., St. Louis, 102-116.

Barb, W., J.B. Park, G.H. Kenner, and A.F. Von Recum (1982). J. Biomed. Mater. Res. to be pub.

Charnley, J. (1960). J. Bone Joint Surg., 42B, 28-30.

Charnley, J. (1970). Acrylic Cement in Orthopaedic Surg. E&S Livingtone, Edinborough and London.

Eftekhar, N.S. (1978). Principles of Total Hip Arthroplasty. C.V. Mosby Co., St. Louis.

Park, J.B., W. Barb, G.H. Kenner, and A.F. Von Recum (1982). J. Biomed. Mater. Res., to be pub.

Piermatter, D.L. and R.G. Greeley (1966). An Atlas of Surgical Approaches to the Bones of the Dog and Cat, W.B. Saunders Co., Philadelphia.

Willert, H.G. and M. Semlitsch (1976). Advances in Artificial Hip and Knee Joint Technology, M. Schaldach and D. Hohmann (ed.), Springer-Verlay, Berlin.

FUNCTIONAL CONSIDERATIONS IN IMPLANT DESIGN

E. E. Frisch

Dow Corning Corporation
12334 Geddes Road
Hemlock, Michigan 48266

ABSTRACT

The normal physiological response to essentially all implanted foreign objects or materials, in both bone and soft tissue, is encapsulation in fibrous (scar) tissue. Thus, most all implants (including bone cement) are eventually fixed or attached within the structures of the body by this biologically dynamic, unstable, fibrous tissue interface. An important functional consideration in the long-term success or failure of any implant design is the response it elicits in the fibrous tissue capsule, and in turn, the response of surrounding structures to the fibrous capsule. Both biocompatibility and biomechanical aspects can be important. Fibrous tissue is known to remodel when subjected to gentle stress or pressure, and hypertrophy can be initiated by either chemical, biological, or mechanical stimulus. Most of the reactions which occur in fibrous tissue and the adjacent structures of the body are predictable, and account for many of the problems and complications which occur in implants. Whenever possible implant design should use the fibrous tissue response it elicits to help support the reconstructive procedure, and force-transfer to adjacent structures should duplicate those encountered in normal anatomy to the greatest extent possible.

Key Words

Implants, Fibrous Tissue, Scar, Encapsulation, Force-Transfer, Design, Failure Modes, Mammary Implants, Hip Joints, Finger and Toe Joints.

Introduction

Success of the silicone elastomer hydrocephalus shunt[1] in 1955 initiated an era, which continues at present, of substantial research and development effort to perfect new implants for use in the correction of human health problems by reconstructive surgery. Medical grade silicone elastomer provided a new soft, flexible, and elastomeric material of construction with excellent biocompatibility and biodurability characteristics and thus allowed researchers to consider the potential for implants where elastomeric materials of construction were needed.

Implant grades of various metals had become quite well established, and implant grades of strong, rigid plastics, such as ultra high molecular weight polyethylene were also being developed during the 1960's. As implants gained increased attention it was hoped, and believed by many that implants fabricated from one or more of these materials would provide mechanical replacement parts to treat many of the unsolved problems encountered in human health. Fortunately, as a result of this research and development many new implants were developed to improve or facilitate reconstructive surgical procedures.

However, the limitations of implants in reconstructive procedures, the problems and complications, and the physiological reactions of bone and soft tissue to an implant also become apparent. Because of the increased potential for problems and complications inherent in any reconstruction done with an implant, use of an implant is justified only when it offers enough benefit to offset the increased risk. The physiology resulting from reconstruction with living tissue can be essentially normal, while use of a man-made implant creates permanently abnormal physiology.

There are many considerations which determine the degree of benefit patients receive from implant reconstructive surgery, including patient selection, surgical technique, health characteristics of the patient, implant design and quality, the nature of the postoperative therapy, the postoperative activities of the patient, and of great importance, the nature of any longer term changes which occur in the tissues of the patient as a result of the presence of the implant.

The benefits from an implant procedure end when failure first occurs in any component of the reconstruction critical to success, whether in the implant or in the tissues of the patient. However, failures which occur because of changes in tissues adjacent to the implant, which may occur slowly over several years, as a result of the presence of the implant can be the most serious as correction may be difficult or impossible. Bone or soft tissue once lost may be irreplaceable leaving the patient and surgeon no reasonable alternative for subsequent reconstruction. In addition to biocompatibility, biodurability, function, and strength an important functional consideration with any implant is that of avoiding designs and structural features which result in progressive destructive changes or loss of living tissue. All implants elicit a predictable response in the host. Whenever possible, either by design or serendipity, implant function should use the expected physiological response elicited by the presence of the implant to the advantage of the reconstruction.

Fibrous Tissue Encapsulation

Relatively early postoperatively, as a normal response to a foreign body, essentially all implants, in either soft tissue or bone become encapsulated in fibrous tissue which then interfaces between the implant and surrounding structures. Thus, understanding the characteristics and reactions of fibrous tissue is important in the functional considerations of implant design as the changes which occur in fibrous tissue, and in the tissues with which the fibrous tissue interface can ultimately determine the long-term success or failure of a reconstruction.

Fibrous tissue which forms around implants is histopathologically the same as the fibrous tissue produced during wound healing, commonly known as scar tissue. Structurally composed of collagen fibers, it has a dynamic biology undergoing constant metabolism and regeneration as mature collagen fibers break down when metabolized and new collagen fibers are formed during regeneration. The regenerated fibers essentially duplicate those metabolized. The rate at which metabolization and regeneration occur is highest in immature fibrous tissue and decreases with time; however, the process is continuous.[2] The shape of fibrous tissue stabilizes when an equilibrium has been established between forces which stretch it, and its natural force of contraction.

Hypertrophy of scar tissue can be initiated by physical or mechanical trauma, stress, and chemical or biological factors. When ruptured or traumatized by mechanical or physical forces hypertrophy of scar tissue is probably a natural biological healing response to initiate the development of fibrous tissue with adequate strength and thickness to withstand the traumatizing force. Hypertrophy of scar tissue is also common in areas of the body where motion and stress occur. Burn scar hypertrophy is generally more of a problem in mobile areas of the body such as the axilla, hand, and anterior cervix of the neck. A bone plate which has become loose will generally develop a hypertrophic fibrous tissue layer between the plate and bone, often with bone resorption to make room for the excess fibrous tissue.

Fibrous tissue production is essential to the mechanism of wound healing. Thus, in situations where the conditions of inadequate wound healing exist or are simulated fibrous tissue hypertrophy is likely to occur.

Fibrous tissue is inherently inelastic with a typical elongation before rupture of 1%. However; fibrous tissue can be remodeled[3] by long acting, relatively low forces which gently hold the scar stretched as collagen fibers are metabolized and regenerated in the elongated shape. The remodeling of scar with prolonged low force has become widely used in the revision of burn scars.[4]

Tissue Reactions to Implants

Implants Used in Breast Reconstruction

The implants used in breast reconstruction (silicone gel, saline inflatable, and silicone gel-saline inflatable combination) have essentially no shape, are soft and pliable, and are designed to feel like breast tissue.[5] When implanted the shape of the implant is determined by the forces which act upon it. These implants become encapsulated in the usual fibrous tissue capsule. The most common complication is a contraction in the fibrous capsule as it matures and shrinks. As the surface area of the fibrous capsule shrinks during contracture the implant is slowly reshaped to a spherical configuration. Because the fibrous capsule is inelastic, and a sphere inherently unyielding, the resulting implants feel hard even when intraimplant pressure is low and the fibrous capsule thus subjected to only weak stretching forces. When equilibrium between stretching and contracting forces has been reached the shape of the implant becomes stable, and measurement of intraimplant pressure at equilibrium conditions has provided a means to determine the contractile strength of the fibrous capsule. With a central venous pressure manometer using a hypodermic needle into the implant pressure measurements were conducted. In-vitro pressures (implant and capsule previously removed enbloc) varied from 12cm/H_2O to 15cm/H_2O. In-vivo pressures (done with patients prior to undergoing surgery) varied from 14cm/H_2O to 18cm/H_2O. These low intraimplant pressures indicate that it is a combination of developing a spherical shape, the non-compressibility of the silicone or saline, and the inelasticity of the fibrous capsule which result in symptoms of capsular contracture, and not strong contracture forces.

Not all patients with breast implants develop symptoms of capsular contracture. Some surgeons advocate a program of daily exercise and implant manipulation to keep the capsule stretched, and avoid capsular contracture symptoms.

In other patients, skin and tissue tension, or activities of daily living may keep the capsule stretched, thus maintaining a flattened shape in the implant and avoiding capsular contracture symptoms. It has also been reported that implants placed beneath the pectoralis muscle have less tendency to develop capsular contracture symptoms.[6] Logically, the pectoralis muscle exerts a compressive force against the implant maintaining a flattened shape and providing the intraimplant pressure needed to maintain a stretched capsule.

Surgical treatment of capsular contracture may consist of either blunt or sharp capsulotomy, thus creating an opening in the capsule so that the implant can flatten and feel soft.[7] However, the advantages of such surgery may be only temporary as the opening in the capsule will heal with additional scar tissue, and the tendency to contract recur.

Many studies have been done in an attempt to find a cause-and-effect relationship for capsular contracture.[8] However, when the biophysics of the typical shapeless mammary implant, and the physiology of fibrous tissue are considered, the development of capsular contracture is the predictable end result.

Hip Joint Implants

The system developed for hip joint reconstruction using implants to replace both the head of the femur and the acetabulum, secured in place with bone cement, has probably been more successful in restoring function and reducing pain than any other bone and joint implant. The development of bone cement has been important to this success since it facilitates immediately strong attachment of the implants to bone. The term "bone cement" is somewhat misleading in that it does not truly adhere to either the implants or bone. Applied in a semi-fluid state the cement flows into the contours of both bone and implant and in a few minutes it polymerizes in-situ and serves as a strong, rigid interpositional material that mechanically interlocks the implant with bone.

However, one of the most frequently occurring complications in hip joint reconstruction with an implant is loosening with loss of bone and eventual need for revisional surgery. The abnormal transfer force across the reconstructed joint to the intramedullary canal, rather than to cortical bone, as a result of bone cement fixation is probably a contributing factor to bone resorption in the proximal femur (Wolff's law). When the fibrous capsule has fully formed the implant is no longer fixed to bone with bone cement, but rather the combination of the implant and bone cement are fixed to bone by the fibrous tissue layer. The high force loading which must now be borne entirely by the fibrous capsule initiates fibrous tissue hypertrophy which occurs simultaneously with loss of bone and the implant becomes progressively looser.

The fixation of weight-bearing implants to bone by ingrowth into a variety of porous materials has also been proposed.[9,10] However, obtaining bone ingrowth into a porous surface requires rigid, prolonged fixation between the porous material and bone as even the slightest movement will result in the development of fibrous tissue ingrowth, rather than bone ingrowth. Prolonged immobilization is not practical with many patients. When fibrous tissue ingrowth occurs with the pores the implants proceed to become loose essentially the same mechanism as with implants fixed by bone cement. Fibrous tissue is inadequate for fixation of implants to the musculoskeletal system..

Finger and Toe Joint Implants

In the smaller joints of both upper and lower extremities the bones appear to be even more reactive to the abnormal physiology established when implants are fixed into the intramedullary canal with either bone cement or by ingrowth into a porous material. The implants become loose rather early from fibrous tissue hypertrophy and bone loss.[11] Particularly with joints of the fingers immobilization for bone ingrowth or biological attachment is contraindicated as unless early motion is started in the reconstructed joints the tendons become adherent, joint capsules contract, and the loss of motion is permanent.

In reconstruction of the small joints of lower and upper extremities a variety of low modulus, flexible, one piece, spacer implants with unfixed intramedullary stems have become widely used.[12,13] "Total" joints have two intramedullary stems joined by a flexible hinge-spacer. Hemi joints contain a single stem and an articulating spacer which is usually either convex or concave depending upon the intended application. The spacer portions of the implants transfer force across the joint to cortical bone. The interface between the fibrous capsule and silicone elastomer is low friction, thus the fibrous tissue is subjected to minimal mechanical stress minimizing the potential for fibrous tissue hypertrophy. In addition to its excellent resistance to fatigue, medical grade high performance silicone elastomer has excellent biocompatibility which further reduces the fibrous tissue reaction. Thus, when the unfixed implant stem design is proper, and the implant has been correctly fitted to bone the fibrous capsule in the intramedullary canal remains thin. When fibrous tissue hypertrophy in the intramedullary canal is avoided, and force transfer and stress simulate normal anatomy the response is thickening and strengthening in cortical bone as predicted by Wolff's law. A review of long term clinical response in patients with finger joint implants has confirmed this prediction.

Conclusions

The biophysical interaction between any implant and the fibrous capsule which eventually forms around essentially all implants is perhaps the most important functional consideration in any implant design. The fibrous capsule provides universal fibrous tissue fixation of implants to the body. When stressed gently the fibrous capsule will remodel, and when stressed traumatically it can hypertrophy. Fibrous tissue is thus inadequate to withstand force transfer fixation of musculoskeletal implants. Fibrous tissue hypertrophy within an intramedullary canal results in loss of bone and implant loosening. Force transfer in reconstructed joints to cortical bone across a low modulus spacer is an important consideration if bone is to be long maintained. The most successful implants are those designed with both mechanical features and materials of construction which minimize the tendency for fibrous tissue hypertrophy.

Bibliography

1. LaFay, H., "A Father's Last Chance Invention Saves His Son", Readers Digest, Vol. 57, Jan. 1957, pp. 29-32.

2. Madden, J.W., "Clinical and Biologic Features in Wound Healing", Sabiston, D., ed., Christopher's Textbook of Surgery, 11th ed., Philadelphia, W.B. Saunders Co., 1977.

3. Arem, A.J., and Madden, J.W., "Effects of Stress on Healing Wounds: Part I - Intermittent Noncyclical Tension", J. Surg. Research, Vol. 20, 1976, pp. 93-102.

4. Malick, M.H., Carr, J.A., "Flexible Elastomer Molds in Burn Scar Control", American Journal of Occupational Therapy, Sept. 1980, Vol. 34, No. 9, pp. 603-608.

5. Cronin, T.D., and Gerow, F.J., "Augmentation Mammoplasty: A New 'Natural Feel' Prosthesis", Excerpta Medica, International Congress Series No. 66, Proc. Third International Congress of Plastic Surgery, Washington, D.C., Oct. 1963.

6. Gruber, R.P., Kahn, R.A., Lash, H., et al., "Breast Reconstruction Following Mastectomy: A Comparison of Submuscular and Subcutaneous Techniques", Plast. & Reconstr. Surg., Vol. 67, No. 3, March 1981, pp. 312-317.

7. Baker, J.L., Jr., Bartels, R.J., and Douglas, W.M., "Closed Compression Technique for Rupturing a Contracted Capsule around Breast Implants", Plast. & Reconstr. Surg., Vol. 58, No. 2, Aug. 1976, pp. 137-141.

8. Rudolph, R., Abraham, J., Vecchione, T., et al., "Myofibroblasts and Free Silicone around Breast Implants", Plast. & Reconstr. Surg., Vol. 62, No. 2, Aug. 1978, pp. 185-196.

9. Homsy, C.A., Cain, T.E., Kessler, F.B., et al., "Porous Implant Systems for Prosthesis Stabilization", Clin. Orthop., Vol. 9, 1972, p. 220.

10. Rhinelander, F.W., and Nelson, C.L., "Experimental Implantation of Porous Materials into Bone", Acta Orthop. Belg., Vol. 40, 1974, p. 771.

11. Goldner, J.L., Gould, J.S. Urbaniak, J.R., et al., "Metacarpophalangeal Joint Arthroplasty with Silicone-Dacron Prostheses (Niebauer Type): Six and a Half Years' Experience, J. Hand Surg., Vol. 2, No. 3, May 1977, pp. 200-211.

12. Swanson, A.B., "Flexible Implant Arthroplasty for Arthritic Finger Joints - Rationale, Technique, and Results of Treatment", J. Bone Joint Surg., Vol. 54A, No. 3, April 1972, pp. 435-455.

13. Swanson, A.B., "Flexible Implant Resection Arthroplasty in the Hand and Extremities", St. Louis, The C. V. Mosby Co., 1973.

Session 14

ORAL SURGERY AND DENTISTRY

IMPROVEMENT OF EXPERIMENTAL ELASTOMERS FOR PERMANENT SOFT DENTURE LINERS: PHYSICAL AND BIOLOGIC PROPERTIES

L. Gettleman, C. Farris, R. LeBoeuf, H. Rawls, and E. Dillingham*

Gulf South Research Institute, P. O. Box 26518, New Orleans, Louisiana 70186
*Materials Science Toxicology Laboratories, University of Tennessee, Memphis, Tennessee

ABSTRACT

The development of improved bonding, tensile, and toxicologic properties of a polyphosphazine fluoroelastomer is documented. This elastomer is used as a soft liner in removable dentures over anatomic areas that pinch the soft tissues against bone, or on atrophic oral mucosa.

KEYWORDS

Removable dentures; permanent soft denture liners; polyphosphazine fluoroelastomer (PNF); adhesive bonding; toxicology of polymers.

INTRODUCTION

Several industrial elastomers have been investigated for use as permanent liners for removable partial and complete dentures. Such liners fulfill many clinical requirements, such as in areas where the oral mucosa is very thin and is pinched between bony ridges and the hard acrylic denture base. They are more comfortable to wear, enable more patients to tolerate removable dentures, require fewer adjustments by the dentist, and overcome anatomical deformities in the patient. Existing products are either self- or heat-curing. They are made of plasticized acrylic, silicone rubber, or polyurethane rubber. Failure usually occurs within months where the liner debonds from the denture base, harbors the growth of fungus, discolors, or dries out and cracks.

MATERIALS

The as-cast physical properties of a baseplate resin and two commercial permanent soft denture liners were compared with three experimental materials using ASTM or ADA tests. Lucitone 199 (grafted PMMA, Caulk) is a modified hard acrylic resin; Super-soft (Coe) is a plasticized acrylic; and Molloplast-B (Köstner, Buffalo) is a silicone rubber. The test materials were chlorinated polyethylene (CPE-3615, Dow), a Kraton-1107 block copolymer (Isoprene-Styrene-Isoprene, Shell), and a polyphosphazine fluoroelastomer (PNF-200, Firestone). The latter is a stable, soft, semiorganic elastomer with a -[-P=N-]- backbone; polymerization is by cross-linking of the fluorocarbon side groups (U.S. Patent No. 4,251,215).

METHODS AND RESULTS

Rubber indentation hardness is most easily tested and provides a measure of the comfort that the patient will experience while wearing the denture. Hardness at 1 mm and at 6 mm (ASTM D 2240, Durometer A) was Ss 77/75, MB 65/50, CPE 68/57, ISI 55/42, and PNF 55/17. As shown in Fig. 1, these materials differ in firmness at thicknesses greater than 3 mm, but the differences in firmness are much less at 1-2 mm, the thickness of most denture liners.

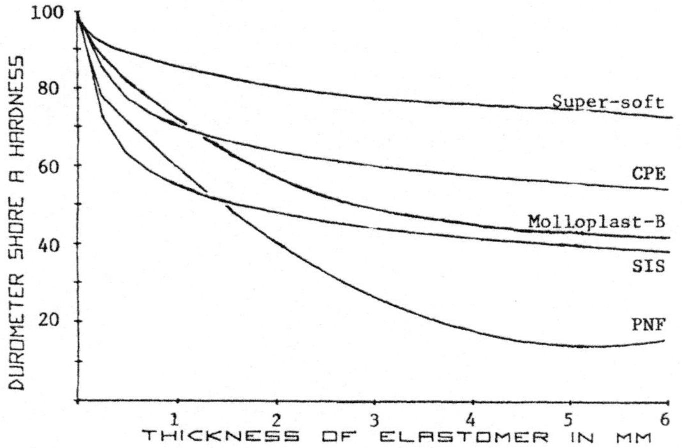

Fig. 1. Durometer hardness as a function of denture liner thickness.

Tensile strength in MPa and % elongation (ASTM D-412) were: PMMA 87/-, Ss 2.8/244, MB 4.3/370, CPE >4.0/>400, ISI >5.3/>400, PNF 1.5/419. Adhesion in N/m using ASTM D-903 (a 180° peel test) was: CPE 16,600, ISI 1660, PNF 2840. For Ss and MB, dry adhesion was greater than cohesion at room temperature.

Contact angle in degrees was: PMMA 79, Ss 73, MB 75, CPE 85, ISI 109, PNF 99. Water sorption and solubility in g/m^2 using ADA 12.4.3.3 and 12.4.3.4 were: Ss 2.3/5.1, MB 0.4/0.02, CPE 8.9/0, ISI 3.6/0.1, PNF 12.8/0. Color stability (ADA 12.4.3.6) indicated no change for all but ISI, which depolymerized after 24 h exposure to UV. Because of the modest bonding properties and degradation of ISI in accelerated aging, it was dropped from further development.

A primary screening was done to assess the acute toxicity of these materials. PMMA was difficult to dissolve, while CPE and ISI dissolved completely in cottonseed oil, so tests were not performed on these extracts. The partial/total Cumulative Toxicity Indices were: PMMA 123/-, Ss 83/108, MB 125/200, CPE 53/-, ISI 72/-, PNF 78/153. Molloplast-B was most toxic, with five tests higher than the minimum toxic response. PNF produced a distinct hemolytic response and was positive in the cottonseed oil intradermal irritation test. CPE appeared to be the least toxic.

The PNF formulation has been selected for its combination of mechanical and toxicologic properties. The use of lauroyl peroxide-magnesium oxide has reduced the curing temperature to 100°C and plasticized the rubber. By milling Lucitone 199 beads and fibers into the rubber and exposing to MMA monomer prior to packing, adhesion to PMMA is improved. Loading of 10 and 40% PMMA yielded soft and firm puttys which can be used selectively at the center and periphery of the denture. By incorporating tetraethylene glycol dimethacrylate as a crosslinker, bond strengths as high as 3590 N/m have been achieved. Combining TEGDMA and PMMA beads is expected to improve bonding even more; it is apparently a combination of polymerization of PNF around the PMMA beads, and fusion of the PMMA beads to the underlying baseplate as shown in Figs. 2 and 3.

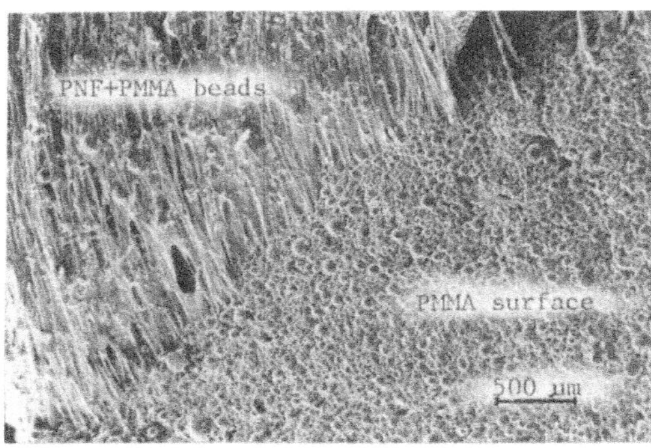

Fig. 2. Peel specimen of PNF on a PMMA substrate (SEM, 70X).

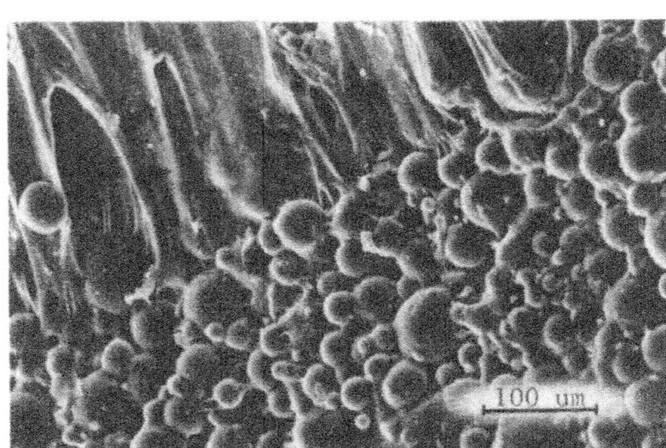

Fig. 3. Higher power view of peel specimen (SEM, 200X).

MB, Ss, and PNF are moldable as a putty; CPE and ISI can be used in sheet form, and CPE can be vacuumformed for use as a stabilized baseplate for waxup. Ten to fifteen grams is required for a liner in a complete denture. MB costs 90¢/g, Ss 8¢/g, raw CPE 0.8¢/g, raw ISI 0.1¢/g, raw PNF 22¢/g. Long term property tests under simulated intraoral conditions of PNF are underway.

Supported by grant No. DE-04814 from the NIDR.

FACIAL AUGMENTATION WITH PROPLAST:
Long Term Retrospective Investigation

John N. Kent, D.D.S., Russell L. Westfall, D.D.S.
and David M. Carlton, D.D.S.
Dept. of Oral and Maxillofacial Surgery
LSU Medical Center
New Orleans, LA

Proplast (Vitek, Inc., Houston) has been successfully used for facial augmentation for over eight years. The long term stability of proplast chin and malar implants has not been examined. A retrospective clinical study was undertaken to determine the long term stability of chin and zygomaticomaxillary augmentation with proplast implants and stability relative to the type of dentofacial deformity and concomitant surgical procedure.

MATERIALS AND METHODS

Most of the patients in this study had zygomatic and chin implants of grey proplast I (polytetrafluoroethylene (PTFE/graphite), and a few had implants of white proplast II (PTFE/aluminum oxide). These implants were placed over an eight year period at the Louisiana State University Medical Center and affiliated hospitals. Proplast I and II are unique, ultraporous, low modulus implant materials designed especially for facial bone augmentation. Both materials are 70% to 90% porous by volume, with pore sizes ranging from approximately 100 to 400 m. Dendritic pore interconnections are greater than 80 m. The materials are resilient, may be bent without regard for memory, and are easily carved. Preformed chin and zygomatic implants have recently been introduced to reduce surgical time and eliminate problems such as palpable edges and asymmetry resulting form carving.

In the past few years, both extraoral and intraoral techniques of implant placement procedure have been refined. Placement protocol for the chin and zygomatic implants in this study primarily involved transoral placement (94%), preplacement infusion with antibiotic solution, development of a subperiosteal pocket, fixation of the implant to underlying bone or overlying soft tissue by suture, and careful layered tissue closure.

Over 150 patients have received chin and facial implants. Fifty-four patients with chin implants and 45 with paired malar implants were available for follow-up.

Chin Implant Radiographic Evaluation - Thirty-three of 54 patients with chin implants had adequate records for long term radiographic follow up. These patients were followed for up to 96 months. Serial cephalograms taken with the patient's head in a natural position and the patient's lips at rest were obtained preoperatively, postoperatively, and at each follow up interval. At the time of surgery, the thickness of the implant was measured and recorded.

Serial cephalograms were measured for implant thickness, changes in soft tissue thickness, superior or inferior implant migration, and posterior implant migration into bone. Soft tissue changes were measured from the lingual cortex of the mandible horizontally to the skin surface at the level of the pogonion. These measurements were done on each cephalogram on two seperate occasions; the first measurements were made by hand directly from the cephalogram, the second measurements were done on the Zeiss MOP-3 from a tracing of the cephalogram. The measurements were averaged to the nearest 0.5mm, and the mean recorded. Long term augmentation was also expressed as percent of the original implant thickness at placement.

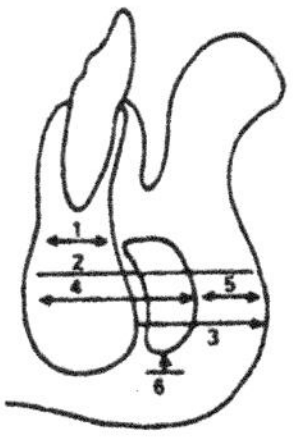

1. Buccal-lingual width of mandible measured at Pogonion, mid-implant level, and superior implant level.
2. Lingual cortex to skin surface, measured at Pogonion
3. Buccal cortex to skin surface at Pogonion
4. Lingual cortex to facial of implant at Pogonion
5. Facial of implant to skin surface at Pogonion
6. Mentonto inferior surface of implant

-fig 1 levels of measurement-

On each cephalogram, superior implant migration was determined by measurement from menton to the inferior edge of the implant. Superior migration of the implant was recorded as a positive value and inferior movement as a negative value.

Posterior implant migration into bone was measured by determining the lingual to buccal cortex width of the mandible at three implant levels: pogonion, midimplant level, and the superior implant level. The mean of the two determinations of posterior implant migration at each of these levels was recorded for each patient.

Evaluation of implant thickness at each time interval was done to determine the long term dimensional stability of the implant.

Patients were separated into six groups according to cephalometric and clinical diagnosis and the type of orthognathic surgery, if any, performed before or concomitant with the chin augmentation surgery (table 1.) descriptions and patient population.

Patients in group 6 had horizontal osteotomies of the chin for advancement as well as onlay augmentation with Proplast block. This group was analyzed separately from the others because of the multiple factors affecting total augmentation and long term augmentation.

Data for the various parameters were compiled with differentiation as to the category of patients. Means for all patients in each category were obtained for each follow up interval. Comparisons of raw data, determinations of means, and statistical calculations were performed using the BMDP Biomedical Computer Programs - 1979 of the University of California Los Angeles.

Zygomaticomaxillary Implant Evaluation -It is not possible to radiographically evaluate zygomatic implants, due to anatomic interferences. Therefore, long term clinical evaluation of zygomatic implants consisted of tabulating postoperative results (infection, gross displacement, or less of implant).

RESULTS

Gross Observation of Implants - Long term evaluation of postoperative complications indicated that removal of only four (7.4%) of the chin implants and two pairs of zygomatico-maxillary implants was necessary.

No. of group	No. of patients	% of total	Description
1	7	22	Patients who had chin augmentation only. No other skeletal deformity or surgical procedure.
2	3	9	Patients who had maxillary or mandibular surgery, excluding correction of vertical maxillary excess and mandibular advancement surgery.
3	6	18	Patients who had mandibular advancement surgery concurrent with chin augmentation.
4	3	9	Patients who had uncorrected vertical maxillary excess or vertical maxillary excess tendency (SN-MP 40°) and had chin augmentation.
5	10	30	Patients who had vertical maxillary excess corrected by maxillary impaction with chin augmentation.
6	4	12	Patients who had osteotomy of the chin for advancement, with overlying implant.

-fig 2 patient groups-

These all occured in the early postoperative period. One pair of malar implants was removed because of infection and the other because of excessive implant displacement. Of the chin implants, three were removed because of infection, and the last due to excessive displacement. There was no instance of extrusion or overlying tissue dehiscence. Although temporary anesthesia was frequently reported after surgery, only one patient had permanent mental nerve anesthesia and one had infraorbital anesthesia.

There was no obliteration of vestibular depth or exacerbation of periodontal conditions, as well as, no evidence of tooth root resorption or devitalization secondary to posterior implant migration.

In three patients with zygomaticomaxillary implants, inferior displacement resulted from the absence of stabilization to the infraorbital rim. In two patients, the displacement was variable and did not require repositioning, but in one, it did require repositioning. Generally, patients receiving zygomatic or chin implants were extremely pleased and often commented that the implants felt "natural" and "a part of them", indicating satisfactory stability.

Radiographic Evaluation of Chin Implants - Over half of the patients who received chin implants were followed for three years or longer.

Implant mean superior migration averaged 0.72 $\pm$1.1mm. Superior migration of the implant was found to be significantly greater in patients with concomitant ramus osteotomies done for mandibular advancement (group 3). Also, patients who had vertical maxillary excess corrected concomitantly with autorotation of the mandible (group 5) were found to have significantly more superior implant migration. Patients who had no concomitant surgical procedures (group 1) had less superior migration than other groups.

Posterior migration at the superior extent of the implant was equal to, or more than, the posterior migration at the pogonion and midimplant when the groups were compared. Statistical analysis of groups for posterior implant migration showed significant differences between groups 2 and 5 at the level of pogonion (P=.052). At the midimplant level, there were statistically significant differences between groups 1 and 2 (P=.023) and groups 2 and 5 (P=.052). Significant differences were found between groups 4 and 5 at the superior implant level (P=.091). It is to be noted that group 4, with uncorrected vertical maxillary excess, and substantially increased posterior implant migration at the pogonion (2.5mm), midimplant (2.5mm), and superior implant (2.66mm) levels.

Average implant thickness for this study was 8.58mm as compared with that of a previous study, in which mean implant thickenss was 5.4mm. The linear augmentation findings in each group at various time intervals are seen in figure . There were significant statistical differences between groups 1 and 2 (P=0.94), groups 1 and 3 (P=.038), groups 1 and 5 (P=.035), and groups 4 and 5 (P=.103) one year postoperatively. Two years postoperatively, there were significant statistical differences between groups 1 and 3 (P=.094) and groups 1 and 5 (P=.052. In the three year postoperative data, significant differences were found between groups 1 and 3 (P =.026) and groups 1 and 5 (P=.040). Only groups 1 and 4 were statistically different four year postoperatively.

Effectiveness of augmentation was also calculated as a percent of the implant thickness at placement. Overall mean percent of augmentation after a year compared with original implant thickness was 57%. Two years post implantation, it was 47%, and at three years, it was 46%. Statistical analysis showed no significant change in percent of augmentation in any group of patients after two years of implantation. It can be inferred from this analysis that for many patients, the amount of long-term augmentation stabilizes after a year, and that for all patients, it stabilizes within a two year period. It also shows that there is noticable better long term augmentation in patients who have no other skeletal deformity or other sugical procedure (group 1).

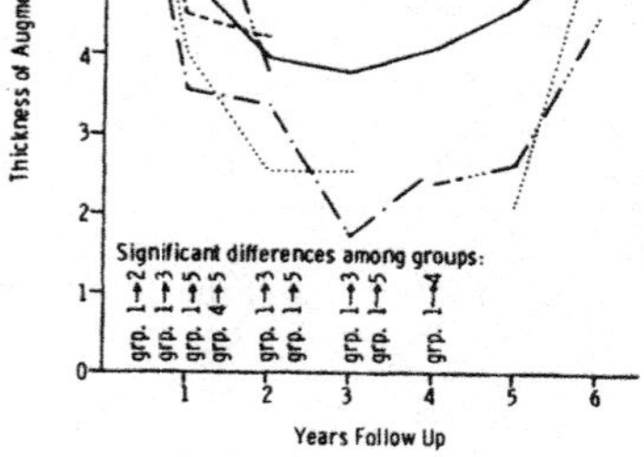

Fig 3

Cephalograms measured to compare original implant thickness with that of the three year postoperative period, showed excellent results. All implants had at least 98% long term stability, and many had 100% dimensional stability over the test period.

DISCUSSION:

The results of this investigation have been very enlightening, in that, dividing the patient population into the various groups has helped to more clearly define skeletal and soft tissue influences.

Superior migration of chin implants was found to be significantly greater in patients who had mandibular advancement surgery, and patients who had corrected vertical maxillary excess. In both of these situations, chin soft

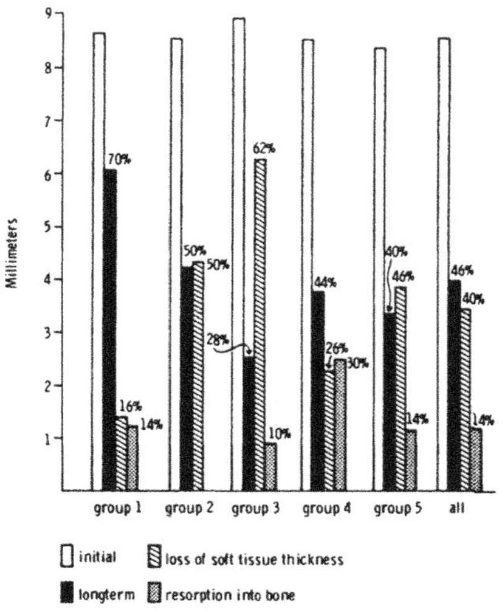

Fig 4

tissue tension increased from forward and upward mandibular repositioning. Superior displacing forces on the implant would be expected from peri-implant soft tisue stresses. Anatomic variation may also have a bearing on superior migration of the implant.

Posterior implant migration into bone was greatest in group 4 (uncorrected vertical maxillary excess) patients, in whom increased mentalis muscle activity is found. Group 5 (corrected vertical maxillary excess) patients also had significant implant migration into bone, probably secondary to increased soft tissue tension from the autorotation of the mandible as well as residual mentalis muscle habituation. Significantly, group 2 (prognathism patients corrected by mandibular set back) had no implant migration into bone. This is explained by the relaxation of peri-implant soft tissue tension and muscular forces secondary to retropositioning of the mandible.

Comparing all patient categories, patients having only a chin contour deficiency, with no other skeletal deformity or associated orthognathic surgical procedure, had significantly better chin augmentation results. Patients having chin augmentation concurrent with mandibular advancement, showed the greatest amount of long term loss of augmentation. It is apparent then, that the more stable postoperative soft tissue environment probably contributes to a better long term augmentation result. Loss of augmentation over the long term seems to be primarily caused by a redistribution and circumferential displacement of soft tissue, rather than to posterior migration of the implant into bone, as previously thought.

Long term chin augmentation has been shown to average 46% of the original implant thickness for all categories. The augmentation stabilizes within two years after implantation; no significant changes in augmentation occurred after two years in this population of patients. Patients in group 1 had significantly better long term augmentation results (70% of original implant thickness), whereas group 3 had significantly inferior long term stability (28% of original implant thickness).

Patients in group 6, with proplast placed over a horizontal advancement genioplasty had findings similar to those of groups 3. A 38% long term augmentation was found when comparison was made with the original implant thickness. It was difficult to determine whether loss of augmentation was caused by posterior implant migration or postoperative resorptive remodeling of the underlying osteotomy segment.

SUMMARY

The results of a long term study of Proplast chin and zygomaticomaxillary implants are presented. The implants are well accepted, and the stability of the augmentation is related to the type of facial deformity and concurrent corrective surgery.

REFERENCES

1. Kent, J.N.; Westfall, R; and Carlton, D. Chin and Zygomaticomaxillary Augmentation with Proplast: long term follow-up. J. Oral Surg. 39:912-919. Oct. 1981.

2. Kent, J.N., and others. Pilot studies of a porous implant in dentistry and oral surgery. J. Oral Surg. 30:608-616, 1972.

3. Homsy, C.A., and Anderson, M.S. Functional stabilization of soft tissue and bone prostheses with a porous low modulus materials system. In biocompatability of implant materials. Tunbridge Wells, England, Sector Publishing Co., 1976.

4. Vitek, Inc., data sheet 81001, March 1981.

Nine Year Report on Proplast Coated Metallic Prostheses for Condyle Replacement.

J.N. Kent* and D.J. Misiek**
*Professor and Head, Dept. of Oral and Maxillofacial Surgery
Louisiana St. Univ., School of Dent. New Orleans, Louisiana
**Senior Resident, Dept. of Oral and Maxillofacial Surgery,
Louisiana St. Univ., School of Dent. New Orleans, Louisiana

INTRODUCTION

Over the years, reconstruction of the temporomandibular joint has presented a challenge to both the surgeon and the bioengineer. The basic problem for reconstruction revolves about two concepts, restoration of condylar function and re-establishment of ramus vertical dimension (Kent, and others 1974). Biologic materials, such as fascia, dermis, cartilage, bone, and hard tissue growth transplants, have all been tried but have limitations, including: inability to restore and maintain ramus vertical dimension, displacement, unpredictable resorption, reduced function from necessary mandibular immobilization, infection, and potential donor site morbidity. Alloplastic materials, reviewed by Kent and James (1978) (acrylic, teflon, silicon rubber, and various metals), have been developed to circumvent these problems but have complications of their own, including: forshortening of the ramus, erosion of adjacent bone, and slippage of the implant material.

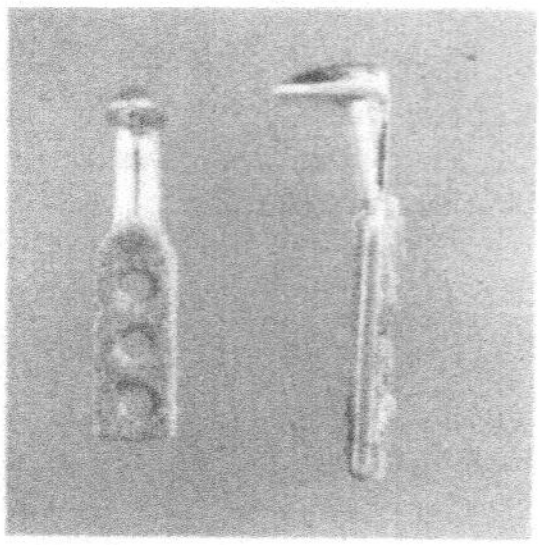

Fig. 1 Flat (TMC) Condyle

The use of Proplast as a biocompatible interface between metallic implants was proposed by Homsy, and others (1970a, 1970b, 1972) and Hinds, and others (1971). Implant design with case studies was presented by Kent, and others (1972) with follow up by Hinds, and others (1974). The original design was a flat shank (Fig. 1) cast from ticonium surgical alloy with a one mm Proplast coating on both medial and lateral surfaces. Forty-five, 50, and 55mm lengths were available, accommodating 3 self-tapping screws. In 1975, Kent modified this design (box type) by adding parallel flanges to the shank (Fig. 2) which fit into recesses cut into the lateral cortex of the mandibular ramus. Since the Proplast coated metallic condyle was introduced, more than two hundred have been placed for ankylosis, with and without skeletal deformity, traumatic condylar loss, tumors or infections requiring condylectomy, and facial deformities (open bite, retrognathia, asymmetry, etc.) associated with the above. Contraindication for use include those situations compromising both placement and acceptance of the implant materials, as well as patient cooperation with postoperative physiotherapy.

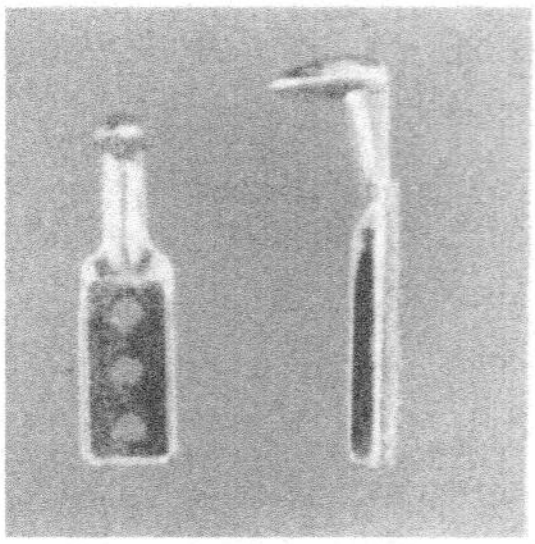

Fig. 2 Box (TMCK) Condyle

Prosthesis placement utilizes a combined retromandibular/submandibular approach. After excising the condyle and performing a coronoidectomy, articular fossa preparation with a bur may be necessary. With the patient in intermaxillary fixation, the ramus is modified to accept the prosthesis, which is then secured to the ramus with bone screws (Fig. 3). With the flat (TMC) condyle, 3 weeks of immobilization were recommended with gradual function thereafter. The box (TMCK) condyle was designed for immediate function thus obviating the need for postoperative immobilization.

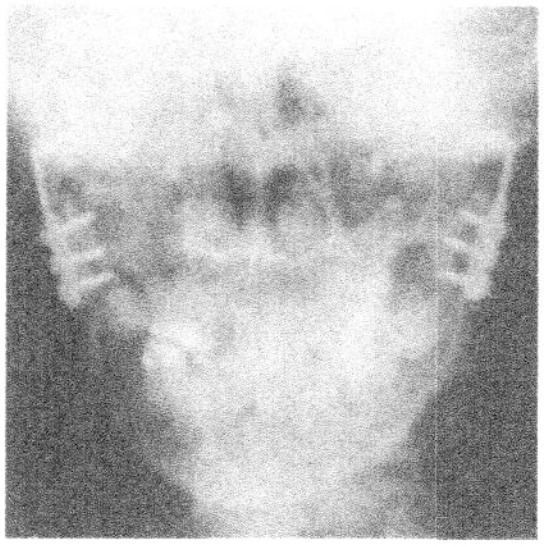

Fig. 3 Prosthesis in place

MATERIALS AND METHODS

In 1980, a retrospective study of those patients who have had Proplast coated metallic condylar reconstruction was undertaken (Misiek, 1981). A comprehensive questionnaire was sent to oral surgeons in teaching centers and in private practice to report their experience. Parameters evaluated included: preoperative historical and clinical findings, prior conservative therapy, positive laboratory and radiographic findings, prior surgical therapy, diagnosis, operative findings, current therapy, and postoperative follow-up. The information reported was tabulated and presented as raw numbers or percentages. The nonparametric nature of the data negated statistical analysis, which proved unwieldly and inappropriate. Failure of the proplast coated metallic condyle, as defined in our study, represents those prostheses which have been removed, or where removal is indicated.

RESULTS

A total of 109 prostheses placed in 80 patients were reported. Sixty percent of these patients represent the efforts of 20 private practitioners and teaching institutions,while the remaining 40% were treated at the University of Texas Health Science Center at Houston, and at the Louisiana State University Medical Center at New Orleans. Females outnumbered males 3:1 or more in the diagnosis categories of ankylosis, rheumatoid arthritis, osteoarthritis, and miscellaneous. The average ages ranged from just over 33 years in ankylosis, to just under 44 years in rheumatoid arthritis, with ages ranging from 15 to 60 years. Seventy-six prostheses were placed for ankylosis, which was the most common diagnosis. The miscellaneous category included diagnoses of condylar hyperplasia, aseptic nerosis, condylar hypoplasia, desmoplastic fibroma, trauma, and osteochondroma. The average follow up was 25.4 months with individual follow-up ranging from one to 120 months.

Preoperative radiographic findings reported ranged from frank bony ankylosis to the subtle changes found in degenerative diseases. Decreased interarticular space, represented clinically by decreased ramus vertical dimension, was noted in 93 of 109 condyles placed. Prior surgical intervention was reported with great frequency, including gap arthroplasty with interpositional materials, silicon rubber (21 cases) and teflon (7 cases). All 21 cases which failed using silicon rubber block as an interpositional material, eventually went on to successful use of the Proplast coated metallic condyle. Mandibular fractures, in particular subcondylar fractures (22 cases), in which immobilization was the only therapy, comprised a surprisingly large group of patients requiring eventual condylar reconstruction indicating possible inadequacy of initial therapy. Success and failure for both flat (TMC) and box (TMCK) shank condyles are represented in Table 1. In each diagnostic category the box (TMCK) condyle had performed better than the flat condyle, particularly in ankylosis where the box (TMCK) condyle was successful in 49 instances, failing in 3, as opposed to the flat (TMC) condyle success/failure ratio which was 14 to 10. Even more impressive is the fact that the box (TMCK) shank condyle, placed twice as often, 74 versus 35, had an overall

TABLE 1

TYPE OF CONDYLE PLACED PER DIAGNOSIS

	FLAT (TMC)		BOX (TMCK)	
	SUCCESS	FAILURE	SUCCESS	FAILURE
ANKYLOSIS	14	10	49	3
RHEUMATOID ARTHRITIS	0	0	8	4
OSTEOARTHRITIS	5	2	5	0
MISCELLANEOUS	2	2	5	0
	21	14	67	7

TABLE 2

POSTOPERATIVE FINDINGS: ALL CASES (109 CONDYLES)

		SUCCESS	FAILURE
I. VERTICAL DIMENSION OF RAMUS	MAINTAINED	59.0%	31.6%
	INCREASED	39.8%	21.2%
	DECREASED	1.2%	47.3%
I. INCISAL OPENING	MAINTAINED	36.6%	36.8%
	INCREASED	59.8%	36.8%
	DECREASED	3.6%	26.4%
III. LATERAL EXCURSIONS	MAINTAINED	59.8%	57.9%
	INCREAESD	26.8%	15.8%
	DECREASED	13.4%	26.3%

success rate of 89.4% while the flat (TMC) shank condyle succeeded just 60% of the time. Postoperative clinical findings, Table 2, represents the functional results of Proplast coated metallic condylar reconstructions. In successful cases, ramus vertical dimension and incisal opening was maintained or increased over 96% of the time. Lateral excursions, difficult to restore because of loss of lateral pterygoid muscle attachment, was maintained or increased in more than 86% of the cases reported.

TABLE 3

CONDYLE FAILURES

	FLAT	BOX
REANKYLOSIS	5	
INFECTION	4	1
INADEQUATE STABILIZATION	4	1
GLENOID FOSSA RESORPTION*		4
PLACEMENT ERROR		1
METAL FATIGUE	1	
	14	7

*RHEUMATOID ARTHRITIS ONLY

Failures listed in Table 3, when addressed individually show reankylosis to have only occurred with the use of the flat (TMC) condyle. Infection and inadequate ramus stabilizations were also primarily seen with flat (TMC) condyle use. Placement error was a surgical problem and metal fatigue was an isolated manufacturing defect. Glenoid fossa resorption was noted in two rheumatoid arthritis patients who had functioned with the prostheses for 2 years or more. Encroachment upon the middle cranial fossa in these cases was imminent, therefore reconstruction of the glenoid fossa, with a custom implant, was perfomed. Function was restored postoperatively but long term follow-up is pending.

DISCUSSION

In the years from 1971 to 1980, at least 109 Proplast coated metallic condyles were placed for diagnoses ranging from ankylosis to aseptic necrosis of the mandibular condyle. Overall, nearly 90% of these reconstructive attempts were successful. Prior to 1975, when only the flat (TMC) shank condyle was available, reankylosis and inadequate ramus stabilization were predominant causes for implant failure. The box (TMCK) design was developed to reduce these problems. The flanges, which inlay into the ramus, were designed to increase ramus stabilization, especially in the immediate postoperative period, allowing for early mobilization thus markedly decreasing the incidence of reankylosis. Refinement of the implant placement technique, has reduced the infection rate as well.

We have not placed the Proplast coated metallic condyle in children or adolescents whose growth potential has not been fully realized. The reasons for this are two-fold: first the anatomy of the ramus and temporomandibular articulation prohibit the use of the implant in its present size. Second, we feel that transplanted bone/cartilage composite grafts do have a place in condylar reconstruction because of their ability to respond to a dynamic growth situation by resorption, apposition, and remodelling.

Rheumatoid arthritis has been a challenging problem, to which we feel the solution is total temporomandibular joint reconstruction incorporating the use of the Proplast coated metallic condyle along with a custom glenoid fossa implant. With the fossa implant, a more mechanically favorable situation is created by distribution of forces over a wider area, thus minimizing further bony resorption in function.

CONCLUSION

1. The Proplast coated metallic condyle was successful in 87.3% of cases requiring functional temporomandibular joint reconstruction.
2. The box (TMCK) design, with a success rate of 89.4%, was found to be clinically superior to the flat (TMC) design, success rate of 60%.
3. Three screws are necessary for adequate ramus stabilization.
4. Immediate postoperative function is essential.
5. Ramus vertical dimension can be predictably restored and maintained.
6. The Proplast coated metallic condyle has been used successfully in all 21 cases of failure of silicon rubber block interpositional arthroplasty.

REFERENCES

Hinds, E.C., Homsy, C.A. and Kent, J.N. (1971) Use of a biocompatible interface for combining tissues and prostheses in oral surgery. In Lester W. Kay (Ed.) Transactions of the IVth International Conference on Oral Surgery, 1972, Munksguard, pp. 210-216.

Hinds, E.C., Homsy, C.A., and Kent, J.N. (1974) Use of a biocompatible interface for binding tissues and prostheses in temporomandibular joint surgery, Oral Surg., 38, 512-519.

Homsy, C.A. (1970a) Biocompatibility in selection of materials for implantation, J. Biomed. Mater. Res., 4, 341-356.

Homsy, C.A., Cain, T.E., Anderson, M.S., and King, J.W. (1970b) Dynamic Stabilization of implanted prostheses, J. Bone Joint Surg., 52A, 604.

Homsy, C.A., Cain, T.E., Kessler, F.B., Anderson, M. Sidney, and King, J.N. (1972) Porous implant studies for prosthesis stabilization, Clin. Orthoped., 89, 220-235.

Kent, J.N., Homsy, C.A., Gross, B.D. and Hinds, E.C. (1972) Pilot studies of a porous implant in dentistry and oral surgery, J. Oral Surg., 30, 608-615.

Kent, J.N., Lavelle, W.E., and Dolan, K.D. (1974(Condylar reconstruction: Treatment Planning, Oral Surg., 37, 489-497.

Kent, J.N., and James, R.B. (1978) Alloplasts in maxillofacial surgery. In Transactions of the 4th Annual Meeting Society for Biomaterials and the 10th Annual International Biomaterials Symposium, Vol. II:9-11, San Antonio, Texas.

Misiek, D.J., (1981) Nine year report on Proplast coated metallic prostheses for condyle replacement. Presentation. 63rd Annual Meeting. American Association of Oral and Maxillofacial Surgeons, Washington, D.C.

STRESS DISTRIBUTION IN RESTORATIONS
A THREE-DIMENSIONAL FINITE ELEMENT STUDY

C. Rubin*, H. Yi*, and E. Capilouto**

*Department of Mechanical and Materials Engineering,
Vanderbilt University, Nashville, TN 37235
**Department of Operative Dentistry, University of Alabama
School of Dentistry, Birmingham, AL 35294

ABSTRACT

Amalgam restorations are introduced into a three-dimensional finite element model of a mandibular right permanent first molar. Stresses in the restorations are obtained and compared to previous two-dimensional, axisymmetric studies.

KEYWORDS

Finite elements; stress; tooth; teeth; restorations; molar.

INTRODUCTION

Development of a three-dimensional finite element model of the human tooth (Rubin and co-workers, 1981) has enabled investigators to evaluate physical stresses due to chewing forces more accurately. The purpose of this paper is to further develop the three-dimensional finite element model of the mandibular right first molar by introducing restorations.

Finite element techniques approximate a structure mathematically by means of "elements" which may be thought of as building blocks. These elements are connected by "nodes". The more elements that are used, the closer the approximation becomes to a description of the actual structure. The finite element model used in this study is shown in Fig. 1. This model is composed of 336 elements and 520 nodes.

The finite element computer program employed was WHEEL, a program developed at Battelle Columbus Laboratories. This program allows the calculation of stresses, strains, and deformations in an arbitrarily shaped three-dimensional finite-element model representing a structure under static loading. The theoretical formulation underlying the program is based on linear elastic material behavior. The element used in WHEEL belongs to the "isoparametric" family. This name derives from use of the same interpolation function to define the element shape as was used to define the displacements within the element. The element within WHEEL is of the "quadratic" type. A detailed description of the theoretical development of WHEEL may be found in the Technical Manual (Sampath, 1978). For a general theoretical formulation, see Bathe and Wilson (1971).

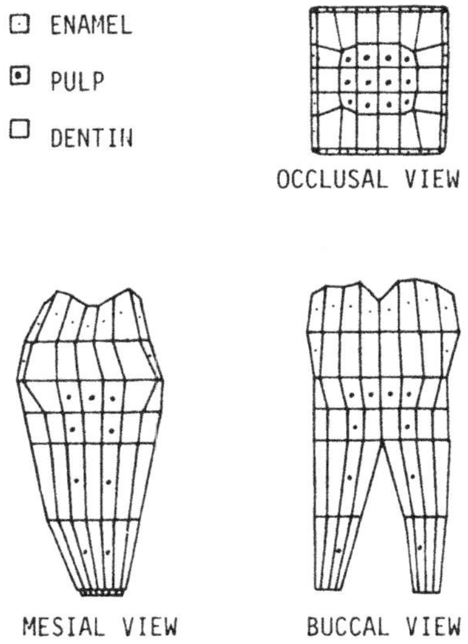

Fig. 1. Three-dimensional Finite Element Model.

Several finite element studies have been made of the stresses in restorations. A good summary of this work with associated bibliography has been done by Craig and Farah (1977). These studies treated the tooth as an axisymmetric structure with axisymmetric loading. Since neither structure nor typical loading of a tooth are actually axisymmetric, it is evident that more accurate predictions of stresses could be made with a three dimensional model in which loads and structures are closer to the actual case. Figure 2 shows a typical loading and restoration shape for the mandibular right permanent first molar.

PROCEDURE AND RESULTS

The three-dimensional finite element model is composed of three materials plus an amalgam restoration. The material properties used are given in Table 1. These properties are the same as those used by Farah, Hood and Craig (1975) in treating the effects of cement bases on stresses in amalgam restorations (Pulp was treated as a void in the 1975 study; this will have no affect on the stresses in the amalgam).

Rubin and co-workers showed that stresses in a mandibular right permanent first molar are lower than predicted by an axisymmetric finite element model. It would be expected, similarly, that the stresses in amalgam restorations would be different from those predicted by the axisymmetric model.

An attempt was made to reproduce the axisymmetric conditions of Farah, Hood and Craig by replacing a layer of enamel elements with amalgam and applying an approximately annularly concentrated load of 440 N to the model in Fig. 1. Although it was not possible to make the loading and restoration completely axisymmetric, agreement with the previous study was good. Figure 3, Case 1, shows the range of compressive vertical stresses in the amalgam which were obtained.

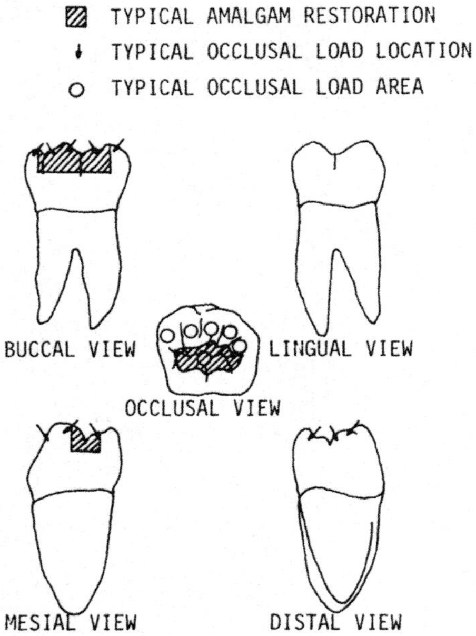

Fig. 2. Typical Occlusal Loading and Restoration.

TABLE 1 Material Properties

	Young's Modulus (MPa)	Poisson's Ratio
Enamel	8.25×10^4	.33
Dentin	1.86×10^4	.37
Pulp	2.07	.45
Amalgam	6.9×10^4	.35

Next, the load was changed to be applied in a manner closer to real chewing forces. The magnitude of the vertical forces remained the same, however, additional lateral loading was introduced. The lateral load components were approximately 40% of the vertical loads. Figure 3, Case 2 shows the results. In Case 3, a less symmetric restoration was used, roughly approximating the restoration shown in Fig. 2. A symmetric load (as in Case 1), was applied. Figure 3 shows the results. Case 4 employed the more realistic restoration and the more realistic chewing loads.

As may be observed from Fig. 3, the restoration shape (provided that a restoration of similar size is treated), has less effect on the stresses than the load distribution. Putting it another way, an assumption of axisymmetric load distribution introduces greater error than does an assumption of axisymmetric restoration shape. The stresses in Cases 2 and 4 are much more evenly distributed over the amalgam than they are for annularly concentrated loads. Thus axisymmetric treatments have consistently overestimated stresses within the amalgam.

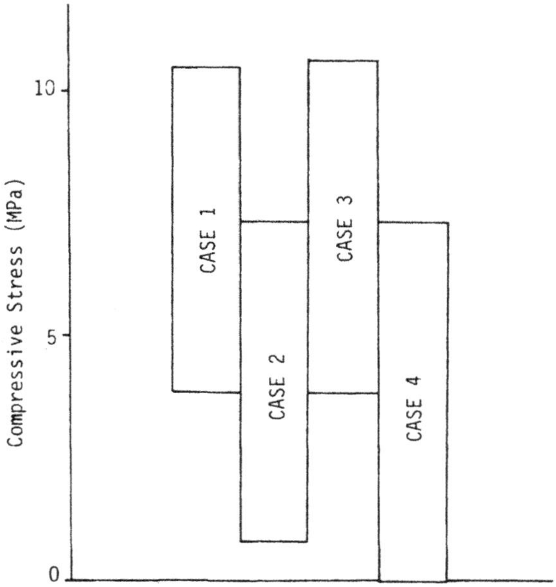

Fig. 3. Range of Longitudinal Stresses in Amalgam Restorations.

Note the slight tensile stress produced in Case 4. This stress occurs in the element furthest removed from the tooth center. It is due to bending moments induced by the loading and would not be observed to occur in the two-dimensional axisymmetric model.

FUTURE WORK

The present three-dimensional model has been used to the fullest extent possible and further refinement must be made to obtain more information. The work in making these refinements has begun. This will make it easier to approximate and compare a wider variety of restoration shapes and materials, and will lead to additional work on other types of artificial tooth replacements.

A new finite element program will be used for the refined mesh. This program can treat anisotropic materials, elastic-plastic stresses, and heat transfer problems, all of which have applications in stress analysis of the human tooth.

REFERENCES

Bathe, K. J., and E. L. Wilson (1971). Numerical Methods in Finite Element Analysis. Prentice-Hall, Inc., Englewood Cliffs, N.J.

Craig, R. G., and J. W. Farah (1977). Stress Analysis and Design of Single Restorations and Fixed Bridges. Oral Science Review, 10, 45-74.

Farah, J. W., J. A. Hood, and R. G. Craig (1975). Effects of Cement Bases on the Stresses in Amalgam Restorations. Journal of Dental Research, 54, N1, 10-15.

Rubin, C., N. Krishnamurthy, E. Capilouto and H. Yi (1981). Stress Analysis of the Human Tooth using a Three-dimensional Finite Element Model. Submitted to Journal of Dental Research.

Sampath, S. G. (1978). Technical Manual for Program WHEEL. Batelle Report to Track Train Dynamics, Association of American Railraods.

Session 15
COMPUTERS IN MEDICINE

COMPUTER IDENTIFICATION OF CLINICAL LABORATORY SPECIMENS

Robert L. Keenan, P.E. (1) Bruce E. Mount (2)
Garland W. Persinger (3) William Stickel (4)
John Melynk Ph.D. (3)

City of Hope Medical Center - Duarte, California 91010

ABSTRACT:

This paper discusses patient I.D. quality control used in a system for processing human blood to produce chromosome preparations on microscope slides, suitable for karyotyping. The system operates semi-automatically, under computer control. Bar codes placed on specimen tubes and microscope slides are read with hand-held and fixed optical scanners and then stored in memory and on a disc. A data table listing names, plus all I.D. numbers is generated. Only the kinematic actions of specimens and specimen hardware are given. A reference is supplied for the mechanical and control details.

KEYWORDS:

Specimen; Bar Code; Identification; Computer; Chromosome.

INTRODUCTION:

This paper discusses the methods used to identify patient specimens in a semi-automatic system which processes human blood to generate chromosome preparations on microscope slides. The system operates under the control of a computer, a DEC PDP-11/34 with dual floppy discs. It consists of two consoles plus a specimen I.D. matrix. Console one operates in two modes. Mode one transfers blood from specimen tubes to culture trays. After processing the culture trays through console two, mode two of console one is used to transfer specimens from the culture trays to microscope slides, (see Fig. 1). Only the kinematic motions of specimens and their associated vessels are shown. The reference by Keenan (1980) gives the mechanical and control details of these consoles.

SPECIMEN I.D. MATRIX:

The operation begins at the teletype keyboard and the specimen tube

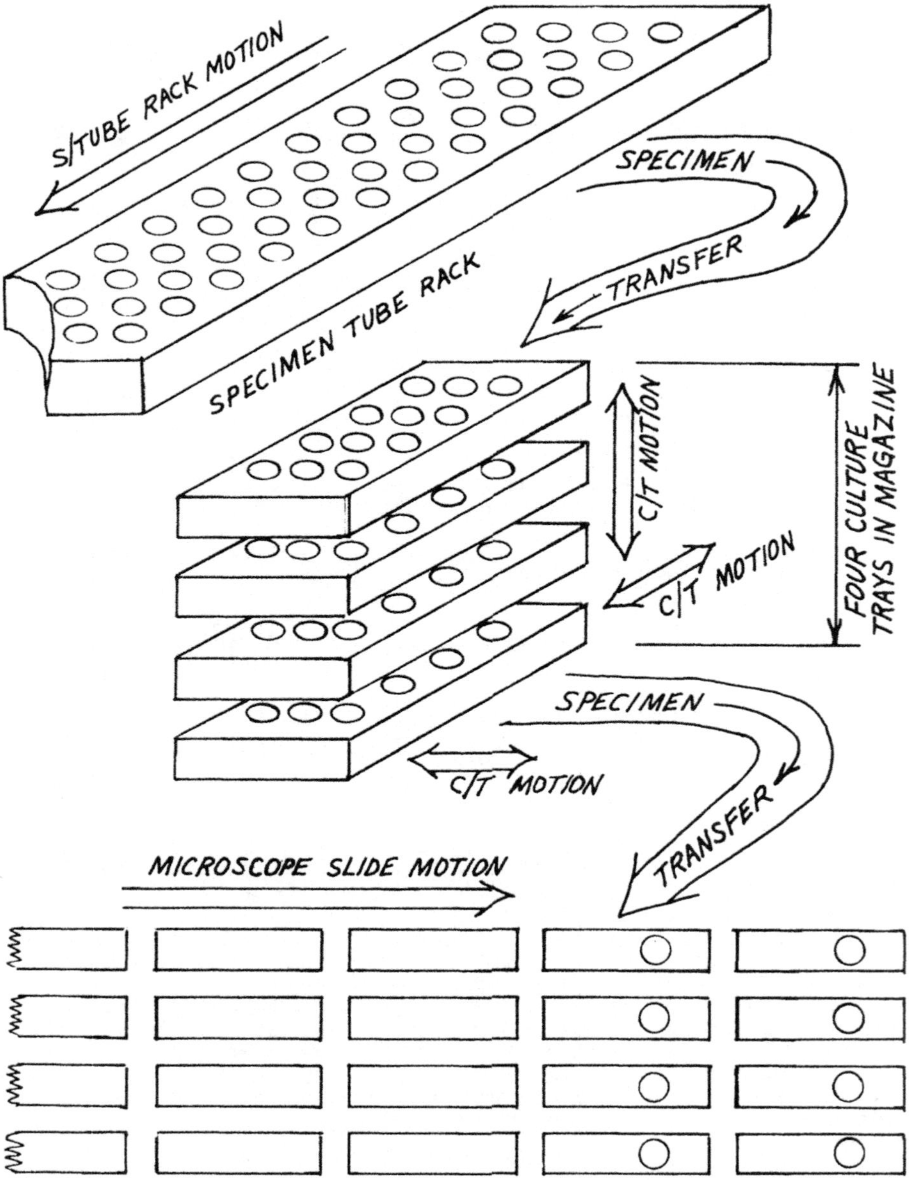

Fig. 1. Kinematic specimen and hardware actions

matrix, upon which the specimen tube rack is placed. The matrix contains a 4 x 12 array of LED/photocell pairs which detect the presence of a specimen tube when it is inserted in the rack. A simplified block diagram of the information flow is given in Fig. 2. The operator is prompted for most data items in a data table heading. This heading includes the batch no, three processing dates, and four culture tray bar codes.

Next, the operator is prompted for patient name, birth date, specimen draw date, and the bar code attached to the specimen tube. The specimen numbers are read by both teletype, and a hand-held optical scanner. Any discrepancy is detected by the computer. After reading each specimen number, the tubes must be placed in the rack in proper sequence. If this is not done, an alarm is sounded, and no further operation is allowed until the tube is properly placed. This process insures that each place in the tube rack has a corresponding place in the main body of the data table. Later, when the bar codes are machine read on each of the two microscope slides, these numbers are associated with the same location in the data table.

CONSOLES ONE AND TWO:

The specimen tube rack is then placed on console one, for the transfer of specimens to the culture trays. Here the bar codes on the trays are read automatically and the numbers read are placed in the data table heading. Later, in console two operation and in mode two operation of console one, the numbers read are checked against those already in the data table. Figure 2 does not show the console one, mode two check on culture tray numbers. The final step is mode two operation of console one. Here the microscope slide bar codes are read and placed with the appropriate patient's name in the data table, which is determined by precise control over the sequencing of specimen transfer actions. Data table four contains everything in data table three plus all slide numbers.

REFERENCES:

Keenan, R.L. and others (1980), A Computer-Controlled Semi-Automated System for Processing Human Blood to Produce Chromosome Slides. Biomed. Sci. Instrum., 16, 117-122.

This work was supported by a contract from the National Institute of Child Health and Human Development.

CURRENT ADDRESSES:

(1) Glendale Adventist Medical Center
1509 Wilson Terrace
Glendale, California 91206

(2) Perkin Elmer Co.
2771 N. Garey
Pomona, CA 91767

(2) Medical Genetics
11160 Warner Ave.
Fountain Valley, CA 92708

(4) T and B Ansley Corp.
4371 Valley Blve.
Los Angeles, CA 90032

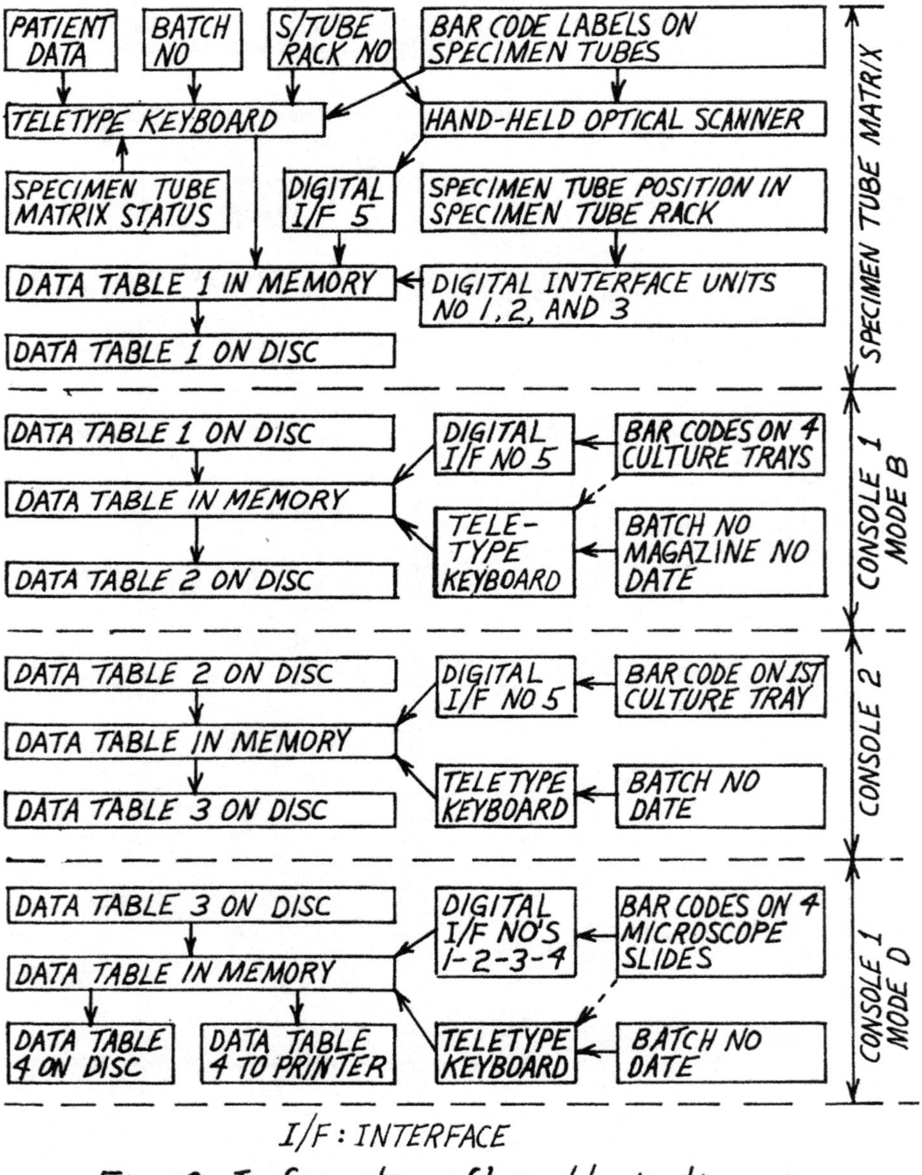

Fig. 2. Information flow block diagram

A MICROCOMPUTER SIMULATION OF THE ISOGRAVIMETRIC DOG LUNG

K. A. Gaar, Jr.

Dept. Physiol. and Biophys., L.S.U. Medical Center
Shreveport, Louisiana

ABSTRACT

An Apple II Plus microcomputer and a mathematical model were used to simulate the estimation of pulmonary capillary pressure with the isogravimetric technique. This model predicted that blood volume shifts into and out of the lung during isogravimetric maneuvers would dominate the weighing behavior of the system and make it impossible to obtain an accurate estimate of capillary pressure.

KEYWORDS

Pulmonary capillary pressure; pulmonary filling pressure; mean pulmonary pressure; pulmonary vascular resistance; dog lung.

Pappenheimer and Soto-Rivera (1948) introduced the isogravimetric technique for determining capillary pressure and arterial and venous resistances in the isolated hindlimb of the cat and dog. We have since used it for similar determinations with the isolated dog lung preparation (Gaar and co-workers, 1967).

There are several obstacles to obtaining an accurate estimate of the capillary pressure in the lung using the isogravimetric technique. First, the intravascular blood pressure must be adjusted in balancing the microvascular forces affecting transcapillary fluid movement to obtain an isogravimetric state, but the blood pressure cannot be changed without changing the lung blood volume which, by itself, causes lung weight to change. Second, lung weight change due to capillary filtration of fluid occurs more slowly and is perhaps hundreds of times smaller than the weight change due to blood volume shift following a change in the intravascular blood pressure. Thus, the isogravimetric state is mainly dependent on lung blood.

It is thought that the large initial weight change that occurs during an isogravimetric measurement is a vascular capacitive effect of blood that lasts for about a minute, and that any subsequent minor weight change is due to an imbalance of the microvascular forces at the capillary. Therefore, once the initial blood capacitive effect subsided and no further weight changes were observed during several minutes of monitoring, it could be assumed that the microvascular forces were in balance. But it has not been possible to verify this experimentally. Because of the greater potential of lung blood volume changes to override lung weight changes caused by capillary fluid shift, an accurate determination of capillary pressure in the lung may not be feasible. This computer simulation study was made to determine if variation in lung blood volume has any effect on the accuracy of estimates of capillary pressure using the isogravimetric technique.

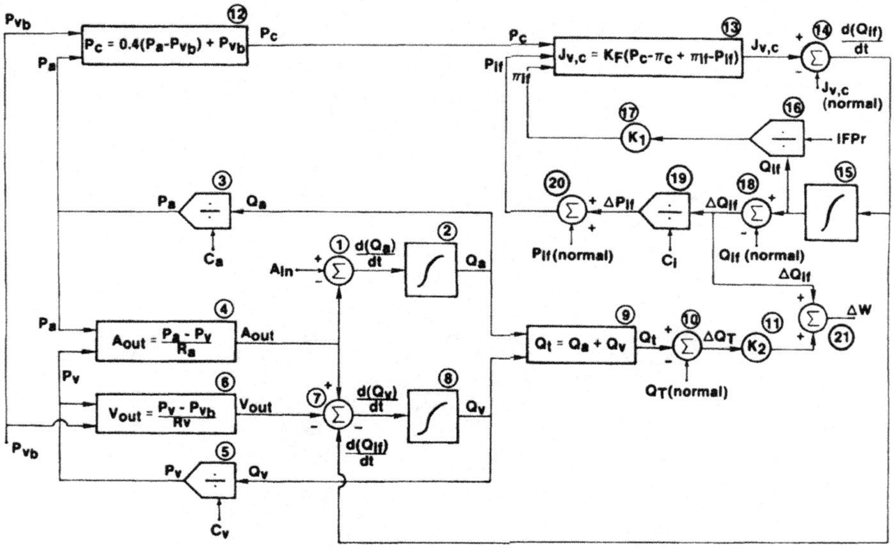

Fig. 1. Block diagram of isogravimetric lung model.

The model used for the simulation studies is shown in Fig. 1. Basically, it consists of a hemodynamic section (blocks 1-12) and a transcapillary fluid dynamics section (blocks 13-20). Block 21 joins these two segments to produce the total lung weight change. In the hemodynamic section blood pressures are represented by P with appropriate subscripts for arterial (a) and venous (v). Other symbols are: R, vascular resistance; Q, blood volume; C, vascular capacitance; Aout, arterial outflow; Vout, venous outflow; Ain, arterial inflow; Pv_b, venous back pressure. Block 13 represents the capillary fluid dynamics. Jv,c is volume flow of fluid through the capillary membrane, and K_F is the capillary membrane filtration coefficient. The Pi symbols represent osmotic forces for the capillary (c) and interstitial fluid (if). IFPr is interstitial fluid protein. The output of block 16 is IFPr concentration which is multiplied by the proportionality constant of block 17 and converted to osmotic pressure. Block 19 converts interstitial fluid volume changes to interstitial fluid pressure changes. Also, any change from normal in the interstitial fluid volume is summed with blood volume changes at block 21 to produce changes in the lung weight. For simplicity, volume and weight changes are considered to be synonomous.

The constant K2 in block 11 can be set to any value between 100% and 0% to vary the effect of blood volume change on total lung weight change. For example, by setting K2 near zero the effect of blood volume change can be dissociated from lung weight change caused by capillary fluid shift. Weight changes as little as 0.01 grams can be detected, irrespective of the K2 setting, which might be somewhat generous considering the practical weighing limitations of some isogravimetric systems.

A unique feature of the model is that a new and different value of capillary pressure is generated each time a new simulation is run. The value of capillary pressure is not revealed until the simulation ends. These "unknown" capillary pressures were determined under several different conditions. Thirty measurements were obtained using step-wise reductions of blood flow (Ain) as shown in Fig. 2(A). Following each step reduction, the venous back pressure, Pv_b, was adjusted until an isogravimetric state was obtained, as indicated by the mostly horizontal segments of the blood pressure tracings.

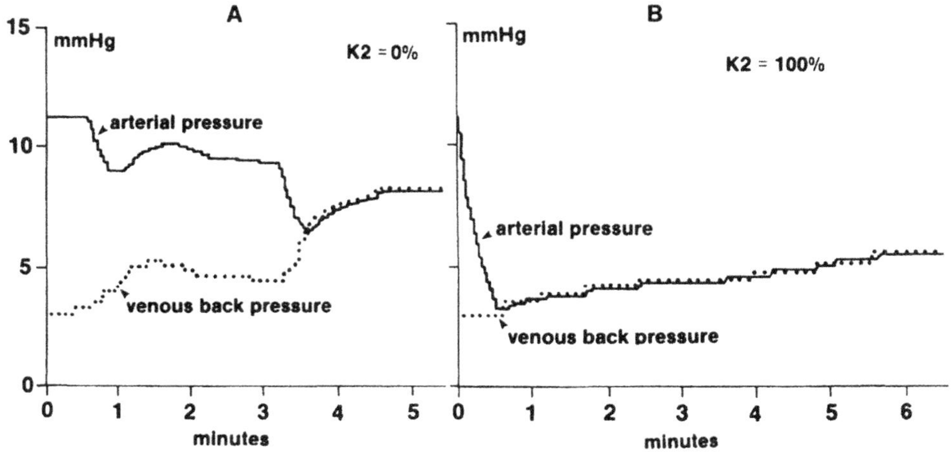

Fig. 2. Isogravimetric simulations using step-wise blood flow reduction (A) and immediate reduction of flow to zero (B).

In thirty additional simulations, instead of a step-wise reduction of blood flow, the flow was reduced all the way to zero at the very beginning. This is shown in Fig. 2(B). In both figures 2(A) and 2(B) the final blood pressures attained were considered to be the isogravimetric capillary pressures.

The last method used studied the effect of measurements made when the total lung weight was maintained constant throughout each simulation, with K2 set to 100%. Under these conditions there was an almost exact correspondence of the isogravimetric measurements to the value of the "mean pulmonary filling pressure", which was always 9.2 mmHg by design. Fourteen estimates were obtained with this method.

In contrast, when K2 was set to near 0% there was almost perfect correspondence every time between the true capillary pressure and the isogravimetric capillary pressure, regardless of the method used for measurement. This is because shifts of blood volume could not interfere with measurements, and the changes in lung weight due to capillary fluid shifts could be observed unimpeded. Thirty-three estimates were obtained and these were about equally divided between each of the three different methods of measurement previously described.

Figure 3 represents a composite summary of all the data. The overriding feature noted is that the isogravimetric estimates of capillary pressure have a tendency to be independent of the true capillary pressure because of the effects of blood volume shifting during the course of measurements. This effect is most pronounced when the total lung weight is not allowed to change during measurements, and always yields a value close to the "mean pulmonary filling pressure" as shown by the curve labeled A. Other means of making isogravimetric measurements yielded altogether different results. Curve B represents data obtained using a step-wise lowering of blood flow, but flow was lowered initially to zero to obtain the data represented by curve C. In either case, however, no attempt was made to keep the total lung weight exactly constant as it was for curve A. Finally, the solid line of identity corresponds to those estimates of capillary pressure that were obtained without any interference from the effect of blood volume changes. In summary, when K2 is set to 100% (curves A, B & C of Fig. 3) blood volume changes tend to dominate the weighing determinations and the isogravimetric estimates of capillary pressure do not exactly correspond to the true capillary pressure. Furthermore, an unexpected discrepancy in results was observed when different methods were used to make the measurements. In conclusion, this study suggests that reliable estimates of pulmonary capillary pressure may not be obtainable with the isogravimetric technique.

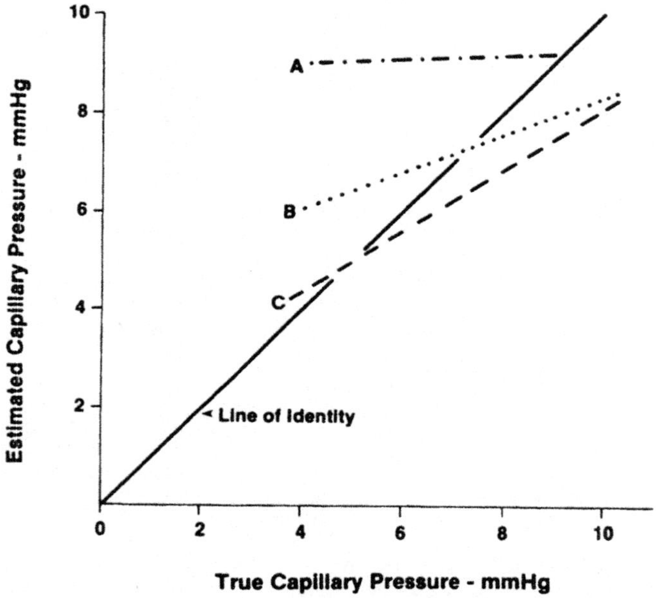

Fig. 3. Composite summary of results (explained in text).

REFERENCES

Gaar, K. A. Jr., A. E. Taylor, L. J. Owens, and A. C. Guyton (1967). Pulmonary capillary pressure and filtration coefficient in the isolated perfused lung. Am. J. Physiol., 213, 910-914.

Pappenheimer, J. R., and A. Soto-Rivera (1948). Effective osmotic pressure of the plasma proteins and other quantities associated with the capillary circulation in the hindlimb of cats and dogs. Am. J. Physiol., 152, 471-491.

SIMULATED MASS SPECTROMETER USING MICROPROCESSOR TECHNOLOGY AND A GAS CHROMATOGRAPH

James J. Hatcher, M.D., James P. Baker, M.D., and Peter A. Koen, Ph.D.

Dept. of Pulmonary Medicine, Eastern Virginia Medical School
Norfolk, Virginia

ABSTRACT

A sampling device which simulates a respiratory mass spectrometer (Begin, 1975; Hall, 1973; Sackner, 1975) using microprocessor technology and a gas chromatograph has been developed. The device has the advantages of being less expensive, readily adapatible for the measurement of multiple gases, and offers the ability to detect concentrations smaller then commerically available mass spectrometers.

KEYWORDS

medical mass spectrometer; gas chromatograph; microprocessor technology; respiratory mass spectrometer; gas analysis techniques.

INTRODUCTION

A sampling device which simulates a respiratory mass spectrometer has been developed. This device offers the advantages of being less expensive then a medical mass spectrometer and is readily adapatible to simultaneously measure multple gases. The device can also measure smaller concentrations then commerically available mass spectrometers. In this paper, we will discuss the device design and several experiments utilizing the device.

METHODS

EQUIPMENT

The device consists of three parts: the sampling port where samples are obtained, the controller, and a three way valve (see fig. 1). The sampling port is made of plexiglass with thirty-two ports each attached to a solenoid switch (Lee Co. miniture 3-way valve Model# LFAX) operated by the controller. Each port contains a vial to contain the gas samples. The controller consists of a Motorola MC6800 microprocessor, support circuity (Heathkit ET-3400 and ETA 3400) and appropriate logic designed to operate the solenoid switches

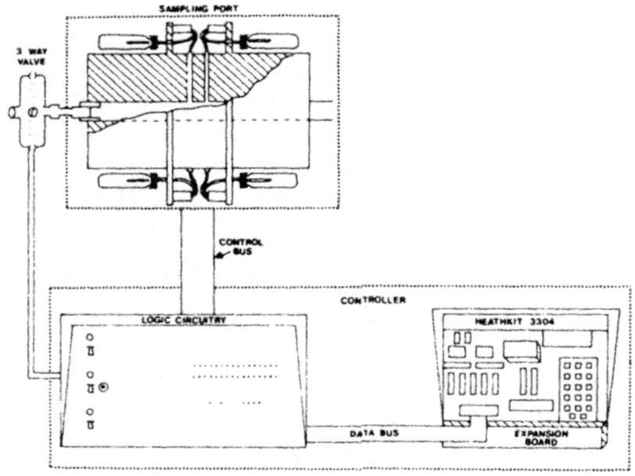

Figure 1

The Sampling Device

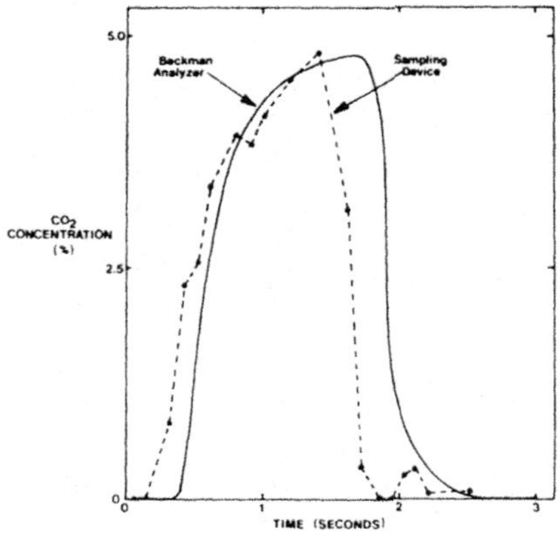

Figure 2

Simultaneous Curves for a Single Breath

of the sampling port. The logic circuity consists of addressing information for a perpherial interface adapater (PIA) which controls two four to sixteen demultiplexers. Inverters and transistors then operate the opening and closing of the solenoid switches. The thirty-two samples are obtained by sequentially opening each vial using the controller to determine the sampling interval and rate. The interval between samples and the time the solenoid switches remain open can be software controlled allowing different combinations of discrete time intervals to be selected. The third component, a three way valve (Fluorocarbon Model # AT3166A), operated by the controller allows the system to begin sampling at any desired time.

PROCEDURE

The procedure for gas analysis involves initial evaculation of the vials on the sampling port to a negative 200 mm of Hg using a vacuum pump. Circuity is available to open all solenoid switches to the vials simulanteously while the three-way valve closes the central channel of the sampling port. After the vials are evaculated, the solenoid switches are closed. (The vials are able to hold a vacuum for several weeks.) The system is now "primed" to collect samples for the event being studied. The interval between samples and the time the solenoid valves remain open (duty time) are software controlled allowing different combinations of discrete time intervals to be selected. The solenoid switches have a response time of six milliseconds which determines the theroetrical lower limit of the timing interval between samples. The events of interest in our work occur over a single respiratory cycle, which, depending on the event being studied, can be from three to thirty seconds. A timing interval of from 0.1 seconds to 1 second between samples is usually chosen which includes a duty time of 50 milliseconds.

After all the vials are evaculated the sequence of sample gathering is started. The controller opens the three-way valve allowing the flow of gas through the sampling port and begins the timed sequential collection of gas samples. The gas is then stored in the vial until it is measured by the gas chromatograph using standard gas chromatography techniques. The system is capable of measuring any number of gases limited only by the chromatographer's ability to choose separation columns and the instrument's detector being used. We have used this device for the simultaneous sampling of five gases. The gas chromatograph in use in our lab is a Varian Model 3700 equiped with an electron capture detector, a thermal conductivity detector, and a flame ionization detector.

RESULTS

The system has been shown to reproducibly measure multiple gases in equal concentrations from each of the thirty-two ports within a five percent experimental error. A partial list of the gases which have been studied include, sulfa-hexafloride, cyclopropane, acetlyene, ether, carbon-dioxide, and halothane. Gas concentrations as low as one hundred parts per billion have been detected.

A typical example showing the ability of the device to sample physiologic events is shown in fig. 2. This graph compares the simultaneous curves produced by a carbon-dioxide analyzer (Beckman

LB-2) and one plotted from measurements obtained by the device for a single breath. As can be seen, the curves follow almost exactly with the exception that the sampling device has a slightly faster analysis time. The carbon-dioxide curve obtained by the Beckman LB-2 analyser is somewhat shower then the sampling device as a result of smapling delay due to the time it takes the sample to traverse the inlet tube of the analyser.

DISCUSSION

The device has been shown to measure small gas concentrations and to concurrently follow physiologic curves, while giving the ability to detect multiple gases simultaneously. The device will allow the development of experimental protocols which use gases beyond the capability of most available respiratory mass spectrometers.

ACKNOWLEDGEMENT

The authors would like to express their gratitude to Ms. Debbie Bishop for her help with the preparation of this manuscript. This project was supported in part by NIH Biomedical Research Support Grant No. 2507 RR 05771.

REFERENCES

Begin, R., A. C. DeGraff, Jr., R. M. Gardner, L. G. Hall, R. P. Howard, R. L. Johnson, Jr., E. D. Michaelson, A. H. Morris, R. W. Olsen, and M. A. Sackner(1975). Test of a Rebreathing Technique for Measuring Cardiac Output, Diffusing Capacity and Pulmonary Tissue Volume. Reprint of a Paper delivered at the XV Meeting of the Scandanavian Society for Clinical Chemistry and Clincal Physiology: Aarhus, Denmark, June 25 to 27.

Hall, L. G., J. B. West(1973). An Improved Respiratory Mass Spectrometer. Abstract given at the Association for the Advancement of Medical Instrumentation, Washington Hilton Hotel, Washington, D.C., March 22.

Sackner, M. A., D. Greeneltch, S. E. Heiman, and N. Atkins(1975). American Review of Respiratory Disease. 111, 157-165.

AN ADAPTIVE COMPUTER-AIDED INDEPENDENT LIVING SYSTEM FOR SEVERELY HANDICAPPED

Stanley R. Cronk and Roy W. Schubert

Department of Biomedical Engineering
Louisiana Tech University
Ruston, Louisiana 71272

ABSTRACT

This paper describes the Biomedical Engineering Department's progress in developing an adaptive microprocessor-based independent living system at Louisiana Tech University. The results from a three week clinical trial are presented and discussed.

KEYWORDS

microprocessor; independent living; communication aid; general purpose interface; handicapped

INTRODUCTION

It has long been known that computers can be used to extend the capabilities of handicapped persons (COMPUTER, 1981). It was the goal of the Biomedical Engineering Department at Louisiana Tech University to develope a computer "orthosis" to interface severely handicapped individuals with their environment. A preliminary report was published in 1980 (Cronk and Schubert, 1980).

APPARATUS

A General Purpose Interface (GPI) to be used as a communication aid/independent living system has been developed principally with off-the-shelf hardware, consisting presently of a microprocessor-based microcomputer, a dual floppy disk drive, a matrix printer, a duophone unit for phone communication, and up to 256 BSR modules for on/off control of lights and appliances. The Biomedical Engineering department developed a custom-made printed circuit board for input/output which plugs directly into the computer's user port without requiring any modifications of the computer as well as the software to allow the user to control the computer's actions. The circuit board accepts control signals from both the user and the computer and implements the computer's signals to control the user's environment. The software allows the user to command the computer through use of either its keyboard or a switch provided to the user. The software presents the options available to the user by presenting individual menus, or lists of selections, each

with a different function. Functions presently available to the user include environmental control, telephone dialing, creation and storage of messages on a highlighted field at the bottom of the CRT screen, and utility functions to alter scanning speed, provide aural feedback for a moving cursor, and transmit messages from the screen to a printer for permanent storage. The menus are:
(1) SYS - This menu contains the names of all other menus which the user can access as well as certain utility commands such as WRITE to printer and ERASE messages.
(2) MODE - This menu allows the user to choose a selection technique.
(3) LETTERS - This menu presents letters and numbers for creation of messages and telephone numbers.
(4) WORDS - This menu presents a list of commonly used words which the user can select for use in messages.
(5) ENVIRONMENT - This menu allows the user to activate and deactivate electrical appliances and lights. The circuit board plugged into the user port sends the appropriate control signal via the house wiring to a commercially available BSR module to which the appliance is attached for the actual activation/deactivation process.
(6) PHONE - This menu allows the user to make telephone calls by creating a phone number using the LETTERS menu or by selecting a number from a user-modified list. Each time the user enters the PHONE menu he is given the options to SAVE a new phone number or DELETE a number currently stored on the phone list as well as to dial a number.

METHODS

A GPI was placed in one of the resident handicapped students' rooms for clinical testing. All selections made by the student and the time at which they were made were recorded on a floppy disk for later analysis. The results for the first three weeks of usage are summarized in Tables 1 and 2. Both tables contain the weekly totals of selections as well as the relative frequencies of selection for the three week period. Table 1 lists the user's selections of menu names (and therefore the numbers of times he moved from one menu to the other) while Table 2 presents the usage of command and communication selections within the menus. The "other" selection in Table 2 represents all selections not individually listed, including the commands which transmit messages to the printer and ERASE messages from or RECALL messages to the highlighted field at the bottom of the screen, among others.

RESULTS

Selection techniques

The subject, a Cx-level quadriplegic college sophmore, was offered both row/column scanning and direct selection and was given the ability to switch back and forth between the two selection techniques at will during program execution. From examination of the raw data listings it was determined that the student, who has well-developed neck movement and is very familiar with a standard typewriter, used direct selection for about 98% of all selections, and seemed to try row/column scanning only out of curiosity.

Communication menu usage

The student had no real need to use the communication menus LETTERS and WORDS because he is able to communicate orally. The exception was using LETTERS to add new numbers to the PHONE menu. Interestingly though, during the first week only he used LETTERS quite frequently to create messages which he would then send to a printer for hard copy.

Table 1 Results of Three Week Clinical Trial of Prototype GPI - Menu Name Selections

Selection	Week 1	Week 2	Week 3	Relative Freq. (%)
SYS	98	75	50	46
MODE	1	3	3	1
ENVIRONMENT	29	28	26	17
PHONE	56	27	25	23
LETTERS	22	9	3	7
WORDS	16	7	2	5
Weekly totals:	222	149	109	99

Table 2 Results of Three Week Clinical Trial of Prototype GPI - Command and Communication Selections

Selection	Week 1	Week 2	Week 3	Relative Freq. (%)
letters selected	-	90	81	22
words selected	2	0	0	1
phone calls	15	19	28	8
SAVE/DELETE (PHONE)	39	7	10	7
devices on/off	112	62	7	32
other	141	53	41	30
Weekly totals:	309	231	235	99

Environmental control usage

Approximately 32% of the student's command selections were to control electrical appliances. At the time of the trial, the student wished to control only two devices, his television set and his stereo, though he has on/off capability for up to 256 electrical appliances and lights. The student has expressed great satisfaction with the environmental control routine.

The phone usage, accounting for 9% of the command selections, is continuously increasing. It is unknown whether the student is still adjusting to the system or whether he is used to only infrequently making telephone calls because of previous reliance on an aid for dialing phone numbers.

It is significant that the student now feels comfortable with the phone number SAVE and phone number DELETE commands. In early preliminary testing, the subject had difficulty following the instructions for each step printed on the CRT. The programming was then modified and the instructions amplified for clarity before the latest clinical trial.

Menu Switching

Somewhat disturbingly, about 38% of all selections are menu names; that is, the student simply jumped from one menu to another. The surprisingly high number of menu name selections downgrades the efficiency of the system. This may not be too serious, as the user has not complained about having to switch back and forth between menus. This does suggest, however, that a well designed system could improve the rate of information transfer between the client and his surroundings.

Useful modifications to the system were made as a result of user comments and requests. In early clinical testing the user was bothered by brightness of the CRT screen when the lights in his room were turned out. The program was modified to blank out the screen, except for a small bar indicating that the system is in its "wait mode", when no selections are made for a period of about 40 seconds. Another request leading to a program modification was that the "noise-making" appliances, such as stereo and TV, be automatically turned off whenever the user enters the PHONE menu to make telephone calls. During the present clinical trial, he also requested the ability to edit more than the two lines of text he is allowed so that he may write full-page letters to his family and friends. A commercially marketed word processing software accessible from the GPI should provide the user with the desired editing capability.

CONCLUSIONS

(1) The system in its present state is a very functional independent living system.
(2) The present method of storing selection data will prove instrumental in analyzing future system changes.
(3) Each system should be provided with sufficient secondary memory storage to allow the user to use other desired software in addition to the GPI. The floppy disk drive seems the only reasonable solution presently available.
(4) The user should be provided with the ability to create an additional menu which will contain a mixture of phone numbers, environmental control functions, and any utility commands the user wishes to include. This menu should eliminate much of the menu switching seen thus far in the clinical trial, but it is not yet clear whether the users' needs can be met with a single menu.
(5) Each user should have a custom-made system created from commercially available hardware and custom software to meet his own needs. If each individual who desires a GPI is tested to determine his abilities and needs, appropriate menus and selection techniques could be drawn from a pool of in-house routines so that each system would be optimized.

FUTURE PLANS

Other changes in the system are viewed as important to take full advantage of the computer as well as advances in technology. They are:
(1) The user should be given the ability to alter any of the menus by adding or deleting selections just as in the PHONE menu, so that the system will adapt to the user rather than the reverse which is commonly the case.
(2) The user should be given a new menu containing timed messages and environmental control functions. The computer's internal clock can be programmed to constantly compare a list of times and actions against its clock and act accordingly when the programmed time is reached.
(3) The program should be divided into smaller segments which can be individually accessed from the disk. This will permit less expensive computers with smaller amounts of memory to be used.
(4) The system should include an RS-232 communication link. Computer networks such as the Source and Compuserve are becoming increasingly important in information transfer, and handicapped persons should be given equal opportunity to access these networks as a step toward gainful employment.
(5) Each user should be provided with a radio-linked keyboard and/or switch mounted on his wheelchair, allowing him to interact with the GPI even when not in direct visual contact with the system.

ACKNOWLEDGMENTS

This work was supported by a grant from the Division of Vocational Rehabilitation, State of Louisiana.

REFERENCES

Cronk, S.R., and Schubert, R.W. (1980). A general purpose interface (GPI) for independent living. 30th Alliance Conference on Engineering in Medicine and Biology, 209.

(1981). Computer, 14, 19-61.

Session 16
BONE MECHANICS

FRACTURE MECHANICS OF CORTICAL BONE - EVALUATION BY THE COMPACT TENSION METHOD

W. Bonfield and J.C. Behiri

Department of Materials, Queen Mary College,
London, E1 4NS, England.

ABSTRACT

The fracture mechanics parameters of critical stress intensity factor (K_c) and critical strain energy release rate (G_c) are determined for longitudinal fracture of bovine tibia cortical bone by the compact tension method. It is demonstrated that, for a given bone density of 1.96 - 1.98 Mgm^{-3}, K_c and G_c depend on the loading rate, and resultant crack velocity, with a maximum in fracture toughness at a crack velocity of $\sim 1.0 \times 10^{-3}$ ms^{-1}.

INTRODUCTION

As cortical bone is a brittle, and hence notch sensitive solid, the presence of surface cracks significantly reduces the energy absorbed during fracture (Bonfield and Li, 1966). Consequently, an evaluation of fracture toughness provides a more appropriate measure of fracture than a determination of a critical stress such as the ultimate tensile strength and, as a result, a variety of fracture mechanics techniques have been utilized in the past decade to assess the fracture toughness of bone specimens with a "characterized" crack, which have included measurements of the critical strain energy release rate (G_c) (or the specific surface energy γ, $= G_c/2$), and the critical stress intensity factor, K_c (Melvin and Evans, 1973; Margel-Robertson, 1973; Bonfield and Datta, 1974, 1976). The specimen geometries used in the earlier experiments in this series all produced rapid crack propagation, with an unknown and variable crack velocity. More recently, in contrast, with the use of the compact tension method (Wright and Hayes, 1977), it has been possible to propagate a crack in bone at a relatively slow and measurable rate and G_c and K_c values for transversely oriented bovine femur and tibia bone specimens were determined by this method for various crack velocities (Bonfield, Grynpas and Young, 1978; Behiri and Bonfield, 1980). The results of these investigations are shown for comparison in Table 1. The general assumptions involved in the application of fracture mechanics concepts to the fracture of bone have been critically reviewed (Bonfield, 1981), with particular reference to the physical implications of the single edge notched, centre notched cylindrical and compact tension test methods. It was suggested that the compact tension method has two particular advantages, namely, first, that G_c can be determined without any prior assumption as to the appropriate Young's modulus (E) and Poisson's ratio (ν) values, and, second, that the imposed rate of deformation, or more fundamentally, the resultant crack velocity, can be varied in a controlled manner.

TABLE 1 Some Previous Investigations of the Fracture Mechanics Parameters for Bone

Investigation	Method	Orientation	Bone type	G_C (Jm^{-2})	K_C ($MNm^{-3/2}$)
Melvin and Evans (1973)	Single-edge notched	Transverse longitudinal	Bovine femur	1388-2557	3.21
		Longitudinal	Bovine femur	3135-5534	5.58
Margel-Robertson (1973)	Three-point bending	Longitudinal	Bovine femur	-	6.56
Bonfield and Datta (1974)	Centre notched (shock tube)	Transverse	Bovine tibia	4	0.23
Bonfield and Datta	Single-edge notched	Longitudinal	Bovine tibia	780-1120	2.2 - 4.6
Wright and Hayes (1977)	Compact tension	Transverse	Bovine femur	820	3.5
Bonfield, Grynpas and Young (1978)	Compact tension	Transverse	Bovine femur	920-2780	2.4 - 5.2
Behiri and Bonfield (1980)	Compact tension	Transverse	Bovine tibia	1736-2796	4.46 - 5.38

In the current investigation, the compact tension specimen approach has been extended to evaluate the fracture of transversely oriented (i.e. with fracture in the longitudinal direction) bovine tibia compact bone specimens, in terms of G_C and K_C, for a range of crack velocities from $\sim 10^{-5}$ to 10^2 m sec^{-1}. As density has an important effect on the fracture mechanics parameters (Wright and Hayes, 1977), the test specimens were grouped within four narrow bands of density, namely 1.92-1.94, 1.94-1.96, 1.96-1.98 and 1.98-Mgm^{-3}. The K_C results obtained for the 1.96-1.98 Mgm^{-3} density group were reported recently (Behiri and Bonfield, 1982), and are extended here, together with new data for G_C.

EXPERIMENTAL PROCEDURE

Details of the compact tension test procedure have been reported previously (Bonfield, Grynpas and Young, 1978). The critical stress intensity factor (K_C) and the critical strain energy release rate (G_C) are defined as follows:-

$$K_C = \frac{P}{B\,w^{\frac{1}{2}}} Y_2 \tag{1}$$

and

$$G_C = \frac{P^2}{2B} \frac{\delta C}{\delta a} \tag{2}$$

where P is the load associated with a crack length a,
B is the specimen thickness,
w is the specimen width,
C is the experimentally determined specimen compliance for any given crack length, a,

and Y_2 is a geometrical factor, given by:-

$$Y_2 = 29.6 \frac{a}{w}^{1/2} - 185.5 \frac{a}{w}^{3/2} + 655.7 \frac{a}{w}^{5/2} - 1017 \frac{a}{w}^{7/2} + 638.9 \frac{a}{w}^{9/2} \quad (3)$$

Transversely oriented compact tension specimens (i.e. with the starter crack introduced parallel to the long axis of the bone) were wet machined from the mid-diaphysis of adult bovine tibiae and were stored in Ringers solution at -20°C until tested (Bonfield, Grynpas and Young, 1978; Behiri and Bonfield, 1980). The test specimens, saturated with Ringers solution, were mounted in an Instron testing machine and subsequently loaded at room temperature at cross-head speeds ranging from 1.7 to 330 x 10^{-7} m sec^{-1}.

As the crack propagated across the test specimen, the progressive decrease of load (P), was monitored on the Instron chart recorder, while the extension of the crack length (a) was determined either with a microscope (for "slow" crack velocities) or with a high speed camera.

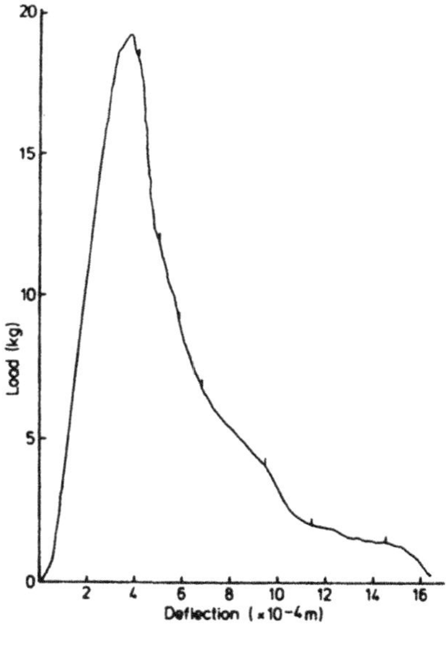

Fig. 1 A typical load-deflection recording showing the progressive decrease in load (P) as the crack length increased. The vertical markers on the curve correspond to the times at which the crack front reached known distance markers on the specimen.

It should be emphasized that K_c and G_c can be obtained in two ways, as follows:-

(1) By measuring the single value of P at which the machined crack length a starts to propagate. Following the ASTM procedure, P is not strictly the maximum value, but is taken as the value corresponding to the intersection of a secant line, reduced by 4% from the elastic slope, with the experimental curve. Hence, by substitution in equations (1), (2) and (3), this procedure gives a single value of K_c and G_c.

(2) By measuring a series of values of P corresponding to various crack lengths (a) produced as the crack propagates across the specimen. Then, by substitution in equations (1) - (3), values of K_c and G_c can be derived for any P - a combination. This procedure allows the validity of the approach to be demonstrated, for if K_c and G_c are unique materials parameters, then, for a given loading rate or crack velocity, each P - a combination should yield similar values of K_c and G_c. Such a result, i.e. that approximately constant values of K_c and G_c are obtained for different P - a combinations, has been demonstrated for values of $a/w < 0.7$ (Bonfield, Grynpas and Young, 1978; Behiri and Bonfield, 1980). In addition both methods (1) and (2) yield similar results for K_c and G_c (Bonfield, Grynpas and Young, 1978; Behiri and Bonfield, 1980), which implies that the radius of curvature of the "actual" starter crack is comparable to that of the propagating crack (Bonfield and Datta, 1974, 1976).

RESULTS AND DISCUSSION

An increase in the cross head speed from 1.7×10^{-7} to 170×10^{-7} ms^{-1} produced an almost proportional increase in the resultant velocity of the propagating crack from $\sim 2 \times 10^{-5}$ to $\sim 1.0 \times 10^{-3}$ ms^{-1}. The increase in cross head speed and, hence, in crack velocity produced a significant increase in both K_c and G_c (derived as average values from the methods (1) and (2) outlined previously) as shown in Fig. 2.

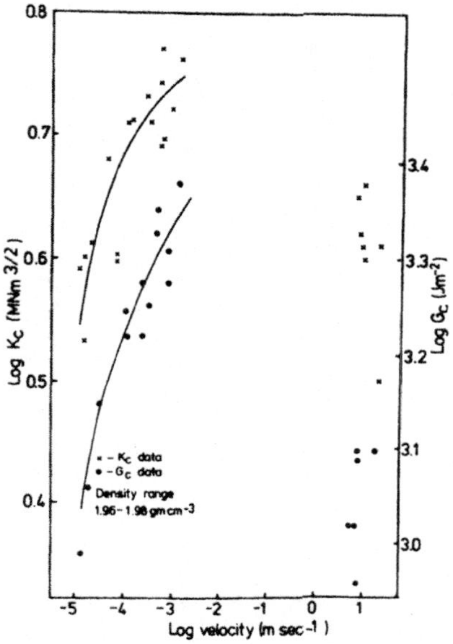

Fig. 2. The transition in K_c and G_c with increasing crack velocity for transversely oriented bovine tibia sections.

However, if the cross head speed was increased to the next available increment (330×10^{-7} ms^{-1}), then a considerably increased crack velocity was produced

($\sim 10^0 - 10^2$ ms^{-1}), which resulted in a decrease in both K_c and G_c as is also demonstrated in Fig. 2.

The transition from relatively slow, controlled crack propagation to rapid, catastrophic crack propagation was also accompanied by a change in the fracture characteristics (Behiri and Bonfield, 1980), with a transition from the relatively rough fracture surfaces produced by the passage of the crack around intersecting osteons or lamellae to relatively smooth fracture surfaces.

The results demonstrate that, for a given bone density, the values of K_c and G_c depend on the cross head speed utilised in the test and, more fundamentally, on the velocity of the propagating crack, with a maximum in fracture toughness occurring at a relatively low loading rate and crack velocity.

ACKNOWLEDGEMENT

The continuing support of the Science and Engineering Research Council for the programme of study on the deformation and fracture of bone is gratefully acknowledged.

REFERENCES

Behiri, J. C., and W. Bonfield (1980). Crack velocity dependence of longitudinal fracture in bone. J. Mater. Sci., 15, 1841-1849.
Behiri, J. C., and W. Bonfield (1982). Fracture mechanics of cortical bone. Third Meeting of the European Society of Biomechanics, Nijmegen, The Netherlands.
Bonfield, W. (1981). Mechanisms of fracture in bone, in Mechanical properties of bone. Ed. S.C. Cowin, AMD Vol. 45, ASME (New York).
Bonfield, W., and P.K. Datta (1974). Impact fracture of compact bone in a shock tube. J. Mater. Sci., 9, 1609-1614.
Bonfield, W., and P.K. Datta (1976). Fracture toughness of compact bone. J. Biomechanics, 9, 131-134.
Bonfield, W., M.D. Grynpas, and R.J. Young (1978). Crack velocity and the fracture of bone. J. Biomechanics, 11, 473-479.
Bonfield, W., and C. H. Li (1966). Deformation and fracture of bone. J. Appl. Phys., 37, 869-875.
Margel-Robertson, D. (1973). Ph. D. Thesis, University of Stanford, CA.
Melvin, J. W., and F. G. Evans (1973). ASME Biomaterials Symposium, Detroit, MI.
Wright, T. M., and W. C. Hayes (1977). Fracture mechanics parameters for compact bone - effects of density and specimen thickness. J. Biomechanics, 10, 419-430.

BENDING STRENGTH AND ASH CONTENT OF BONES DURING EMBRYONIC GROWTH

S. Saha, A. Pal, S. Pal and J.A. Albright
Biomechanics Laboratory
Department of Orthopaedic Surgery
L. S. U. Medical Center
Shreveport, Louisiana 71130

ABSTRACT

Mechanical properties and mineral content of embryonic bone obtained from growing chick embryos were measured. The weight of the embryo, bending strength and ash content of the tibia increased linearly with the age of the embryo. A significant positive correlation between the bending strength and the mineral content of tibia was observed.

KEYWORDS

Bending strength; chick embryo; ash content; long bone.

INTRODUCTION

Mechanical properties of bone as a material have been studied in great detail by various authors. However little is known regarding the in-vivo properties of bone during its formation stage. Although it is well documented that the mechanical strength of both compact and cancellous bones from adults are highly correlated to mineral content (Evans, 1973), no similar study has been done with embryonic bone. In this study, we used the model system of growing chick embryos to determine the mechanical properties of embryonic bone. Both the ash content and the strength increased with increasing age of the embryo. We also found a significant positive correlation between the bending strength of these bones and their mineral content.

MATERIALS AND METHODS

Fertilized chicken eggs of high breed quality white leghorn of Indian River breed were collected from a local hatchery. Each egg was weighed and numbered and subsequently incubated in a wooden incubator (Leahy Mfg. Co., Model 416E). The temperature and the humidity of the incubator were maintained constant at 37° and 80% respectively. The eggs were placed in paper crates approximately 30° to the vertical with their blunt face on top. They were turned back and forth three times daily. In this study, four batches of eggs, each batch containing 60 eggs, were used. In each batch several embryos were sacrificed on alternate days beginning from the 12th day of incubation. After removal from the egg shells the embryos were numbered, weighed, and their length measured. They were then fixed in alcohol for 48 hours and their bones were stained with Alizarin red S. The embryos were then fixed in formalin (10%) until used. The long bones were subsequently excised and their length was measured under a magnifying glass with an accuracy of 0.5 mm. The tibiae and femora were dried for 24 hours at 40°C in a vacuum oven (Isotemp) and the dry weights were determined using an electrical microbalance (Sartorius model 2434).

The long bones were too short to be tested in tension mainly due to the difficulty in gripping the specimens. Initially we attempted to test the bones fresh under three point bending, but the failure loads were too small for accurate measurement of the load cell output. Therefore the bone samples were dried in a vacuum oven and then tested mechanically. A special fixture for supporting and loading the sample was fabricated. The supporting end was fixed to a two component dynamometer (KIAG-9271A) which was then installed on the loading piston of an Instron Servohydraulic material testing machine (model 1321). The loading blade was fixed to a pneumatic grip of the Instron as shown in Fig. 1. The load was monitored by the dynamometer, the output of which was amplified by a matched charge amplifier and fed to a strip chart recorder (Cole Parmer Inc.). The deformation rate for each test was maintained constant at 0.5 mm/sec by the servohydraulic system of the Instron testing machine. The loading system was checked for calibration with dead weight after each batch of experiments.

(a) (b)

Figure 1. Experimental set-up for bending test of the embryonic bone. (a) the specimen held on a jig on top of the load cell; (b) enlarged view of the specimen and the loading configuration.

To determine the percentage of mineral content, the dried bones were heated in a muffle furnace (Lindberg model 51894) at 580°C for 24 hours and the ash weight was measured.

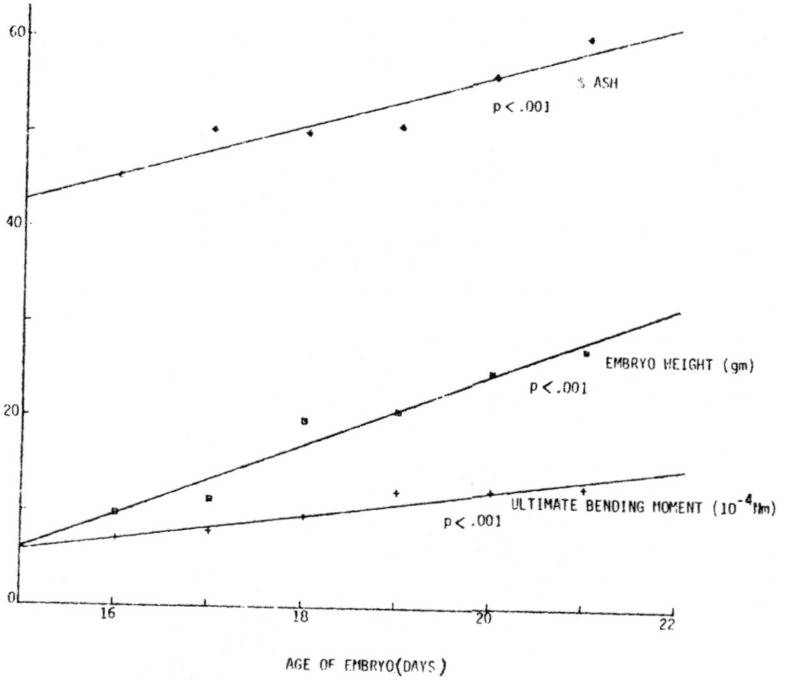

Figure 2. The mean embryo weight (gm), ultimate bending moment of tibia and the ash content of tibia as a function of the age of the incubating eggs.

RESULTS

Figure 2 shows graphically the linear relationships between all the growth parameters measured, and the age of the embryo. Similar increase in the embryo weight has also been reported by Romanoff (1967). As shown in Fig. 2, the embryo weight increased at an average rate of 15% per day for this growth period compared to this, the bending moment increased at an average rate of 17.6% per day while the ash content increased at a much lower rate of 6% per day. Table 1 shows that the increasing weight of the embryo also accompanies a similar enlargement in the size and weight of the tibia.

Table 1

Incubation period (days)	16	17	18	19	20	21
Tibia length (mm)	14.16±0.76	16.80±1.12	18.75±1.15	19.25±.54	21.75±2.79	22.30±1.56
Tibia dry weight(mgm)	8.01±2.32	8.90±2.23	17.76±3.32	19.09±3.19	22.37±6.91	28.15±2.52

DISCUSSION

We have used this model system of chick embryo development to quantatively determine the effect of electromagnetic stimulation on the skeletal growth of such embryos (Saha, Pal and Albright, 1981, 1982). This model system can also be used to evaluate the effect of other external stimulations or drugs on the strength or calcification of bones during embryonic development.

ACKNOWLEDGEMENT

This research was supported by a USPHS Research Grant No. 5 R01 26020 and funds from Shriners Hospital for Crippled Children and by a RCDA Award No. 5 K04 AM 00756 and from USPHS to Dr. Saha.

REFERENCES

Evans, F.G. (1973) Mechanical Properties of Bone. Charles C. Thomas, Springfield, Ill.

Romanoff, A.L. (1967) Biochemistry of the Avian Embryo. Interscience Publishers, New York, pp. 5.

Saha, S., Pal, A., Albright, J.A. (1981) Proc. 9th Northeast Bioeng. Conf., Pergamon Press, pp. 351-354.

Saha, S., Pal, A., Albright, J.A., and Misra, R.P. (1982) Trans. 28th Ann. Mtg., Orthop. Res. Soc., 7, pp. 29.

Frequency Dependence of Ultrasonic Characteristics of Cancellous Bone

S. Pal, S. Saha and G. N. Reddy
Biomechanics Laboratory, Dept. of Orthopaedic Surgery
L.S.U. Medical Center
Shreveport, Louisiana 71130

ABSTRACT

Ultrasonic properties of cancellous bone, namely the velocity and attentuation coefficient, were measured. The velocity increased slightly with increasing frequency. The compressive strength showed significant positive correlation with wet density and mineral content.

KEYWORDS

Cancellous bone, ultrasonic velocity, attentuation coefficient, frequency dependence, compressive strength, density, ash content, elastic properties.

INTRODUCTION

Compared to soft tissues, ultrasonic properties of bone have remained unexplored. Ultrasonic technique has been used to study the elastic properties of compact bone (Lang, 1970; Yoon and Katz, 1976), but almost no information is available on the ultrasonic properties of cancellous bone. This prompted us to study the ultrasonic behavior of cancellous bone. Previously we have shown the anisotropic nature of the ultrasonic characteristics of cancellous bone (Saha, Pal and Rao, 1981). In this study we investigated the frequency dependence of the ultrasonic properties of cancellous bone. The attenuation coefficient, an important acoustical parameter, was also measured. Attempts were also made to correlate the ultrasonic properties with the compressive strength and modulus of elasticity as measured by mechanical testing. These data on the acoustical and mechanical properties of cancellous bone are needed for diagnostic use of ultrasound as well for mathematical modelling of various artificial human joint replacements where spongy bone structure forms the interface with bone cement or prosthesis.

MATERIALS AND METHODS

A fresh tibia was obtained from the lower limb amputation of a 52 year old black male. The proximal end of the cleaned tibia was sliced into approximately 9 mm thick layers using a band saw. Then individual samples were obtained by using a sharp punch to cut out 12.7 mm diameter cylindrical specimens. Then dimensions of each sample were measured with an accuracy of 0.01 mm. Immediately afterwards the wet weight of each sample was measured using a microbalance (Sartorius - Model

2434) and the wet density was calculated. The samples were kept moist with normal saline and subsequently kept frozen at -20°C until used for further test.

The ultrasonic properties were measured as describerd below.

Velocity measurements: In this study we have used a single frequency gated pulse to measure the ultrasonic velocity in test samples. Such type of excitations is expected to avoid errors due to velocity dispersion with frequency (Bhagat et al, 1980). The gated pulse was generated using a pulser (Tetronix, PG505) and a gated function generator (Tektronix PG 502) as shown in Figure 1.

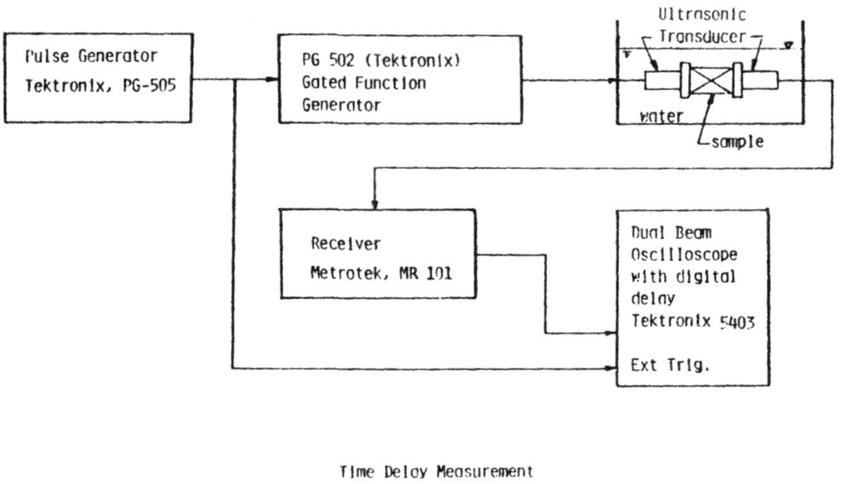

Figure 1. Experimental set-up for ultrasonic velocity measurement

The ultrasonic transducers (Dapco and Aerotech) were excited near their resonance frequency. The signal received after transmission through bone samples was amplified by the receiver (Metrotek, MR101). The time delay was measured accurately with a dual beam oscilloscope with digital time delay to an accuracy of 1 nanosecond (ns). The velocity of ultrasound in cancellous bone samples were then calculated from the measured length (accuracy of .01 mm) of the specimens and corresponding time delay.

Measurement of attenuation coefficient:

The experimental set-up used for the measurement of attenuation coefficient is similar to one used for velocity measurements except that a storage oscilloscope (Nicolet explorer III) was used to store the reference signal and a spectrum analyzer was used to obtain signal spectrum as shown in Fig. 2 (using Metrotek MG701) to avoid the initial electrical noise. The gated pulse was stored as a 1024 point trace on one quarter of a track. A second reading was then obtained and stored with the sample. Now the reference signal spectrum (without sample) and the signal spectrum (with sample) were obtained using the spectrum analyzer. The two spectrums were displayed on A and B channels of the analyzer. The attenuation at a frequency was then obtained by noting the corresponding points on reference spectrum and the signal spectrum. If x volts is the reading on reference spectrum and y volts is the reading on the signal spectrum at a frequency (f) and with sample thickness of t, then the attenuation is given by α and coeff. by α^*:

$$\alpha = 20\ (\log_{10}x - \log_{10}y)/t \quad \text{db/cm} \qquad (1)$$

$$\alpha^* = 20\ (\log_{10}x - \log_{10}y)/t.f \quad \text{db/cm/MH}_z \qquad (2)$$

After measuring the ultrasonic properties, the samples were tested in compression using an Instron servohydraulic material testing machine (Model 1321). The load deformation diagram obtained from the recorder of the Instron was used to calculate ultimate strength and static modulus of elasticity. Subsequently the samples were washed in running water, then cleaned of all marrow and fat content by filling their individual container with ethyl alcohol and subjecting them to vibration in an ultrasound cleaner for two hours. This procedure was repeated three times to get the clean cancellous bone matrix. The samples were then dried at 100°C for 24 hours in a vacuum oven and their dry weights determined. The samples were then burned to ash in a muffle furnace for 24 hours to determine mineral content.

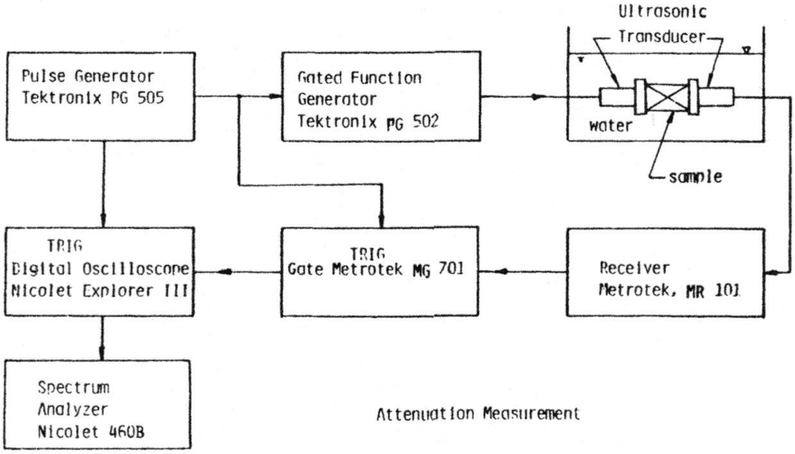

Figure 2. Set-up for ultrasonic attenuation measurement.

RESULTS AND DISCUSSION

There was a small but measurable increase in the velocity of ultrasound in human cancellous bone with increasing frequency. The mean velocity increase ranged from 2.5 to 3.8% for increase of frequency from 0.3MHz to 4.0MHz as shown in Fig. 3. The least square parabolic fit from the mean velocity gives a correlation coefficient (r) of 0.99 whereas straight line fit gives r=0.98 at $p<.001$. This agreed with similar findings for soft tissues (Bhagat, Kadaba and co-workers, 1980).

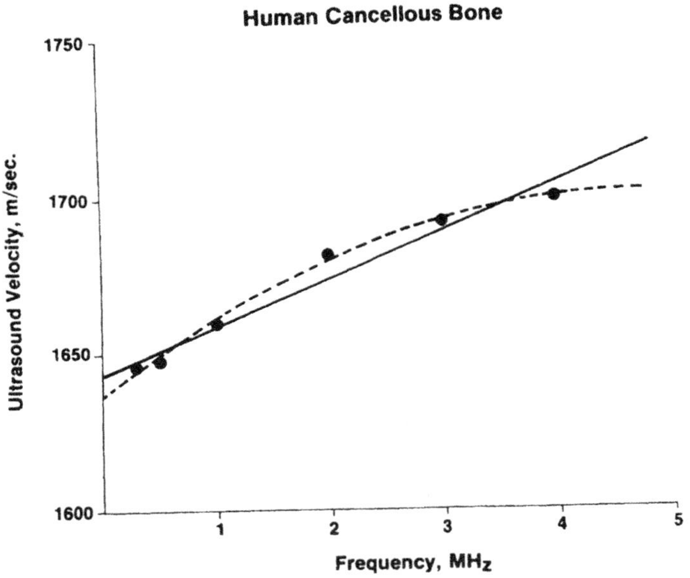

Figure 3. Mean velocity of ultrasound in cancellous bone vs. the frequency of gated ultrasonic pulse. The best fit parabola ($r=.99, p<.001$) and straight line ($r=.98$ $p<.001$) are shown. Each point represents mean for 16 points.

The mean attentuation coefficients for the cancellous bone specimens tested were 13.3 (range 7.7 to 24.2) db/cm/MHz measured at a frequency of 2.25 MHz. The attenuation coefficient showed a significant positive correlation with the wet density. This relationship is shown in Fig. 4.

The ultimate strength and elastic modulii showed significant positive correlation with wet density and mineral content of bone. This was in agreement with the similar findings for shear strength of cancellous bone (Saha and Gorman, 1979). The ultrasonic velocity did not show any correlation with mineral content. This finding, though surprising, agreed with the results observed by Lees and Davison, (1979).

Table 1 shows the mean values of the properties of the cancellous bone samples tested. However, the compressive strength and modulus of elasticity are somewhat less than the other published data indicating the osteoporotic nature of these specimens.

Table 1: Mean (1 Std) Values of Various Properties of Cancellous Bone (n=17)

Density Kg/m^3 $x10^3$	Mod. Elas. MN/m^2	Ult. Comp. Stg. MN/m^2	Vel. of Ultrasound m/sec ($1MH_z$)
0.964(032)	39 (6.25)	1.28 (.90)	1660(43)

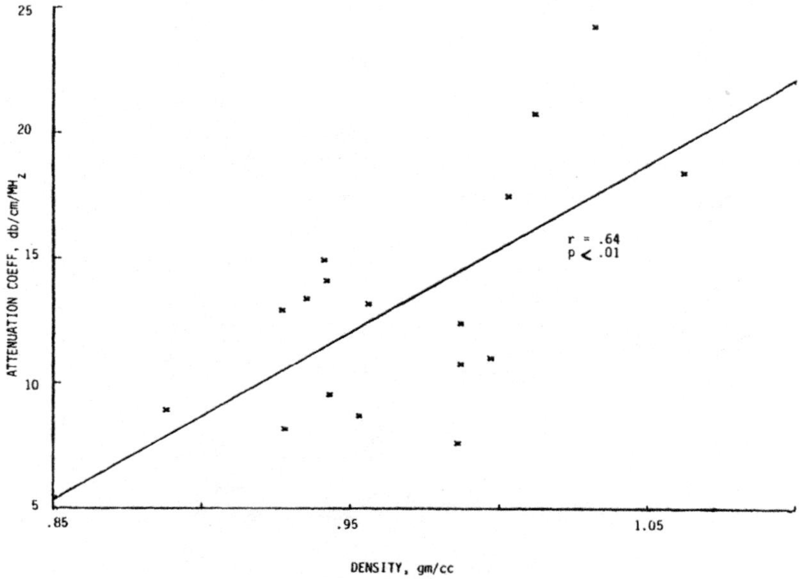

Figure 4. Mean attenuation coefficients of ultrasound (frequency 2.25 MH_z) in ultrasound (frequency 2.25 MH_z) in human cancellous bone vs wet density. Each point represents mean for 16 points.

REFERENCES

Bhagat, P.K., Kadaba, M.P., Gupta, V.N. and Wu, V. (1980) Med. Instrumentation 14: 220-224.

Lang, S.B. (1970) IEEE Trans. Bio-Med. Engg. 17: 101-105.

Lees, S. and Davison, C.L. (1979) Ultrasonic Tissue Characterization II, NBS. Spec. Pub. 525 (U.S. Govt., Washington, D.C.), 179-187.

Saha, S. and Gorman, P.H. (1979) Trans. 5th Ann. Mtg. Soc. of Biomat., III, 174.

Saha, S. Pal, S. and Rao, V. (1981) 1981 Biomechanics Symposium, ASME, 279-281.

DIELECTRIC PROPERTIES OF DOPED BONE MATERIALS

S. Singh

Department of Physics, U.P. (P/G) College,
Varanasi-221002, India

ABSTRACT

In recent years, the study on dielectric properties of bone has become a subject of considerable interest. This is because, firstly, it will determine the relationship between the current and electric field and even the current paths and energy-loss during electrical stimulation. Secondly, the knowledge of dielectric properties is of potential importance in physical characterization of bone materials as ultrasonic transducer. The work undertaken in the present investigation is an attempt in this direction wherein the data on dielectric properties (dielectric-constant, dielectric-loss) at 1 KHz and their variation with temperature in the range (25-225°C) have been reported on test pieces of different bone compositions. Bone compositions have been obtained by mixing bone powder with different foreign additives namely $AlBr_3$, ZnO, Sb_2O_3, $SrCO_3$, Li_2CO_3, Na_2CO_3, Nb_2O_5, PZT, $Ba(OH)_2$ and $Pb(NO_3)_2$ in 5% to 10% proportions by weight. Thus obtained bone materials have been used to prepare test pieces in disc shape of typical dimensions: diameter = 12 mm and thickness = 2.0 mm. Measurements have been carried out with the help of standard techniques and data have been found to be consistent and reproducible. Results on dielectric-constant and its variation with temperature show that 10% addition of ZnO and $Ba(OH)_2$ has desired effect. It is expected that the present study will determine as to how the dielectric properties are related to bone structure. The data on the dielectric properties of bone is of immense use in clinical applications i.e. cardiology, therapy (chemotherapy and physiotherapy) and bone surgery etc. besides providing knowledge needed in design of biocompatible transducer material.

KEYWORDS

Bone materials; foreign additives; milling; grinding; mesh; poling; test pieces; dielectric constant; loss-factor; Curie-temperature.

INTRODUCTION

In the last decade, dielectric properties of bone have been studied extensively. Part of the reason for this growing interest has been to understand the charge and current distribution in bone, subjected to electric potential difference in biomedical research, medical diagnosis and therapy. Secondly, the data on dielectric properties are also relevant and significant for physical characterization of bone materials as ultrasonic transducer. There have been several studies on the dielectric properties of bone (Shamos and Lavine, 1968; Reinish and Nowick, 1976; Marino, Becker and Bachman, 1967; Lakes, Harper and Katz, 1977). Shamos and Lavine (1968) reported results on dielectric constant of bone, collagen and apatite at room temperature and 100 KHz but did not mention about humidity. Marino, Becker and Bachman (1967) and Lakes, Harper and Katz (1977) measured the dielectric constant at humidity below 90 percent relative humidity (RH) while Reinish and Nowick (1976) reported measurements on samples equilibrated at 98 percent RH. In the following years, Maeda, Tsuda and Fukada (1976) reported the dependence of dielectric properties of bone temperature and hydration. These investigations further indicate that the dielectric properties of bone were anisotropic. In conformity with it, present author (Singh, 1981) conducted a systematic experiment to measure various dielectric (dielectric-constant, dielectric-loss and their variation with temperature), physical (density, resistivity, Curie-temperature), piezoelectric (charge-constant, voltage constant) and electromechanical (frequency constant, coupling coefficient) properties on bone, apatite and collagen. It has been found that the value of dielectric constant is 32, 29 and 6 for apatite, bone and collagen respectively. Furthermore, the study of variation of dielectric constant and phase-angle for these materials with frequency in the range (0.5 - 108 MHz) show that these bone materials are highly capacitive in nature. The data on dielectric properties have been found sensitive to the proportions of the mineral content and trace elements present in it which is suggestive of the fact that biophysical properties may be effected by chemical manipulation in bone composition by adding certain foreign materials. In the light of this, it is imperative to study the effect of foreign additives namely $AlBr_3$, ZnO, Sb_2O_3, $SrCO_3$, Li_2CO_3, Na_2CO_3, Nb_2O_5, PZT, $Ba(OH)_2$ and $Pb(NO_3)_2$ in 5% to 10% proportions by weight. The present paper is an attempt in this direction wherein data on dielectric properties of various bone compositions have been measured on test pieces in disc shape of dimensions: diameter = 12 mm and thickness = 2.0 mm. It is expected that these results have two fold significance, one in regard to characterization of electrical properties of '<u>in-vivo</u>' bone and secondly their relevance in development of a suitable biocompatible transducer material of '<u>in-vitro</u>' bone.

MATERIAL AND METHODS

Specimen Preparation

Full bone has been obtained in powdered form. Thus obtained bone powder was subjected to grinding with the help of dry ball-milling. Grinding is being done to obtain particle size between 150-200 mesh for attainment of chemical equilibrium. Various foreign additives namely $AlBr_3$, ZnO, Sb_2O_3, $SrCO_3$, Li_2CO_3, Na_2CO_3, Nb_2O_5, PZT, $Ba(OH)_2$ and $Pb(NO_3)_2$ in 5% to 10% proportions by weight have been mixed in the powder bone. These ingredients were homogenized by ball-milling in the dry polythene jar and have been dissolved in a common solvent (5% Formic acid and Ethyldiamine in equal volume). Solution has been subjected to magnetic stirring for proper chemical mixing. Thus treated solution was evaporated and solute was dried in an electric oven at temperature 100 - 120°C. The dried materials were mixed with 8% solution of polyvinyl Alcohol in a ratio of 2 -3% by weight of the materials and after that it was compressed into disc shape at a pressure of 11.04 tons/cm^2. The disc is of the typical dimension: diameter = 12 mm and thickness = 2.0 mm. The discs were then dried and electrical contacts were provided by painting colloidal silver paste on either sides.

Poling

Poling has been done by applying a strong electric field to the electroded test pieces. To establish desired properties under optimum poling conditions for various bone compositions, electric field of different strength have been applied for stipulated period. It has been found that for optimum poling, electric field has been applied in the range (3.5 KV/mm - 4.5 KV/mm) for 45 minutes at an elevated temperature. Test pieces were immersed in silicon oil-bath to sustain high electric field applied across them.

Measurements

Dielectric constant (K^T) and dielectric-loss have been measured on unpoled and poled electroded test pieces of bone compositions with the help of radiofrequency bridge (Model B601 Wynekerr, England) at 1 KHz. Variation of capacitance and loss-factor with temperature has been measured and is performed by immersing the mounted test pieces in silicon oil. The oil-bath has been heated to various temperature in the range (25 - 225°C). The test piece is connected to the radiofrequency bridge through the leads to note changes in capacitance and dielectric-loss at 1 KHz due to changing oil-bath temperature. The temperature corresponding to the maximum value of capacitance is noted separately for unpoled and poled pieces of bone materials and is termed as Curie-temperature for these materials.

RESULT AND DISCUSSION

Results on dielectric-constant and loss-factor and maximum value of capacitance corresponding to Curie-temperature have been given for unpoled test pieces in Table 1.

TABLE 1 Results on The Dielectric Properties of doped Bone Materials for unpoled Test Pieces at 30°C

Bone Materials B = Bone	Dielectric-Constant (K^T)	Dielectric-Loss ($\tan\delta$)	Capacitance at Curie-Temperature (pF)	Curie-Temperature (°C)
90%B+10%ZnO	29	0.080	1500	130
90%B+10%PZT	5	0.037	440	180
90%B+10%Ba(OH)2	14	0.06	6000	165
90%B+10%Pb(NO3)2	2	0.035	350	175
95%B+5%AlBr3	5	0.038	145	120
95%B+5%ZnO	4	0.033	60	135
95%B+5%Sb2O3	5	0.035	68	130
95%B+5%Li2CO3	3	0.030	60	165
95%B+5%SrCO3	3	0.035	50	150
95%B+5%Na2CO3	9	0.034	120	150

For poled test pieces, there is no significant variation and so data on poled samples are not mentioned here. Only those observations have been discussed here for which the effect of foreign additives are significant. From Table 1, it is clear that bone compositions (90% Bone + 10% ZnO) and (90% Bone + 10% Ba(OH)2) have higher dielectric constant and loss-factor. It has been further observed that the values of dielectric-constant and loss-factor are found higher in the case of intact bone as compared to the sample prepared from powdered bone material (Singh, 1981). The apparent fall in the values of dielectric properties may be attributed to the pores present in the test pieces as also observed by Katz (1971). A similar observation has been reported in case of multiphase ceramic materials wherein rupture strength decreases as the porosity increases (Ryshkewitch, 1953; Coble and Kingery, 1950; McAdam, 1951; Balshin and Fedotor, 1964).

Fig. 1 shows the variation of capacitance with the temperature in the range (25 - 225°C) for those additives which have maximum effect. Further it shows that 10% addition of ZnO and Ba(OH)2 greatly enhances the value of dielectric constant and curve for the variation of dielectric constant with temperature is very sharp with a maximum value of capacitacnce in the range (5000 - 6000pF) at Curie-temperature. This shows that large structural changes are taking place under temperature variation. Curie-temperature has been found around 160°C for most of the materials. The increase in Curie-temperature may be due to larger lattice constant as reported for other dielectric materials (Dekker, 1975). An identical pattern has been noted in studying the variation of loss-factor with temperature as in Fig. 2. It is expected that these data will shed light to det-

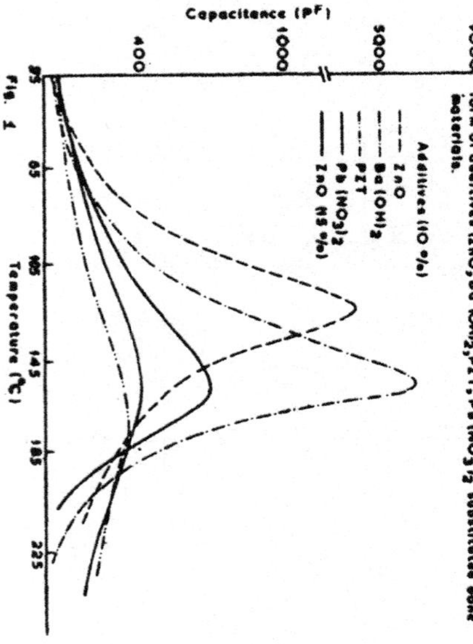

Fig. 1 Capacitance vs temperature for sample in disc shape of dimensions of dia = 1.2 cm thickness, t = 2.0 mm of compositions mixed in powdered form of additives (ZnO, Ba (OH)2, PZT, Pb (NO3)2 substituted bone materials.

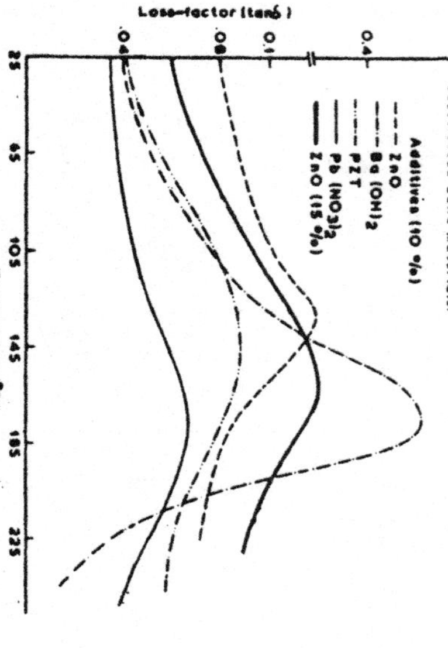

Fig. 2 Loss-factor (tanδ) vs temperature for sample in disc shape of dimensions of dia = 1.2 cm, thickness, t = 2.0 mm of compositions mixed in powdered form of additives ZnO, Ba(OH)2, PZT, Pb(NO3)2 substituted bone materials.

ermine how the bone dielectric properties are related to bone structure. It will help in studying the distribution of the current and charge in the subject material understudy.

ACKNOWLEDGEMENT

The author would like to express his gratitude to the Piezoelectric Ceramic Limited, New Delhi for help in preparation of test pieces. Thanks are also due to Prof. Krishnaji, Allahabad University, India for his valuable suggestions. Also, author wishes to put on record the love and care showered on him by Meena.

REFERENCES

Balshin, M. Yu. and Fedotov, S.G. (1968). Sov. Powd. Metall. Met. Ceramic, 4, 310-312.
Coble, R.L. and Kingery, W.D. (1956). J. Am. Ceram. Soc. 39, 377-385.
Dekker, A.J. (1975). Solid State Physics, Chap. 8. McMillan Press N.Y. 184-207.
Kartz, J. Lawrence (1971). J. Biomechanics 4, 455-473.
Lakes, Roderie, S.; Harper, Richard A. and Katz, Lawrence, J. (1977). J. Biomechanics 12, 652-678.
Lakes, R.S.; Harper, R.A. and Katz, J.L. (1977). J. Appl. Phys. 48, 808-811.
Maeda, M.; Tsuda, K. and Fukada, E. (1976). Japanese J. Appl. Phys. 15(12), 2333-2336.
Marino, A.A.; Beeker, R.O. and Bachman, C.H. (1967). Phys. Med. Biol. 12, 367-368.
McAdam, G.D. (1951). J. Iron Steel Inst. 168, 346-358.
Reinish, G.B. and Nowick, A.S. (1976). J. Electrochem. Soc. Vol.123 1451-1455.
Ryshkewitch, E.J. (1953). J. Am. Ceramic Soc. 36, 56-68.
Shamos, M.H. and Lavine, L.S. (1964). Clin. Ortho. Rel. Res. 35, 177-188.
Singh, S. (1981). Ph. D. Thesis, Submitted to J.N.U., New Delhi, India.

A NOTE ON AN ASPECT OF ELECTROMECHANICAL CONVERSION IN A BIASED THERMOPIEZOELECTRIC BONE MEDIUM

RUPA SEN CHATTOPADHYAY AND D. K. SINHA

DEPARTMENT OF MATHEMATICS, JADAVPUR UNIVERSITY, CALCUTTA - 700 032, INDIA.

ABSTRACT

This paper is a theoretical model of a piezoelectric biological element such as bone constitutively endowed with thermal attributes. The expression for the electromechanical conversion efficiency factor has been derived in the light of the theory of Adler. It has been found that the energy conversion efficiency factor brings about a time decaying characteristics to the overall increase in the energy conversion efficiency. Finally from the available numerical values of piezoelectric constants the conversion efficiency factor is evaluated.

KEYWORDS

Delay-rod combination, biopiezoelectricity, conversion efficiency factor, thermopiezoelectricity.

INTRODUCTION

Studies in electromechanical behaviour particularly piezoelectric behaviour of biological elements undertaken in recent times, are considered to be important because of their utilitarian purposes. While there have been many experimental studies vide Lakes & Katz(5), Yoon & Katz(8) there do not seem to have similar investigation from a theoretical stand point particularly in regard to problems of biopiezoelectricity, one of the trends in theoretical studies in piezoelectricity has been to make use of principles of classical mechanics of continua and of electricity in problems of piezoelectric devices such as transducers, in which the main concern is to obtain electromechanical conversion efficiency. The complexity of the problems is increased because of possible thermal attributes of piezoelectric bone. Hence the need for a theoretical model of a piezoelectric biological element constitutively endowed with thermal attributes. This is essentially the interest of this paper. For the purpose of obtaining the corresponding conversion efficiency factor, the thermopiezoelectric material for a bone is taken in the form of a transducer, on the line of Yoon & Katz(8) Guzelsu & Saha(4). A special configuration of the transducer has been assumed as in Adler(2). To determine the conversion efficiency factor numerical computations are also done. The expression for this factor, obviously complicated, shows that the thermal attributes qualitatively alter the efficiency to the extent of of increase of terms aligning the thermal parameters with wave number which of course remain as in a non-thermal case dominantly dispersive in character.

PROBLEM AND FUNDAMENTAL EQUATIONS

As stated earlier the biological element is considered in the form of a transducer with a delay rod configuration as in Adler(2) and consists of a thin film transducer deposited into a delay rod, where the transducer and the delay rod are at many wave lengths transverse to the direction of the acoustic wave propagation and the transducer electrodes are much thinner than the acoustic wave length. The transducer with the x-axis as the thickness direction is loaded with a delay rod of mechanical impedance z_R. The transducer delay rod combination is assumed to be under the action of some sort of heat flow, and is excited by an a.c. field E_0.

The fundamental equations of the problem are as follows :

$$\frac{\partial T}{\partial x} = \rho \frac{\partial^2 u}{\partial t^2}, \quad \text{mechanical equation of motion} \tag{2.1}$$

$$\frac{\partial D}{\partial x} = 0 \quad \text{divergence equation} \tag{2.2}$$

$$\frac{\partial q}{\partial t} = -k\frac{\partial \theta}{\partial x} \quad \text{Fourier Law of heat conduction} \tag{2.3}$$

$$\frac{\partial^2 q}{\partial x \partial t} = -\theta_0 \frac{\partial \sigma}{\partial t} \quad \text{entropy relation} \tag{2.4}$$

together with the constitutive equations of piezoelectricity taking into account the effect of thermal coupling vide Mindlin(7).

$$T = cS - eE - \lambda\theta \tag{2.5}$$

$$D = eS + \epsilon E + p\theta \tag{2.6}$$

$$\sigma = \lambda S + pE + a\theta \tag{2.7}$$

Where the notations have usual meaning as in Redwood(4) and Mindlin

SOLUTION

The piezotransducer-delay rod combination which is at temperature θ_0 is excited by an a.c. field E_0 and we seek the solution of the form

$$u = U\exp\{i(kx-\omega t)\} \tag{3.1}$$

$$E = E_0 + E\exp\{i(kx-\omega t)\} \tag{3.2}$$

$$\theta = \theta_0 + \theta\exp\{i(kx-\omega t)\} \tag{3.3}$$

Eliminating T, D, θ from the fundamental and constitutive equations the guiding equations of the problem are found which when substituted by the solutions (3.1) to (3.3) give the dispersion equation in wave number K as

$$\begin{vmatrix} ck^2-\rho\omega^2 & ike & i\lambda k \\ -ek^2 & i\epsilon k & ipk \\ -\lambda k\omega & ip\omega & i(a\omega + i\frac{k}{\theta_0}k^2) \end{vmatrix} = 0 \tag{3.4}$$

For four non-zero values of K_j (j = 1,2,3,4) the expressions for E_j and θ_j (in terms of U_j) become

$$E_j = \frac{\{k k_j^2/\theta_o \cdot (c k_j^2 - \rho\omega^2) + i(a\rho\omega^3 - \lambda^2 k_j^2\omega - ac k_j^2\omega)}{k_j\omega(ae + p\lambda) + i k k_j^3 e/\theta_o} \cdot U_j$$

$$\theta_j = \frac{\rho p\omega^3 + k_j^2\omega(e\lambda - cp)}{\{k_j\omega(ae + p\lambda) + i k k_j^3 e/\theta_o} \cdot U_j$$

Since the film thickness is finite, all waves will be present when the film is driven by an a.c. field E_o and the expressions for total displacement, strain, electric field, stress, velocity and temperature are

$$U_F = \sum_{j=1}^{4} U_j \exp(ik_j x) \quad ; \quad S_F = \sum_{j=1}^{4} ik_j U_j \exp(ik_j x)$$

$$E_F = E_o + c \cdot \sum_{j=1}^{4} \frac{k/\theta_o \cdot k_j^4(1 - \omega^2/v_s^2 k_j^2) + i(a\omega^3 v_s^2 - \lambda k_j^2 \frac{\omega}{c} - a k_j^2\omega)}{k_j\omega(ae + p\lambda) + i k_j^3 k e/\theta_o} U_j \exp(ik_j x)$$

$$T_F = -eE_o - \lambda\theta_o + \sum_{j=1}^{4}\Big[ick_j - ec\frac{k/\theta_o \cdot k_j^4(1 - \omega^2/v_s^2 k_j^2) + i(a\omega^3 v_s^2 - \lambda^2 k_j^2 \frac{\omega}{c} - a k_j^2\omega)}{k_j\omega(ae + p\lambda) + i k_j^3 k e/\theta_o}$$

$$- \lambda \frac{\rho p\omega^3 + k_j^2\omega(e\lambda - cp)}{k/\theta_o \cdot k_j^3 + i k_j\omega(\lambda p + ae)}\Big] U_j \exp(ik_j x)$$

$$v_F = -i\omega \sum_{j=1}^{4} U_j \exp(ik_j x)$$

$$\theta_F = \theta_o + \sum_{j=1}^{4} \frac{\rho p\omega^3 + k_j^2\omega(e\lambda - cp)}{k/\theta_o \cdot k_j^3 + i k_j\omega(\lambda p + ae)} U_j \exp(ik_j x)$$

In all the expressions above exp(-iwt) dependence is understood. The a.c. voltage (V) across the film, is given by

$$V = \int_0^L E_F dx = E_o L + c\sum_{j=1}^{4} \frac{k/\theta_o \, U_j k_j^4(1 - \omega^2/v_s^2 k_j^2) + i(a\omega^3 v_s^2 - \lambda k_j^2\omega/c - a k_j^2\omega)}{k_j\omega(ae + p\lambda) + i k_j^3 k e/\theta_o}[\exp(ik_j L) - 1]$$

The power input to the transducer is $P_{IN} = \frac{1}{2}\mathrm{Re}(V I^*)$

$$= \frac{\lambda\theta_o}{2}\mathrm{Re}\Big[\frac{i\omega\epsilon A E_o^2}{\lambda\theta_o} + i\omega\epsilon A E_o \sum_{j=1}^{4} \frac{k/\theta_o k_j^4(1 - \omega^2/v_s^2 k_j^2) + i(a\omega^3 v_s^2 - \lambda k_j^2\omega/c - a k_j^2\omega)}{k_j\omega(ae + p\lambda) + i k_j^3 k e/\theta_o} \cdot v_j\{\exp(ik_j L) - 1\}\Big]$$

The acoustic power into the delay rod is

$$P_{OUT} = \frac{1}{2}\mathrm{Re}[-A T_R v_R^*]$$

Since stress and velocity continuity must hold at the film rod interface

$$P_{OUT} = \frac{1}{2}\mathrm{Re}\Big[-Ai\omega\sum_{j=1}^{4} U_j \exp(ik_j L)\Big\{\sum_{j=1}^{4} ik_j - e\frac{k/\theta_o \cdot k_j^4(1 - \omega^2/v_s^2 k_j^2) + i(a\omega^3 v_s^2 - \lambda^2 k_j^2 \frac{\omega}{c} - a k_j^2\omega)}{\{k_j\omega(ae + p\lambda) + i k_j^3 k e/\theta_o\}}$$

$$- \frac{\lambda^2}{c} \cdot \frac{\rho p\omega^3 + k_j^2\omega(e\lambda - cp)}{k/\theta_o \cdot k_j^3 + i k_j\omega(\lambda p + ae)} - 1 - \frac{eE_o}{\lambda\theta_o}\Big\} v_j \exp(ik_j L)\Big]$$

The electromechanical energy conversion efficiency G is

$$G = \frac{P_{OUT}}{P_{IN}} = \mathrm{Re}\Big[-i\omega \sum_{j=1}^{4} U_j \exp(ik_jL) \Big\{ \sum_{j=1}^{4} ik_j - \frac{k/\theta_0\, k_j^4 (1-\omega^2/v_s^2k_j^2) + i(a\omega^3 v_s^2 - \lambda k_j^2 \frac{\omega}{c} - ak_j^2\omega)}{\{k_j\omega(ae+p\lambda) + ik_j^3 ke/\theta_0\}}$$

$$- \frac{\lambda^2}{c} \cdot \frac{\rho p\omega^3 + k_j^2\omega(e\lambda - cp)}{(k/\theta_0) k_j^3 + ik_j\omega(\lambda p + ae)} - 1 - \frac{eE_0}{\lambda\theta_0}\Big\} \nu_j \exp(ik_jL)\Big] \Big/ \lambda\theta_0 \cdot \mathrm{Re}\Big[\frac{i\omega\epsilon E_0^2 L}{\lambda\theta_0}$$

$$+ i\omega\epsilon E_0 \Big\{ \frac{k/\theta_0 \cdot k_j^4 (1 - \frac{\omega^2}{v_s^2 k_j^2}) + i(a\omega^3 v_s^2 - \lambda k_j^2 \frac{\omega}{c} - ak_j^2\omega)}{\{k_j\omega(ae+p\lambda) + ik_j^3 ke/\theta_0\}} \cdot \nu_j \{\exp(ik_jL) - 1\}\Big]$$

Continuity of stress of the interface and of zerostress at one extremity are the boundary conditions for this problem.
i.e. $T_F(0) = 0$; $T_F(L) = T_R$
Velocity continuity across the interface implies

$$v_R = v_F(L) = -i\omega \sum_{j=1}^{4} U_j \exp(ik_jL)$$

In the delay rod

$$T_R = -Z_R v_R = T_F(L) = i\omega Z_R \sum_{j=1}^{4} U_j \exp(ik_jL)$$

With these boundary conditions the values of ν_j are obtained and thus G is completely solved.

The energy conversion efficiency factor G, involving complex K_j brings about a time decaying characteristics and the thermopiezoelectric parameter contributes to the overall increase in the energy conversion efficiency. From the values given by Lakes & Katz(5), Lang(6), Yoon & Katz(8) and assuming the thermopiezoelectric constants to be of the same order as piezoelectric constants the numerical value of energy conversion efficiency factor is obtained supporting the theoretical model fairly well.

REFERENCES

Adler, E.L. (1967) J. Acoust. Soc. Amer., 43, 842
Adler, E.L. (1969) IEEE Trans. on Sonics & Ultrasonics SU-16, No. 2.
Redwood, M (1961) J. Acous. Soc. Amer. 33, 527.
Guzelsu, N. and Saha, S. (1981) J. Biomech. 14, 19-33.
Lakes, R.S. and Katz, J. L. et.al (1979) J. Biomech. 12, 657-678.
Lang, S. B. (1969) Science, N. Y. 165, 281
Mindlin, R. D. (1961) Proc. Soc. for Industrial and Applied Maths. Pennsylvania, 382-392.
Yoon, H.S. and Katz, J.L. (1976) J. Biomech. 9, 459-464.

ACKNOWLEDGEMENT

The authors would like to express their gratitude to the Research Project conducted by University Grants Commission, Government of India. Thanks are due to Dr. Subrata Saha who assisted us with his valuable advices.

Session 17

CLINICAL ENGINEERING AND BIOENGINEERING EDUCATION

THE NEED OF BIOMEDICAL ETHICS TRAINING IN BIOENGINEERING

Pamela Saha* and Subrata Saha**
*Faith Bible College, 8412 Kingston Rd., Shreveport, LA 71108
**Dept. of Orthopaedic Surgery, LSU Medical Center, Shreveport, LA 71130

ABSTRACT

The need for biomedical ethics training for bioengineers is discussed. Such training will be helpful to: 1) guide conducting human and animal experimentation; 2) meet the legal responsibilities associated with device or implant design; 3) reduce burnout of clinical engineers and 4) decide priorities in research and development. The need for adopting a Code of Ethics for the biomedical engineering profession is also pointed out.

KEYWORDS

Biomedical ethics, ethics education, code of ethics, biomedical engineering, burnout.

INTRODUCTION

In this age of growing ethical concern on the part of physicians, there yet exists no literature addressing the ethical problems of the bioengineer. And while medical schools are beginning to offer courses in medical ethics, no such provision is being made for the training of biomedical engineers.

Unlike the traditional engineer, the bioengineer might experiment on animals, human tissue, or even human volunteers and patients. Questions not relevant to work restricted to metals or plastics in traditional engineering studies become crucial to work dealing with animals or humans.

The training of a bioengineer usually begins with intensive exposure to nonbiologically oriented sciences with emphasis on utility. The engineer's emphasis on utility is expected since engineering is, after all, an <u>applied</u> science or rather a science of application. Often the bioengineer is almost a fully trained scientist in some other field of pure engineering only later to be exposed to a specific area of medicine. But often the transition from pure engineering into engineering for medicine isn't so smooth. The trouble arises when a bioengineer must abruptly and sometimes painfully discover that pure engineering principles of utility and application might sometimes have to be sacrificed for what might appear to be less practical, less useful yet most necessary considerations. Such a change in orientation may actually mean a mild "culture shock" for some scientists who are struggling to meet a serious medical need and yet are impaired by frustrating moral issues they would not have faced as nonmedical engineers. Some bioengineers have actually been forced to discontinue their research because they had failed to accommodate humane or moral considerations in the conduction of their work.

Not only must bioengineers face ethical problems of their own but they must also be sensitive to the ethical situation of the physician who must clinically put to use the devices developed in the bioengineering laboratory. An engineer who is not aware of ethical problems faced by doctors is more likely to develop some application that may be sound technically yet not usable on ethical grounds. This becomes a disappointment to both the clinician and the scientist.

We shall now examine some specific areas that present ethically related problems to the bioscientist. First we will look upon the phenomenon of "burnout" which arises out of trouble with adjustment to new moral or ethical situations in the field of bioengineering. We will then study the effect of law on the ethics and professionalism of the field. We will introduce the ethical issues a biomedical engineer must face in conducting research. We will study the role of economics in making ethical choices between projects to pursue. Finally, we will raise the question of need for a code of ethics for the profession of bioengineering.

BURNOUT OF CLINICAL ENGINEERS

Fennigkoh (1981) warned that "within clinical engineering, burnout may very well become one of the greatest threats to the profession's survival within the hospital." He noted that hospitals are 'people' oriented while clinical engineers are 'thing' oriented. This difference follows directly from the educational background of doctors and engineers. Observe, also, that medical science relies on an acquired art absent in the hard science of pure engineering. Hence, Fennigkoh observes that the black-and-white teachings of engineering education are strained under the grey realities of the hospital environment and this frustrates the engineer in medicine. Therefore, engineers who intend to enter the more people oriented field of biomedical engineering must have enjoyed some exposure to relevant social sciences, particularly medical ethics in their education. An engineering or even a medical curriculum narrowed to a strictly physical science way of defining and attacking problems leaves the student wholly unprepared for the "greys" of biomedical ethics. Years of conditioning to think and function in a limited way, could make the adjustment to a broader scope too great a task. Thus many clinical engineers burn out due to the strain arising from the moral and psychological demands of the bioengineering profession.

Karger (1981) sees burnout as a larger social issue and views it as alienation. This involves an objectification of worker and client making "inanimate objects of production" out of both. Karger's alienation is very like Fennigkoh's phase of "Job Disillusionment" in the burnout process whereby the clinical engineer detaches, disengages, alienates his or herself from the job, (Finnigkoh, 1981 and Karger, 1981). This sort of objectification, abstraction of social or medically related work in burnout flows naturally from an education that teaches from an abstract mechanical perspective. But work should bring a personal involvement and fulfillment impossible if the workers alienate themselves from their services. Without involvement, we lack commitment and grow apathetic. Without fulfillment, we lose self-esteem, become depressed, emotionally exhausted, unable to work, and we "burn out".

Education in ethics would predispose bioengineers as well as strengthen them against the moral frustrations they must face in their work. This preparation would provide a safeguard against the problem of "burnout".

THE PROBLEM OF LAW

Today, professionals are being held more and more accountable for their actions by the society. The number of lawsuits have increased beyond what professionals can withstand in terms of liability insurance and risks in public service. If we consider the examples of doctors, we see how excessive litigation can hurt a profession.

Law and burdensome regulations have caused some biomedical researchers to change careers. Engineers in medicine must be prepared to participate in the process of public regulation of their own profession. They can only do this, however, if they themselves are educated as to their moral and legal obligations.

Of course, law alone cannot ensure quality in the service of bioengineers or in that of any profession. Therefore, ethics has a role and a purpose that cannot be filled by legal regulation alone.

Ethics, however, is not new to even the traditional engineer. The field of engineering has always been governed by law, principles, and standards. A designer, for example, is legally and ethically bound to anticipate all possible forms of misuse or accident with a product (Buck, 1963 and Holder, 1978). These ethical standards for the traditional engineer are more significant in the work of the biomedical engineer who is designing and testing in order to safety-proof a product for patients and physicians. Furthermore, the bioengineer must be very knowledgeable of the medical condition for which he or she is designing.

The consequences due to negligence on the part of traditional engineers may be serious, but they are more likely to be critical or even deadly in the case of improperly designed pacemakers; wheelchairs; surgical equipment; heart, lung, and kidney machines; and incubators for premature infants. Not only must bioengineers design and test carefully such products, but clinical engineers must carefully maintain them. A small undetectable flaw can have dire consequences. And when expensive biomedical equipment fails it cannot be easily replaced.

Negligence in design, research, or maintenance can lead to litigation, (Holder, 1978). The best safeguard against this is education. Bioengineers who are trained in their ethical obligations to those they serve are more protected against lawsuits. But avoidance of lawsuit is not the sole benefit of biomedical ethics education. The law enforces only minimal responsibility. The truly responsible bioengineer will seek to minimize needed caution on the part of others who might be held liable if equipment fails.

Law can never ensure high professional standards unless professionals themselves have a strong understanding of their moral obligations and respect the law as a tool that upholds the standards of their performance. Regulations which are merely obstructive, risk encouraging a service that is legalistic, not ethical. As law cannot enforce internal virtues, we might turn to education to instill them.

HUMAN AND ANIMAL EXPERIMENTATION

In the section on the problem of law we noted the importance of research and adequate testing of a product. While both the traditional engineer and the bioengineer are bound equally under the law, the bioengineer has far greater difficulty satisfying the requirement to fully test a product. The bioengineers face a double bind. They must test completely, says the law, yet the same law restricts the extent to which they can go to completely test.

One stage of the testing procedure is likely to involve animal subjects and regulation of animal research is becoming more specific. The Recommendations from the Declaration of Helsinki (1975) on conduction of animal research include use of anesthetics during operative procedures, and care to minimize pain where recovery from anesthesia is necessary for the investigation. On completion of a study, animals should be humanely sacrificed or cared for according to the standards of veterinary medicine. The Committee for Research and Ethical Issues of the International Association for the study of Pain (IASP) has issued a list of guidelines as well, which included a suggestion that the investigator test the pain stimulus on his or herself when possible. These are examples of some precise regulations of which biomedical engineers must be aware as they begin animal research. The biomedical engineers are entering into an arena of debate as they plan research involving animals.

The bioengineer enters an even more complex and far more serious debate as he or she advances to the next and final stage of biomedical research -- human experimentation. Ultimately, if the bioengineer's work is to be applied to humans, a test will sometime have to involve a human being. One of the most highly debated problems of human experimentation is the matter of informed consent. A human subject is said to have given informed consent to an experimenter, when he or she agrees to be a subject after having been informed what anyone would reasonably expect to know about an experiment. Much difficulty surrounds the matter of determining what potential subjects should reasonably expect to know about an experiment. Some authors feel that human research should be more carefully reviewed by ethics committees, (Barber, 1976) while others blast such efforts to further control human research as counter productive and unnecessary, (Hein, 1980). Thus bioengineers enter this whole area of debate as they design their experiments using human subjects.

In order to be able to conduct research using human and animal subjects in this age of moral enlightenment, the biomedical engineer must be knowledgeable as well as have an understanding of the issues such research involves. If bioengineers are not educated in this area, they will be poorly equipped to plan and defend their research before those who will judge the ethical standard of their work.

THE RISE IN MEDICAL COST AND THE QUESTION OF PRIORITY

Rewards for impressive research and dramatic solutions are great, yet authors like Black (1972) and Wolff (1970) encourage bioengineers to opt for the unspectacular research that makes good care available to many instead of performing miracles for the few. That is, research should seek to offer health care within the financial means of most people. As funds for research are limited, there is some moral imperative in promoting work that may not be eye-catching but may provide low cost solutions to many. If medical technology were to remain solely focused on highly expensive forms of health care, soon the products of biomedical research would become inexcessible to all except for journalists to publicize.

CODE OF ETHICS

Most professions have formed their own code of ethics. The medical profession has been respected for such a code since the time of its inception. The American Medical Association is now revising the old code in order to meet new ethical concerns in modern medicine, (Veatch, 1980). The engineering profession, likewise, upholds a code of ethics and has a tradition of maintaining a standard of professionalism. A code of ethics helps professionals understand their position in relation to those they serve. Such understanding encourages respect between professionals and society alike. Finally, this type of code protects a profession from dishonor. Bioengineering is growing into a new profession all its own, distinct from both medicine and engineering. It must now begin formulating its own code of ethics.

CONCLUSION

We have looked at some areas of concern for bioengineers that establish a need for biomedical ethics education for the engineer in medicine. As biotechnology advances and as engineering grows more significant in the medical field, the bioengineer is sure to face greater dilemmas and face them with increasing frequency. The bioengineer is sure to share in some of the same ethical issues physicians now face and are now wisely trying to deal with thorough education in medical ethics. Bioengineers should be prepared as well. Bioengineers assume more than the burden of medical technology. They take on the load of ethical problems in medicine as well.

REFERENCES

Barber, B. (1976) Sci. Am., 234, 25-31.

Black, M.M. (1972) Developments in Biomedical Engineering. Sussex University Press, London.

Buck, C.H. (1963) Problems of Product Design and Development. The Macmillan Co., New York.

Covino, B.G., R. Dubner, J. Gybels, H.W. Kosterlitz, J. C. Liebeskind, R. A. Sternbach, L. VyKlicky, H. Yamamura, and M. Zimmerman (1980) Pain, 9, 141-143.

Fennigkoh, L. (1981) Med. Instru., 15, 375-379.

Hein, J.W. (1980) J. Dent. Res., 59C, 1353-1363.

Holder, A.R. (1967) Medical Malpractice Law, 2nd ed. John Wiley & Sons, New York.

Karger, H.J. (1981) Soc. Ser. Rev., 55, 271-183.

Recommendations from the Declaration of Helsinki (1975). Transactions of ASME, 102, 118.

Veatch, R.M. (1980) Hastings Center Report, June, 16-19.

Wolff, H.S. (1970) Biomedical Engineering. McGraw-Hill Book Co., New York.

REVISION OF A BMET CURRICULUM

J.O. Wear, L.D. Brown, B.K. Clark, W.E. Graham,
F.E. VanNoy, and F.J. Weber

College of Health Related Professions, University of Arkansas,
Engineering Training Center, Veterans Administration Medical
Center, Little Rock, Arkansas

ABSTRACT

An established associate degree program for BMET's has been revised to take into account changes in technology. Microprocessors and additional electronics laboratory time have been introduced into the curriculum. A combination Chemistry and Physics course has been developed for BMET's. The changes were made without the need for increased staff, equipment, or space.

KEYWORDS

BMET; Curriculum; Microprocessor; Electronics; Associate Degree; Chemistry.

INTRODUCTION

As medical equipment incorporates newer technology, the training of BMET's to repair and maintain the equipment must be updated. Some revisions are made each year by the instructor of each course, but at some time new courses must be added and some courses dropped or shortened. The Associate Degree Biomedical Instrumentation Technology Program in the College of Health Related Professions, University of Arkansas was established in 1972 in cooperation with the Veterans Administration Medical Center, Little Rock, AR. The entire curriculum of the program was reviewed and major revisions made to update the curriculum.

The main thrust of a BMET program is directed at producing competent, analytical repair-persons, who have a proven ability to troubleshoot and repair biomedical instrumentation from each of the major clinical areas. The logical sequence of teaching instrumentation to the BMET is the following four phases:

1. Detailed Principles and Mechanism of Operation.
2. Proper Operation and Routine Maintenance.
3. Troubleshooting.
4. Repair.

Each of these phases require theoretical and practical prerequisites as shown in Table I. Most of the time, some of these prerequisites must be included in the instrumentation courses which reduces the time that can be allocated to "hands-on" experience.

At first, the curriculum was reviewed by the staff with input from former students considering no constraints. The revisions were to include prerequisites in the basic science courses and free up time in instrumentation courses. Then, the following resource constraints were imposed on the plan:

1. No new faculty could be added.
2. No more than 68 total hours could be required.
3. No substantial additions in equipment could be made.
4. There had to be a smooth transition to the new curriculum.

It was found that basic courses, such as physiology and chemistry, were not relevant to the students need for prerequisite theory; and some basic skills with hand tools and test equipment were weak. The students were also entering the second year instrument courses without any reinforcement of electronics.

To address these needs, the following changes were made:

1. A Basic Electronics Skills lab course for one credit hour was added to the first semester.
2. A Chemistry and Physics of Biomedical Instrumentation course (4 hours) was developed to replace the freshman chemistry course in the first semester.
3. A Laboratory course (1 hour) was developed as a co-requisite for the first semester Physiology and Anatomy course with emphasis on instrument measurements.
4. Courses in Hospital Organization and Maintenance Management were combined and reduced from 4 hours to 2 hours.
5. A 3 hour Electronic Test Equipment course was developed for inclusion in the first block of the second year to provide an electronics refresher.
6. A 1 hour Microprocessor course was developed for the second year.
7. The Nuclear Medicine and Radiological Instrumentation courses were combined with a reduction of one hour.

These changes increased the curriculum by 3 hours for a total of 68 hours and met all the needs within the constraints. The new curriculum is shown, by semester, in Fig. 1.

The changes were implemented in the Fall, 1981. At the end of the first year, the revised program appears to have met its objectives of making the material more relevant as well as better preparing the students.

TABLE I Logical Sequence of an Instrumentation Course

Phase	[1]Pre-requisites: Theoretical	Practical
1. Detailed Principles and Mechanisms of Operation.	Knowledge of applicable principles of chemistry, physics, physiology, and electronics.	Ability to read service manuals, to read and understand schematics, and to relate schematics to actual devices or circuits.
2. Proper Operation and Correct Use.		Familiarity with common clinical practices, rules and regulations in the pertinent departments or units of the hospital.
Routine Maintenance Procedures.	Knowledge of applicable hospital safety principles and regulations.	Familiarity with applicable tools and test equipment.
3. Troubleshooting (both hypothetical and practical).	(Additional knowledge of basic systems theory and communications principles may be valuable.)	Familiarity with numerous types of electronic and/or mechanical test equipment.
4. Repair (sub-assembly replacement and individual component replacement).		Ability to use sophisticated test equipment and test procedures. Use of manufacturer's data sheets for replacement, testing, and substitution of components; basic mechanical and electronics skills for disassembly, and reassembly.

[1]An unstated pre-requisite for each phase is the successful completion of the previous phase.

First Semester

DH-2321	-	Physiology and Anatomy	3
BIT-2113	-	Physiology and Anatomy Applications	1
BIT-2101	-	Biomedical Terminology	1
BIT-2412	-	Chemistry and Physics of Biomedical Instrumentation	4
BIT-2501	-	Basic Mathematics for Electronics	5
BIT-2115	-	Basic Electronic and Mechanical Skills	1
			15 hrs

Second Semester

BIT-2401	-	Solid State Electronics	4
BIT-2417	-	Digital Electronics	4
BIT-2302	-	Medical Equipment	3
SPEECH 1300	-	Speech Communication	3
			14 hrs

Summer

ENG-1311	-	Freshman Composition I	3
ENG-1312	-	Freshman Composition II	3
*HIST-1311	-	History of Civilization I	3
*HIST-1312	-	History of Civilization II	3
			12 hrs

*Economics, Political Science, Psychology or Sociology are acceptable alternative courses.

Third Semester

BIT-2331	-	Electronic Test Equipment	3
BIT-2132	-	Microprocessors	1
BIT-2310	-	Clinical Chemistry Instrumentation	3
BIT-2311	-	Clinical Laboratory Instrumentation	3
BIT-2512	-	Physiological Measuring Techniques	5
			15 hrs

Fourth Semester

BIT-2402	-	Practicum	4
BIT-2342	-	Nuclear Medicine and Radiological Instrumentation	3
BIT-2243	-	Hospital Organization and Maintenance Management	2
HIST-2311	-	History of American Civilization	3
			12 hrs
		Total Hours:	68

Fig. 1. Biomedical Instrumentation Technology course schedule.

BIOMEDICAL EDUCATION AND THE PHILOSOPHY OF ORIGINS

A. B. Broderson,* Ph.D., and D. C. Randall,** Ph.D.

*Watkins and Assoc., Inc., Engineers-Architects-Planners, P. O. Box 951, Lexington, Ky. 40588

**Dept. of Physiology and Biophysics, University of Kentucky Medical School, Lexington, Ky. 40506

ABSTRACT

This paper discusses the mixed scientific and philosophical character of the origins issue, the ways it challenges a student's philosophy, and ways the biomedical educator may use the origins issue constructively.

KEYWORDS

Origins; biomedical education; philosophy; science; values.

INTRODUCTION

Questions about the origin and nature of life have strongly shaped the history of reflective thought, and are today commanding a remarkable degree of attention from science, philosophy, and education. The origins question has remained open, many believe, because those for whom the issue is most profound do not consider current scientific hypotheses to be philosophically cogent.

The treatment accorded this issue by many respected journals of science and education does educators a disservice for often it is intellectually shallow. They mistakenly recast the issue as one of science versus religion. This overlooks the conviction of many thoughtful people that the issue is more properly one of the unity of science and human values.

Because the origins issue is so complex, it poses both problems and opportunities for biomedical education. The purposes of this paper are first, to suggest ways to overcome the problems by constructively handling the issue's philosophical dimension, and second, to encourage educators to seize the opportunity this issue presents for the nurture of positive attitudes toward science and human values.

BACKGROUND

The relevance of the origins issue for biomedical education lies in the philosophical character of science. Biomedical subjects seem inherently to convolve scientific facts with a philosophical interpretation of those facts, and the latter is generally grounded in some concept of origins. Since the educator's role is to present scientific facts, but allow students to make their own philosophical decisions, care is required to avoid unintentionally challenging the student's philosophical perspective.

The Student's Philosophy.

Each student of the biomedical sciences possesses some synthetic philosophy or world-view regarding the character of nature, life, and man and how they came into being. Some students may view all things as having originated through purely mechanistic or naturalistic processes, while others may believe that the beauty, complexity, and orderliness of nature speak more of purpose and intelligent design. Such beliefs may, in turn, either reinforce or deny a sense of meaning and value in individual living things.

Students holding differing philosophies about origins may, as a result, differ in their view of science and the scientific method. Some students may view the scientific method as the only epistemology capable of providing reliable knowledge about objective reality. Other students, perhaps most, appreciate science as only one "window on the universe", and acknowledge the existence of other valid windows, such as philosophy and theology. While these other windows encompass a wider circle of knowledge, they are nevertheless expected to work harmoniously with science in the quest for truth. Such students may expect educators not only to respect their personal philosophy, but also to uphold the unity of science and human values.

Biosciences Contain Both Science and Philosophy.

A biomedical education asks students to probe progressively deeper into the inner workings of nature, exposing them to both scientific facts and philosophical syntheses of those facts. For example, biophysics students studying life at the molecular level may encounter not only established laws of chemistry and probability, but also philosophic speculations on whether the universe is inherently random or determinate.

The Science and Philosophy of Biosciences are Convolved.

In most contemporary speculation about life, nature, and their origin, the boundaries separating science and philosophy have largely disappeared, causing these two epistemologies to become heavily intertwined (Barbour, 1966). This makes it difficult for the educator to teach science without concomitantly advancing a particular philosophy.

The Challenge This Poses to Education.

The educator may have difficulty, for example, in analysing biogenesis experiments for some students (see Dickerson, 1978) without appearing to attack their personal philosophy. He may consider his approach to be responsible if he limits his analysis to the recognized chemical kinetics involved, i.e., that one may synthesize certain macromolecules found in living cells by passing simulated lightning through a closed system of reagents. If, however, he discussed whether such experiments demonstrate the inevitability of life's chemical evolution, he has crossed the boundary into philosophy where he assumes the obligation to treat other competing views with respect.

The convolution of science and philosophy may be unavoidable in the biomedical curriculum. The biomedical educator does have a choice, however, in determining whether the philosophical challenge to the student will be directed toward constructive or harmful ends.

DISCUSSION

Biomedical educators should recognize and avoid three varieties of destructive philosophical challenge: blurring the distinction between scientific and philosophic content, elevating the educator's philosophy, and disparaging the student's philosophy.

Blurring the Distinction.

Convolving science with philosophy creates a potential for abuse not unlike that of an advertisement which contains a "hidden persuader". While both the origins lecture and the magazine ad transmit a factual message, they also have the capacity to influence the recipient's philosophical moorings at an intensity below his threshold of apprehension. The conscientious educator will avoid the subliminal advancement of a philosophy by distinguishing the scientific facts from the philosophical model he uses to integrate those facts into a coherent framework. To assert that a human fetus may be scientifically reduced to nothing more than physics and chemistry would be to confuse science's legitimate use of simplifying models with the philosophical doctrine of reductionism.

Elevating the Educator's Philosophy.

The biomedical educator should especially avoid the overt advancement of his personal world-view. Some journal writers do this when asserting that molecules-to-man evolution is a FACT and that science only disagrees over HOW it occurred. It would be more precise to acknowledge such evolution as a philosophical paradigm accepted by many scientists and rejected by others, based on the same scientific data.

Degrading the Student's Philosophy.

Biomedical educators should not inadvertently degrade a student's theological view of origins by identifying it with "religion". The term "religion" is not only poorly defined, but may also be used as a pejorative to imply an illogical, undisciplined, or unenlightened view. Theology, on the other hand, may be considered a rational, disciplined mode of inquiry which, though different in method, is as capable as science of giving objective, empirical knowledge about man and nature (Torrance, 1969). The student who gives rational assent to a theological affirmation about origins may feel the educator has wrongfully degraded his personal philosophy if his view of origins is labeled "religion". The responsible educator will not degrade the distinction between "religion" and theology.

CONSTRUCTIVE TREATMENT OF ORIGINS

The inherent challenge biomedical education presents to the student's synthetic philosophy may be directed, on the other hand, toward constructive purposes. Three approaches may be utilized: teaching by contrasts, investigating by teleology, and examining the nature of knowledge.

Teaching by Contrasts.

The educator may stimulate interest and provoke thought at a deeper level by comparing and contrasting alternate theories of origins (Bergman, 1979). One could, for example, present both a creationist and an evolutionary paradigm for investigating a physiological feedback system. By examining the relative validity of competing theories, the student may enhance his understanding of the concepts involved.

Investigation by Teleology.

If the educator has been teaching primarily from a naturalistic model, and recognizes that some of his students are uncomfortable with an exclusive reliance on that philosophical perspective, he may stimulate the student's attention by careful use of a frankly teleological approach. An excellent example may be found in a text chapter by Burton (1972) on the "design" or "purpose" of blood circulation.

Examining the Nature of Knowledge.

In a more general sense, the origins issue can serve as an excellent springboard from which to launch discussions of the nature of knowledge, helping the student to better understand logic and truth, and to distinguish between inductive and deductive reasoning, law and theory, fact and presupposition, and proof and belief. Properly directed, such discussions can help student and teacher alike to deepen their own intellectual grasp of concepts, enhance research skills, sharpen insight, and promote tolerance for competing ideas.

If the biomedical educator can discuss such issues with a measure of relaxed sincerity, he can inculate students with the healthy attitude that life remains a mysterious phenomenon whose investigation is open to other than purely empirical diagnostics, and whose order, beauty, and functional complexity should never fail to evoke in us what Einstein termed "awe". Properly taught, the biomedical sciences can nurture within us a healthy respect for both positive human values and the excitement of scientific enterprise.

CONCLUSION

The origin of man, life, and the universe remains a controversial, though highly significant issue whose discussion can challenge the student's synthetic philosophy. Biomedical educators should therefore increase their intellectual depth in origins-related issues. This can enable them to avoid unintentionally assaulting the student's philosophical foundation and develop constructive ways to use the origins issue for positive educational ends. These can enable the educator to strengthen concepts, enhance understanding of knowledge, and nurture positive attitudes toward science and human values.

REFERENCES

Barbour, Ian G. (1966). Issues in Science and Religion, Harper Torchbooks, Harper and Row.

Bergman, Jerry (1979). Teaching About the Creation/Evolution Issue, Fastback No. 134, Phi Delta Kappa Educational Foundation, Bloomington, Indiana.

Burton, Alan C. (1972). Why Have Circulation?, Chapter 1 in Physiology and Biophysics of the Circulation, 2nd Edition, Year Book Medical Publishers, Inc., Chicago.

Dickerson, Richard D. (1978). Chemical Evolution and the Origin of Life, Scientific American, 239, 70-86.

Torrance, Thomas F. (1969). Theological Science, Oxford University Press.

CLINICAL ENGINEERING IN THE SOUTHEAST: ACHIEVING ITS POTENTIAL IN THE LARGE COMMUNITY HOSPITAL

Thomas M. Judd, M.S., P.E.
Director, Clinical Engineering Services
Baptist Medical Center
800 Prudential Drive
Jacksonville, Florida 32007

ABSTRACT

A fully developed in-house clinical engineering program has much to offer to a large community hospital. In the southeast U.S., many community hospitals-600 beds or less-have implemented partial programs and, as a result, have received only a partial return on their investment. A comprehensive program extends maintenance support to all patient-care devices, is intimately involved in the new device specification, evaluation, and selection process with the medical staff-standardizing whenever appropriate, and is significantly involved in the continuing education program for in-house users of devices.

KEYWORDS

Clinical engineering; patient-care equipment; medical equipment specification and selection; medical equipment user training; medical equipment maintenance savings.

INTRODUCTION

Two major problems in health care relating to patient care devices got significant recognition and reaction in the mid 1970's. First was the acknowledgement of the fallibility of hospitals and physicians to prevent every forseeable and unforeseeable event or condition which might result in harm or accident to a patient. The reaction was a flood of laws and standards with which health care institutions and health providers must comply. The second major problem has been nearly run-away costs in virtually every area of health delivery, particularly for high technology devices. Here again the problem was met with laws, record systems, paper work and controls; some imposed by legislation; some by third party payment agencies and organizations, and some by the hospitals themselves (Simmons, Wear, 1977). Clinical engineering programs have been shown to be capable of assisting in the cost effective management of both of these conditions while assuring hospital administration of several areas of control and compliance.

Medical devices have proliferated in hospitals in the last two decades to the extent that today there are on the order of 10,000 devices made by approximately 1500 manufacturers in use in health care facilities. This proliferation of medical devices is accompanied by a corresponding expansion of diagnostic, therapeutic, and test equipment. This equipment demands different management and operational skills than those normally exercised by the well established biomedical engineers involved in industry and research and plant facilities/maintenance engineers. To supply these skills, a new specialty, clinical engineering, was created in 1969 to "apply engineering technology for the improvement and delivery of health services" (Shaffer, 1977). In the years that followed, criteria were established such that a clinical engineer, per 250-300 beds, was needed for hospitals or approximately 5,000 engineers nationwide. Currently, roughly 10-15% of those requirements have been realized (Arthur, Boxerman, 1981). As the profession has operationally developed in the past twelve years, it has become clearer what the medical center's range of technical needs are and how clinical engineers can meet them. The Joint Commission on Accreditation of Hospitals (JCAH) has a strong influence on the clinical engineering role by providing definite guidelines for support of patient care equipment. These guidelines are briefly summarized below. As a result of input from the regulatory agencies, the accreditation bodies, and experience, a job description for the clinical engineering role has been developed. It is available in Ostrander's article (1979) and in the Pacela and Forbes article (1977).

Of prime importance is the understanding that clinical engineering deals with all aspects of patient care instrumentation: (1) helping decide what to buy; (2) deciding how to best take care of the devices purchased: and (3) ensuring that users are able to provide the maximum care to patients that the equipment allows. That is, the overall purpose of clinical engineering is to provide the <u>highest quality care at the lowest possible cost for the patient.</u>

The objective of this paper is twofold: <u>first,</u> to examine the full spectrum of management and engineering services that a hospital clinical engineering department can provide and <u>second,</u> to demonstrate how, with proper administrative and medical staff support, the potential benefits of establishing such a department can be achieved.

JCAH SUMMARY, 1981

Guidelines relating to clinical engineering support of patient-care equipment deal with hospital responsibilities for documentation, safety, continuing education for employees and established programs for preventative and corrective maintenance. Performance verification, calibration, and corrective maintenance records are required within Anesthesia, Emergency, Nuclear Medicine, Pathology and Medical Laboratory, Radiology, Respiratory Care, and Special Care components of the hospital. Within the Functional Safety and Sanitation section of the Accreditation Manual for Hospitals (AMH), references are made to clinical engineering safety committee participation - with related continuing education responsibility - and to electrical safety considerations. These considerations include education of device users, assurance of device performance, new device evaluation, and appropriate record systems. All patient care devices are to be tested no less than every six months.

JOB DESCRIPTION

In 1979, Baptist Medical Center recognized that significant economies could be gained through an in-house clinical engineering program. Faced with the complexities of existing equipment management and the future equipment requirements of a large community medical center, nationally recognized biomedical engineering consultants were hired to evaluate and propose a sound plan of action. The plan of action included hiring a clinical engineer with sufficient technical, administrative, and educational background to organize an in-house department and implement the plans with appropriate hospital administrative support. Key elements in the plan which structured it for success included: (1) creating a separate department which could function independently yet complementary to the existing plant facilities/ hospital engineering department; and (2) having the department director/clinical engineer report directly to the medical center Associate Executive Director, which ensured that patient-care equipment priorities were well represented to administration. A clinical engineer was hired in the summer of 1980 and the new department was first implemented in October 1980, guided by the job description as listed in Ostrander (1979) and Pacela (1977) and utilizing three biomedical technicians (BMET'S).

Clinical engineering responsibilities can be summarized into four main areas: technical, administrative, educational, and research. It is not a new idea to attempt to provide these kinds of services in hospitals. Large teaching hospitals and the Veterans Administration have for years implemented similar programs with some amount of success. A key element in developing comprehensive in-house programs is the cost justification of such a department, which relates to the size of the hospital, its range of patient-care services, and its attempts to be a resource to the local community for the widest spectrum of diagnostic and treatment-oriented devices.

In the southeast United States, many community hospitals -- 600 beds or smaller - have implemented a partial clinical engineering program, but have failed to achieve the potential it offers and received only a partial return of their investment. The majority of community hospitals which have biomedical/clinical engineering programs only provide for equipment preventive maintenance, modification, repair and inspection of new devices. Typically, these partial programs are also limited to particular classes of devices. On the other hand, a comprehensive program extends the technical functions listed above to all patient-care devices, is intimately involved in the new device specification, evaluation, and selection process with the medical staff-standardizing whenever appropriate, and is significantly involved in inservice programs for users of the devices. Thus, purchasing support, malpractice avoidance through education, and maintenance support are the hallmarks of a fully developed clinical engineering program. (Note: A relatively complete summary of patient-care devices is included in the AHA/ASHE publication (1978). Patient-care devices are defined here as equipment used in the treatment of diagnosis of a patients condition where it is in the best interest of the patient and medical center that the equipment be periodically scheduled for testing to determine its accuracy, reliability, and safety (Judd, Lalmond, Grant, 1980)).

IMPLEMENTATION OF COMPREHENSIVE IN-HOUSE CLINICAL ENGINEERING PROGRAM

Baptist Medical Center - located in Jacksonville, Florida is the largest community hospital - 563 beds - in northeast Florida. It offers a wide range of tertiary care, employing some 2,000 diagnostic and acute-care treatment devices, valued at over seven million dollars. Included within the medical center is an adult hospital and a children's hospital. Major equipment-intensive areas include: (1) over sixty monitored intensive and progressive care beds for neonate, children and adults - including telemetry and arrhythmia analysis capabilities; (2) a Cardiovascular Center with wide-ranging diagnostic capabilities; (3) extensive operating room services with emphasis on open-heart surgery; (4) extensive emergency services /trauma unit with the region's only jet helicopter ambulance service; (5) radiology and nuclear medicine services offering the full range of diagnostic modalities including CT scanning, nuclear cardiology devices and state-of-the-art ultrasonic imaging devices; (6) radiation therapy offering 4 MEV and 20 MEV linear accelerators and a cobalt therapy device; (7) and strong departments of pulmonary care, neuro diagnostics, labor and delivery, and pathology and clinical lab.

A three year plan was developed by the clinical engineering department to fully provide services to all above mentioned medical center departments. The emphasis of the three year plan has been to progressively take on specific categories of patient-care devices developing the full range of services -- purchasing, education, and maintenance - for these devices. As areas to which full services are offered increase, increases in staff will occur appropriate to and cost-justified for the tasks being added. The three year strategy is to pursue the GENERAL BIOMEDICAL DEVICES - some 1,500 items -- in Year 1 and then the more specialized devices. Year 2, this includes radiology and nuclear medicine devices, i.e. x-ray machines and the CT scanner; and in Year 3, the radiation therapy and laboratory medicine devices.

YEAR 1 ACTIVITIES

1. Maintenance Savings - Providing fast response to daily problems, eliminating costly service contracts and reducing equipment downtime are some of the return-on-investment factors of an in-house department. Within three years, these and other factors such as physician/nursing confidence in devices and hospital reputation for quality care at reasonable prices will yield a return more than offsetting department operating costs. Utilizing the staff of four, some $60,000 in general biomedical device service contracts have been eliminated in Year 1.
2. Equipment Acquisition - This past year, clinical engineering has had a primary role in coordinating the selection of $1 million worth of high technology devices and a secondary role in designing the various facilities - $2 million - to properly utilize the new devices. Examples include : (1) selection of intensive care monitoring devices for 37 ICU beds and 8 central stations: (2) selection and interfacing of x-ray, patient monitoring, cath lab computer analysis, and audiovisual devices for a new heart center; (3) selection of patient monitors for 12 recovery beds; (4) outfitting a new cardiac rehab unit; and (5) selection of a computerized ECG stress test system. Some $30,000 has been saved in the selection and purchasing process in Year 1, directly attributable to c.e. input, either through prudent device selection or timely ordering.

CONCLUSIONS

The large community hospital, if it is to continue to function through the 80's, must manage and operate its high technology patient-care equipment in a very efficient manner. With the soaring costs of new device technologies - in original purchase and support dollars, clinical engineering expertise must be available to administration to assist them in optimizing use of financial and personnel resources. Clinical engineering can serve as an effective bridge between sometimes conflicting requirements of hospital physician, nursing, and administrative staffs.

The full spectrum of management and engineering services that a hospital clinical engineering department can provide has been shown. Proper administrative and medical staff support has caused this to be essentially a "break-even" operation for the medical center in its first year. Device user preparedness and confidence have increased. It is time for the large community hospitals in the southeast to become aware of the full range of benefits and capabilities of an in-house clinical engineering staff, as they develop their strategies for cost-effective delivery of patient care in this decade.

REFERENCES

American Society of Hospital Engineering of the American Hospital Association (AHA/ASHE-1978), "Medical Equipment Management in Hospitals", Chicago.
Arthur, R.M. and S.B. Boxerman (April-June, 1981), "Achieving the Promise of Clinical Engineering", Journal of Clinical Engineering, Vol. 6, No. 2, 105-109.
Judd, T.M., J.T. Lalmond, and D.G. Grant (August, 1980), "The Johns Hopkins Hospital High Technology Equipment (HTE) Quality Assurance Program", Johns Hopkins University Applied Physics Laboratory publication SR 80-4.

Ostrander, L.E. (January-March, 1979), "Presentation of the Clinical Engineering Role to Hospital Administrators", Journal of Clinical Engineering, Vol. 4, No. 1, 11-19.

Pacela, A.F. and E.G. Forbes (July-September, 1977), "Job descriptions & Work Objectives in Clinical Engineering: Part 2", Journal of Clinical Engineering, Vol. 2, No. 3 264-272.

Schaffer, M.J. (1977), "Clinical engineering: an in-depth review of its economic potential", Medical Care, XV: 552-567.

Simmons, D.A. and J.O. Wear (1977), Clinical Engineering Manual, Scientific Enterprises, Inc., Little Rock, Arkansas

APPLICATION OF THE DESIGN PROCESS TO A REHABILITATION ENGINEERING PROJECT: A WHEELCHAIR ADAPTATION

Cecil H. Ramage, Ph.D.
Associate Professor
Mechanical Engineering
University of South Alabama
Mobile, Alabama 36688

ABSTRACT

The design process is a well established sequence of events that must occur to achieve an optimal design. The ideas behind this process have been successfully applied to problems in many fields. This paper will describe the application of the design process to a wheelchair adaptation for a cerebral palsied child.

KEYWORDS

Cerebral Palsy, wheelchair adaptation, postural control design process, handicapped, body support, rolled seat.

INTRODUCTION

The design process will be described and its application to the design of a wheelchair adaptation discussed. The adaptation was designed and constructed for a young boy with cerebral palsy. The final design criteria resulted from interaction between the design team and physical therapists who were cognizant of the boy's seating requirements. The design and construction of a wheelchair adaptation is most successfully accomplished on an individual basis. This is especially true when the user is a handicapped child where body support and postural control are the important design criteria. For the cerebral palsied child, correct posture is imperative to reduce the occurrence of pressure sores and prevent the development of permanent deformities.

Commercial attempts to provide body support for the individual user lack the advantages of a custom-designed support and are very expensive. The basic wheelchair design itself provides a poor mounting surface for postural controlling pads and support straps.

DESIGN PROCESS

The design of the wheelchair adaptation was accomplished following a definite sequence of events. This sequence, known as the design process, consists of seven or more steps. A listing of the design steps follows:

Recognition of need
Problem definition
Literature search
Ideation and concept selection
Preliminary design and evaluation
Detail design
Final evaluation

The design could also include steps such as market analysis, production planning, packaging design, etc. If the design described in this paper were to be manufactured on a commercial basis, these steps would be a necessary part of the process.

RECOGNITION OF NEED

The Mechanical Engineering Department at the University of South Alabama was approached by the Mobile Rotary Rehabilitation Center and United Cerebral Palsy of Mobile for engineering assistance on problems they had experienced with a number of their clients. A design team of two senior students was assigned to work with the physical therapists at the center. A client was selected for evaluation, a young boy with postural and body control problems. The therapists outlined the boy's physical problems and expressed their ideas concerning his needs.

PROBLEM DEFINITION

The therapists described the characteristics of the boy's particular type of palsy as follows:

1. Severe spasticity in his trunk, arms and legs,
2. Moderate lack of control, or athetosis of his trunk and arms,
3. Scissoring, or adduction, of his lower limbs,
4. Moderate lack of head control,
5. Downward rotation of his feet.

The boy's wheelchair did not have seat, chest and head restraints located to properly position his body for prevention of permanent skeletal deformations. The therapists asked that the sling back of the chair seat be modified to allow adjustment of the positioning straps.

After the trip to the rehabilitation center and the discussion with the therapists, the design team spent some very important library time researching not only the problem of supporting a person in a wheelchair but also the new language they found themselves speaking. The literature search and subsequent discussions with the therapists led to a complete redesign of the seating and support system rather than the simple adjustment of straps and pads that was first thought to be the desired solution. One cannot depend on the apparently obvious route to solving a problem giving the best solution.

LITERATURE SEARCH

Once the problem was defined as a need for complete seat redesign, an extensive search was made for literature relating to wheelchair seating designs. Many seating systems were compared and several designs selected for further evaluation.

IDEATION AND CONCEPT SELECTION

As the literature search progressed, several idea-producing or brainstorming sessions were held. Ideas generated at these sessions ranged from conservative run-of-the mill solutions to highly imaginative approaches, some of which showed promise for highly successful designs. In this type session, it was very important that constraints not be considered so that the designer's imagination could work without restriction.

After several of these sessions were held, the design team made a list of the ideas generated. Those having obvious problems were discarded. The ideas that appeared to be most promising were included with approaches found in the literature search and compared. The comparison used a numerical ranking technique that rated each design's characteristics relative to criteria including support, comfort, adjustability, ease of manufacture and cost. These criteria were selected by the design team and the therapists.

A rolled seat configuration was selected over other designs based on this evaluation. The geometry of the rolled seat helps keep the child well back in the chair and inhibits extensor thrust, a common characteristic of palsied children.

PRELIMINARY DESIGN AND EVALUATION

Once the design configuration was selected a preliminary design effort was initiated. This phase of design consists of a conceptual study of the design requirements and how these requirements are to be met by the selected design.

First, the rolled seat was given a plywood back to furnish the required support for positioning pads and straps. Next, an adjustable knee abductor was added at the front of the seat to prevent leg scissoring. Foot and ankle control was provided by an adjustable footrest and toe straps. Pads were specified where appropriate. The preliminary design was reviewed by the physical therapist prior to the team's proceeding into the detail design phase.

DETAIL DESIGN

The detail design phase started with a careful measurement of the wheelchair that was to be modified and of its occupant. The boy's size and weight were used, along with the wheelchair dimensions, to size the seat, back and footrest. Scale drawings were made to define the overall layout and to specify the materials and hardware necessary for construction. In addition a step-by-step construction sequence was developed. Next, material properties and expected loads were checked to ensure the structural integrity of the seating system. Tie-downs and straps were located to support the boy properly in the seat. Pads were sized and materials specified to furnish a high-friction surface that would reduce the boy's tendency to slip around in the seat. A vest-like upper trunk support was designed to support the boy in the chair. A neck pad was added to support his head. The entire design was rechecked by the team and then submitted to the therapists for final approval.

CONCLUSION

This paper has presented a step-by-step correlation of the design process to the design of a wheelchair adaptation for a young boy. In this case the seating system was constructed by the design team and mounted on the frame of a commercial wheelchair. The system was designed to enable its construction by a person having average woodworking skills and basic tools. This makes possible utilization of the seat design by low-income families and gives the parent who constructs the seat the satisfaction of being able to help his or her child.

At this time, the seat has been in use for approximately one year and has met the design criteria. Although the chair is in reasonably good condition, it will soon be replaced to accommodate the boy's growth. A similar seating system will be constructed for the new chair.

REFERENCE

Pietrzyk, J., and Runkel, C., Wheelchair Positioning Aids, Mechanical Engineering Design Project, May 1981.

Session 18

MEDICAL INSTRUMENTATION

BURST PACEMAKER THERAPY FOR CONTROL OF REENTRANT TACHYCARDIA

Michael L. Hardage and Michael B. Sweeney

Intermedics, Inc.
P.O. Box 617
Freeport, Texas 77541

ABSTRACT

The Intermedics CyberTach 60 is a multiprogrammable, automatic antitachycardia "burst" pacemaker designed to deliver a rapid burst of pulses at a preprogrammed rate and duration upon detecting tachycardia. This paper summarizes the history of pacemaker therapy for reentrant tachyarrhythmias, and describes the CyberTach 60. Additionally, we report on the clinical evaluation of the CyberTach 60 in both the atrial and ventricular applications. Automatic "burst" pacing appears to be a safe and effective method of treating patients with sustained reentrant tachycardia.

KEYWORDS

Antitachycardia burst pulse generator; reentrant tachycardia; clinical evaluation.

The available treatments for tachycardia include medical, surgical, or device (pacemaker) therapy, or a combination of these. Historically, device therapy has constituted a very small aspect of this regimen. Because of the ongoing maturation of cardiac electrophysiology, partially due to advanced techniques and instrumentation, there has been increased interest in and development of sophisticated pacemakers to treat tachycardias. This paper deals specifically with one of these recently developed devices, the Intermedics CyberTach 60, which is a multiprogrammable, automatic antitachycardia "burst" pulse generator.

Traditionally, tachycardias have been classified in two basic categories: those resulting from abnormal automaticity, and those produced by a rapidly circulating impulse in an anatomically or functionally determined pathway (reentrant tachycardia). The arrhythmias in the second category, reentrant tachycardias, are readily initiated and terminated by pacing.[1] The mechanism that enables this termination depends upon the delivery of a pacing pulse of sufficient prematurity to "collide" with the tachycardia wavefront, thereby interrupting the circus pathway. Success in termination depends on several factors, including the type and nature of the tachycardia, the proximity of the pacing electrode to the tachycardia circuit, the reliable detection of the tachycardia, and the delivery of impulses at the proper time.

Accepted forms of pacing for tachycardias include overdrive pacing,[2,3] underdrive pacing, single or double extrastimuli,

---

[1]Josephson, M.E., and S.F. Seides (1979). *Clinical Cardiac Electrophysiology: Techniques and Interpretations.* Lea & Febiger, Philadelphia. Chap. 9, p. 148.

[2]Fisher, J.D., R. Mehra, and S. Furman (1978). Termination of ventricular tachycardia with bursts of rapid ventricular pacing. *American Journal of Cardiology*, 41, 94-102.

[3]Hyman, A.L. (1979). Permanent programmable pacemakers in the management of recurrent tachycardias. *PACE*, 2, 28-39.

train pacing, and simultaneous pacing at two (or more) sites.[4]

The mechanism involved in tachycardia overdrive pacing (often referred to as burst pacing) requires pacing at a rate in excess of the tachycardia with 1:1 capture for a few impulses, followed by abrupt cessation of pacing. Underdrive pacing consists of asynchronous pacing at rates comparable to the spontaneous normal sinus rhythm, resulting in spikes occurring randomly throughout the tachycardia cycle. In theory, a spike will ultimately occur at a critical point in the cycle to terminate the tachycardia (termination zone).

A somewhat more effective technique for pacing during the termination zone is to deliver a premature single stimulus or double stimuli, coupled to a preceding intrinsic beat. Upon sensing the intrinsic depolarization, one or two stimuli are delivered after an appropriate delay. This delay may be programmable, may be automatically altered until an effective interval is achieved, or may be automatically calculated as a percentage of the tachycardia interval.

Some tachycardias have a termination zone immediately following the myocardial refractory period. An effective pacing therapy for these patients consists of high frequency trains of stimuli, beginning during refractory period and lasting until a pacing-induced depolarization occurs. This therapy, referred to as train pacing, requires very rapid pacing (on the order of 1000 PPM or more) for a brief period (a few hundred milliseconds, typically).[5]

In many cases pacing the atrium and ventricle simultaneously with a predetermined AV sequential time does not appear more efficacious than single chambered pacing for the underdrive termination of supraventricular tachycardia. It is conceivable that in a small proportion of patients, simultaneous atrial and ventricular pacing may terminate the tachycardia more promptly and in rare instances, simultaneous underdrive pacing may indeed be required for actual termination.

Although experimentation with pacing as a treatment for tachyarrhythmias began back in the mid 1960's, the overwhelming majority of the industry's efforts has been devoted towards refinement of bradycardia-treating pacemakers. Before a description of the Intermedics CyberTach 60 is presented, a brief outline of the devices that preceded it will be given.

Perhaps the most primitive method employed for the treatment of tachyarrhythmias was the patient-activated underdrive units. These units were conventional demand pulse generators that were activated upon the patient's perceiving the onset of the arrhythmia. Activation consisted of applying a magnet over the pulse generator, to convert it to a fixed rate generator. The resulting series of random stimuli would eventually terminate the reentrant tachycardia.[6,7] An outgrowth of this pacing mode led to pulse generators that would double in pacing rate when a magnet was applied; for example, would increase from perhaps 75 PPM to 150 PPM.[8] Another approach was the rate scanning system that upon application of the magnet, would steadily increase its rate from 165 to 300 PPM until the magnet was removed.[9] Other scanning pulse generators evolved that would deliver single or double stimuli at a programmed interval coupled to the intrinsic tachycardia events. The coupling interval decremented by several millisecond steps after each impulse until the magnet was removed. Another patient-activated system was the radio frequency rapid stimulator developed

---

[4]Akhtar, M., C.J. Gilbert, M. Al-Nouri, and D.H. Schmidt (1979). Electrophysiologic mechanisms for modification and abolition of atrioventricular junctional tachycardia with simultaneous and sequential atrial and ventricular pacing. *Circulation*, <u>60</u>, 1443-1454.

[5]Fisher, J.D., S.G. Kim, and E. Ostrow (1981). Ultra-rapid single capture train stimulation for termination of ventricular tachycardia. Second European Symposium on Cardiac Pacing, May 4-6 (Abstract).

[6]Ryan, G.F., R.M. Easly, L.J. Zaroff, and S. Goldstein (1968). Paradoxical use of a demand pacemaker in treatment of supraventricular tachycardia due to Wolff-Parkinson-White syndrome. *Circulation*, <u>38</u>, 1037-1043.

[7]Cooper, T.B., W.A.H. MacLean, and A.L. Waldo (1978). Overdrive pacing for supraventricular tachycardia: A review of theoretical implications and therapeutic techniques. *PACE*, <u>1</u>, 196-221.

[8]Peters, R.W., E. Shafton, S. Frank, A.N. Thomas, and M.M. Scheinmann (1978). Radiofrequency-triggered pacemakers: Uses and limitations. *Annals of Internal Medicine*, <u>88</u>, 17-22.

[9]Mandel, W.J., I. Yamaguchi, M. Laks, and B. Berkovits (1977). The use of overdrive pacing for termination of tachycardia episodes in the Wolff-Parkinson-White (WPW) syndrome. In Yoshio Watanabe (Ed.), *Cardiac Pacing - Proceedings of the Vth International Symposium on Cardiac Pacing, Tokyo, March 14-18, 1976.* Excerpta Medical, Amsterdam. pp. 162-165.

in the late 1960's.[10,11,12,13] This system incorporated a small implantable receiver/stimulator (connected to a standard permanent pacing lead) and a hand-held transmitter with rate adjustability from 50 to 400 PPM. The patient, upon detecting the arrhythmia, placed the transmitter over the implanted receiver, thus activating the receiver for several seconds. This radio frequency system has been used clinically for more than ten years, and although marketed as a rapid atrial stimulator, has demonstrated its success in both the atrium and ventricle.[14,15]

The previously discussed stimulators were often implanted in conjunction with a conventional demand pulse generator to control the patient with bradycardia pacing, since most of the stimulators did not provide this function. One feature common to all those devices was their inability to automatically detect tachycardia. The ability of a pacemaker to detect tachycardia and subsequently deliver the pacing therapy is important in that it does not rely on the patient to perceive the arrhythmia and activate the pacemaker. Patient-activated systems require direct patient interaction that may be psychologically disturbing to the patient and rule out implantation in patients whose tachyarrhythmia induces syncope. This constraint often eliminates pacing therapy for patients who have proven refractory to pharmacologic therapy and are not candidates for arrhythmia surgery.[16]

The CyberTach 60 circumvents the problems associated with patient-activated systems. The device, after automatically detecting the tachycardia, delivers a rapid burst of pulses at a programmable rate and duration. Tachycardia detection is based upon the presence of seven consecutive intrinsic intervals with a cycle length shorter than the programmable tachycardia criterion.

The device is a multiprogrammable bipolar pulse generator with overall dimensions of 6.0 x 4.6 x 1.5 centimeters and a mass of 92 grams. The power source is a 3.2 ampere-hour lithium iodine cell. The circuitry and programming system of the CyberTach 60 are based on the Intermedics CyberLith multiprogrammable technology that has been utilized in pacemaker implants since early 1978. The system is implemented with custom complementary metal-oxide semiconductor large scale integrated circuitry (CMOS LSI) that performs all timing and logic functions of the CyberTach 60.

The circuitry supports a bi-directional communications link between pacemaker and programmer. The programming system is the same as that in the CyberLith series of pacemakers, which includes a high energy, inductively coupled energy link that incorporates four independent security features. A fail-safe programming system is imperative for antitachycardia pacemakers, since misprogramming such devices could prove disastrous. The programming link enables independent adjustability of six parameters, including bradycardia rate, output pulse width, demand sensitivity, tachycardia criterion, burst rate, and burst duration, and also allows activation of telemetry transmission.

There are two programmable modes: standard AAI/VVI mode with programmable bradycardia pacing rates of 60 and 80 PPM, and AAI/VVI mode with tachycardia response. The criterion for detection of tachycardia is based on sensed intrinsic rate and is programmable to one of two values. CyberTach 60 pulse generators nominally incorporate tachycardia criterion rates of 137 and 180 PPM; however, units incorporating other criterion rates are available from Intermedics.

Upon detection of seven consecutive intervals which satisfy the tachycardia criterion rate (when programmed to the tachycardia response mode), the CyberTach 60 emits a series of pulses at a programmable rate (7 values from 180 to 720 PPM) and duration (4 values from 1.3 to 5.3 seconds). A special selection is available at a rate of 1440 PPM for 0.33 seconds.

---

[10]Todo, K., S. Kaneko, T. Fujiwara, K. Sugiki, N. Tanaka, and S. Komatsu (1979). Patient-controlled rapid atrial pacing in the long-term management of recurrent supraventricular tachycardias. In Claude Meere (Ed.), *Cardiac Pacing-Proceedings of the VIth World Symposium on Cardiac Pacing, Montreal, October 2-5, 1979.* PACESYMP, Montreal. Chap. 9-6.

[11]Wald, R.W., M.B. Waxman, J.F. Bonet, A.D. Sharma, D.C. MacGregor, B.S. Goldman (1979). Patient initiated rapid atrial stimulation for treatment of paroxysmal supraventricular tachycardia. In Claude Meere (Ed.), *Cardiac Pacing-Proceedings of the VIth World Symposium on Cardiac Pacing, Montreal, October 2-5, 1979.* PACESYMP, Montreal. Chap. 9-9.

[12]Lister, J.W., A.J. Gosselin, D.A. Nathan, and S.S. Barold (1973). Rapid atrial stimulation in the treatment of supraventricular tachycardia. *Chest*, 63, 995-1001.

[13]Waxman, M.B., R.W. Wald, J.F. Bonet, D.C. MacGregor, and B.S. Goldman (1978). Self-conversion of supraventricular tachycardia by rapid atrial pacing. *PACE*, 1, 35-48.

[14]Hartzler, G.O., D.R. Holmes, Jr., and M.J. Osborn (1981). Patient-activated transvenous cardiac stimulation for the treatment of supraventricular and ventricular tachycardia. *American Journal of Cardiology*, 47, 903-909.

[15]Wiener, I. (1980). Pacing techniques in the treatment of tachycardias. *Annals of Internal Medicine*, 93, 326-329.

[16]Lister, W. (1978). A new technology: Patient-triggered pacemakers, *Annals of Internal Medicine*, 88, 120-121.

The automatic tachycardia burst function of CyberTach is easily disabled by programming to the AAI/VVI mode. If a programmer is unavailable, the burst function can be disabled simply by placing a magnet over the pacemaker which results in asynchronous pacing at the programmed bradycardia rate.

As described previously, the major categories for tachycardia pacing therapy include (1) overdrive (burst) pacing, (2) diastolic scanning (either by underdrive or single/paired extrastimuli), (3) train pacing, or (4) atrioventricular sequential pacing. Because of the diverse characteristics of tachycardias, it would be desirable to provide a single pacemaker that could function in each of the four categories. Although no such device has been developed to date, the CyberTach 60 can perform to some extent in the first three categories. It is primarily designed as a burst pacer (category 1), but by proper selection of burst rate, the function of category 2 can be performed. Because the first pulse of the burst occurs synchronous with a tachycardia beat, by altering the programmed burst rate one is also changing the point in diastole when the second pulse in the burst occurs. The third category (train pacing) is provided by the program selection of 1440 PPM burst for 0.33 seconds.

The clinical evaluation study of the CyberTach 60 has been limited to a relatively small number of institutions with the capability of performing the thorough electrophysiologic examination required prior to implantation. Through the use of intracardiac recording and programmed electrical stimulation, the mechanism of an arrhythmia is identified and localized. The tachycardia is then initiated and terminated with programmed stimulation. These initial studies are of paramount importance in screening out patients in which burst pacing is ineffective or potentially dangerous. In atrial application of programmed stimulation this screening process includes discovery of the presence of a concealed accessory pathway with a short effective refractory period which can result in transmission of the rapid atrial stimulation to the ventricle. In ventricular applications, acceleration of the tachycardia or a degeneration into ventricular fibrillation obviously precludes the use of an automatic burst pulse generator. These electrophysiologic studies are often performed concurrently with anti-arrhythmic drug therapy used to supplement the pacemaker therapy.

Once the patient has been studied and burst pacing has proven to be successful in controlling the tachycardia, the efficacy of the CyberTach 60 system must be thoroughly evaluated. The most effective way to evaluate the function of the CyberTach 60 is to connect an external, non-sterile CyberTach 60 to a temporary lead electrode positioned in the site to be employed if the system is eventually implanted.

A switchable junction box enables the investigator to connect the programmed stimulator and the external CyberTach 60 to the patient. The arrhythmia is induced by programmed stimulation and terminated by the external CyberTach 60. This offers a reasonable and reliable way to duplicate the actions of the implantable system, and arrive at the optimal parameters prior to implantation. At this point in the electrophysiologic study, numerous successful initiations/terminations of the arrhythmia are required. Prior to discharging the patient post implant, numerous successful terminations are again required.

The CyberTach 60 has been in clinical evaluation for 30 months at the time of this writing (February 1982). In these 30 months, there have been 58 pulse generators implanted; 36 atrially and 22 in ventricular applications. Two CyberTach 60's, one atrial and one ventricular application, were removed due to slowing of the arrhythmia below the minimum standard tachycardia criteria of 137 ppm. Both these units were replaced with special parameter CyberTach 60 pulse generators. Therefore, there were 58 implanted and 56 patients. Five patient deaths have been reported. These deaths have been attributed to non-pacemaker related causes and the returned devices found to be within specifications.

The CyberTach 60 implants have been performed in 23 investigational centers by 24 clinical investigators implanting from one to nine units each. Extremely encouraging results have been observed in this study. Burst pacing to terminate tachyarrhythmias appears very efficacious.[17,18,19]

The authors would like to thank K.M. Guerrant , S.C. Hendrix and P.R. Gragert for their invaluable assistance in the preparation of this paper.

---

[17]Griffin, J.C., J.W. Mason, and R.V. Calfee (1980). Clinical use of an implantable automatic tachycardia - terminating pacemaker. *American Heart Journal*, 100, 1093-1096.

[18]Falkoff, M.D., L.S. Ong, R.A. Heinle, and S.S. Barold (1981). Implantable multiprogrammable automatic burst antitachycardia pacemakers for refractory ventricular tachycardia. Second European Symposium on Cardiac Pacing, May 4-6 (Abstract).

[19]Griffin, J.C., J.W. Mason, D.L. Ross, and R.V. Calfee (1981). The treatment of ventricular tachycardia using an automatic tachycardia terminating pacemaker. *PACE*, 4, 582-588.

MICROPROCESSOR MONITORS HEART RATE

Russell E. Puckett, P.E., and Albert B. Grubbs

Department of Engineering Technology
Texas A&M University
College Station, TX 77843

ABSTRACT

Modern medical instrumentation uses technology that resulted from space-age electronics. Integrated circuit (IC) components are fast replacing hard-wired electronic logic circuits. The IC chip called a microprocessor makes possible sophisticated instruments that can measure, compute, and display data for recurring physiological changes such as heart rate. These instruments can operate faster than changes occur, permitting calculations of pulse rate almost instantly. Because the rate is known quickly, corrective action can respond to the measurement.

This paper describes a pulse rate monitor that uses a microprocessor to measure heart rate and compare it to a preset limit. It detects the rate from a measurement of the electrophysiological signal, and activates alarms as upper and lower limits are exceeded. The measured rate is displayed at all times, until alarm conditions are reached for the limits. Then the display shows the rate that caused an alarm, until an operator resets the instrument. Upper and lower limits also can be manually reset by an operator to suit the application.

KEYWORDS

Microprocessor; heart rate monitor; pulse rate display; biomedical instrumentation; electronic instruments; ECG monitor.

INTRODUCTION

Applications of recent advances in electronics can be found in many diverse fields. Biomedical instrumentation is but one of the areas that use integrated (IC) technology to replace bulky devices of only a few years ago. In addition to the advantage of small size, a very significant inprovement in reliability can be realized from large-scale (LSI) and very-large-scale (VLSI) technology. Systems using such components can be packaged in tabletop enclosures, in contrast to multiple cabinets of circuitry that used early versions of the components. With the advent of IC technology many electronic measurement systems became obsolete, and better instruments have resulted for use in the health care field.

Probably the most important recent development is the IC chip called a microprocessor. This tiny device makes it possible to have "smart" medical instruments that can measure and compute rapidly changing data, and generate visual displays for operator use. Recurring changes such as pulse rate can be detected easily, and because of the speed of such instruments, a catastrophic change can be detected almost as soon as it happens. Remedial action can be undertaken quickly in response to the measured pulse rate. Thus, electronic technology that is applied for this purpose is a major asset in biomedical instrumentation.

The microprocessor-based pulse rate monitor described in this paper features the Motorola 6800. This microprocessor was selected for its proven use in countless applications, and because it is supported by a vast array of components and software. Because of these features future expansion of this system is a viable option.

This monitoring system extracts the electrophysiological intervals of cardiac rhythms, and uses them to calculate pulse rate in beats per minutes (BPM) for display as a three-digit number. Whenever upper or lower rate limits are exceeded, an audible alarm is set off to alert an operator of the system. The pulse rate is displayed continuously as it is recalculated for successive beats, for the convenience of the operator. Based on the optical display and alarms, an operator can take whatever action may be necessary for an individual patient.

DESIGN OF THE SYSTEM

In the design of this equipment, it was necessary to consider the problems associated with medical monitoring systems. In particular the measured pulse rate results in a continuous visual display, beat by beat, for an accurate and fast indication of cardiac activity. In addition, alarms are needed whenever either low or high preset limit is exceeded. The rate causing an alarm condition then is "frozen" on the display panel for operator recognition. Otherwise the display shows the last measured and computed pulse rate. The system design incorporates features usually found on medical equipment for monitoring patient vital signs.

In order to obtain clean ECG signals, electrodes can be attached to the chest of a patient. Figure 1 shows a typical ECG signal that can be detected using such an electrode configuration. This waveform is exaggerated to identify various artifacts according to the usual nomenclature for ECG signals.

The pulse rate monitor uses the time intervals between successive R-waves to determine a number for display. Since the R-wave is always the largest amplitude

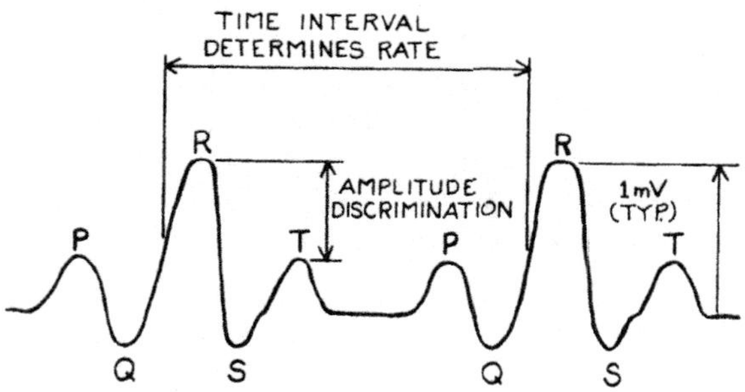

Fig. 1 Features of the ECG waveform.

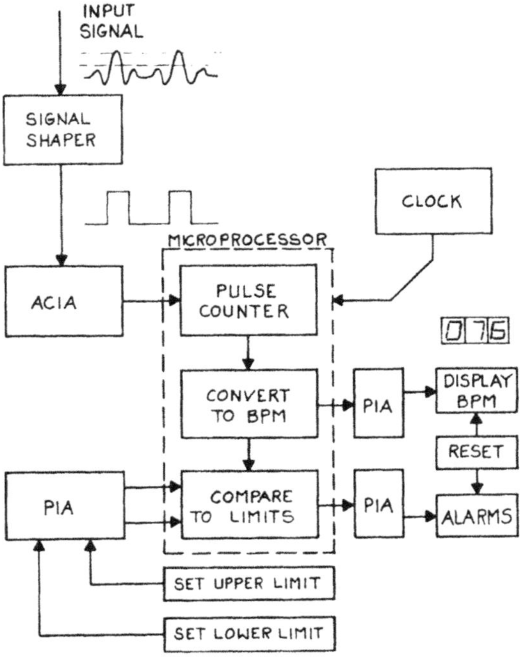

Fig. 2 Functional layout of the pulse rate monitor.

signal present in the recurring ECG pattern for most patients, the time associated with the R-R interval can be obtained by amplitude discrimination in electronic circuits. Lower amplitude P-wave or T-wave signals can be prevented from activating the circuits, and only have a response to R-waves. The measurement of the R-R interval is the basis for the pulse rate monitor.

The instrument includes several distinct operating circuits, as indicated by the functional layout in Fig. 2. The input signal from the patient electrodes is passed through a signal shaper, which will generate pulses corresponding to the time interval of the R-R period. The resulting pulse enters the microprocessor together with pulses from a clock source. Within the microprocessor these two pulse trains produce a measurement of time interval, from action of the pulse counter. The resulting time measurement is then converted to a pulse rate in beats per minute, which is compared to upper and lower limits. Conversion and comparison lead to display of the rate as a three-digit number (expressing BPM) and to activation of any alarm required from the preset limits. A description of the instrument will illustrate use of the microprocessor for real-time ECG analysis.

The signal shaper circuit detects the ECG signal from a patient, amplifies it from the millivolt range to the order of volts, then converts the signal into a usable form compatible with voltage levels needed by the microprocessor. A level detector can discriminate between P-wave and R-wave amplitudes, so that only successive R-waves will trigger calculation of a time interval.

A pulse shaper circuit limits the resulting amplitude of pulses that bound the R-R interval, so that the microprocessor will receive signals that are within its amplitude specifications.

Resident programs in the microprocessor calculate the time interval and convert it to its corresponding pulse rate in BPM. Subroutines in these programs also condition the conversion for activating alarms and the visual display of the pulse rate. Clock pulses are used for the calculations, comparing their known rate to the incoming signals. The software is stored in a read-only memory (ROM), part of the microprocessor 64k memory capability. Additional memory may be added in any expansion of the instrument for other purposes. A 128 x 8 random-access memory (RAM) stores input pulses for the counting and conversion process.

Calculation processes and conversion of data for display may take only about 60 milliseconds for completion. Thus it is possible for a new rate to be available every 60 ms or so. If the three digit number were changed that fast, it would be visible with an obvious flicker. A 60 ms rate would correspond to successive ECG pulses every 60 ms, indicating patient heartbeat at 1,000 beats per minute. A more realistic rate from 60 to 80 BPM also could cause the display to flicker if nothing prevented it. Circuitry has been included in the instrument to synchronize rate calculation and display period for that rate. Every input R-wave triggers the microprocessor to calculate a rate for the last R-R interval and convert it to a rate in BPM. However, only every third R-wave will change the displayed number of beats per minute. The time delay between displays is short enough to capture significant changes in the rate, but is long enough to stabilize the display for operator recognition.

SUMMARY

Modern electronic circuitry such as the microprocessor makes possible versatile and sophisticated instruments for applications in the biomedical field. A pulse rate monitor has been described, which turns irregular ECG pulses into suitable signals that trigger a high-speed calculation of electrophysiological activity for display in a clinical environment.

The instrument can be packaged in a small volume because of compactness of electronic components used with the system. Expansion capability includes additional memory and inherent computing power to store the patient's pulse rate history, and furnish a hard copy record for use by health care personnel in analyzing and prescribing corrective actions. A printout could be continuous, periodic (the time selected by an operator), or on demand. The amounts of data in storage would be a function of memory installed in the system.

Other uses of this circuitry include data collection from other physiological activity, for instance, respiration rate and temperature. Other options include visual display of ECG waveforms on a CRT monitor, as well as associated strip chart recorder for ECG records. In addition this system could be interconnected with a host computer to allow multiple monitoring of patients, as in a large hospital. Certainly this instrument can be put to beneficial use in diverse applications. Its use as a pulse rate monitor illustrates its applicability in real-time analysis of biomedical data.

CUTTING TOOL TECHNOLOGY APPLIED TO INTRA-ARTICULAR ARTHROSCOPIC SURGICAL INSTRUMENTS

W. H. Gondring* and C. R. Wheeler**

*St. Joseph Orthopedic Associates, Inc.,
St. Joseph, Missouri, U.S.A.
**A-R Engineering, Inc.,
St. Joseph, Missouri, U.S.A.

ABSTRACT

With the introduction of powered rotating instruments to arthroscopic surgery, cutting tool engineering and concepts commonly used in other disciplines may be applied. Rotating cutting surfaces require a different engineering approach than that used for straight cutting blades.

The problem addressed is how to design the "window" or opening in a rotating instrument to duplicate as nearly as possible the cutting action of a scissors. The method and motion by which the surgeon accomplishes the cutting task is equally important.

KEYWORDS

Intra-articular shaver; arthroscopic surgical techniques; intra-articular rotating surgical instruments; rotating cutter surface design and analysis; rotating arthroscopic surgical instrument systems.

INTRODUCTION

Since the mid 1970's, powered intra-articular shavers have become more and more commonplace alongside hand powered arthroscopic instruments. Since the shavers have departed from the normal cutting action employed by blades and scissors, a more detailed engineering analysis of the design and application is needed.

This paper will deal in detail with a new window design as well as a cutting technique to develop what is believed to be the first true systems approach to rotating intra-articular arthroscopic surgical instruments.

HISTORY

While arthroscopy was first introduced in 1918, it was not until the 1970's and the perfection of fiber optics that it met with wide acceptance. With increased usage the limitations of manual tools became apparent and the development of powered intra-articular surgical instruments began.

INTRA-ARTICULAR SHAVERS

Three of the more popular shavers available are built by Karl Storz, Stryker and Dyonics, Inc. Since only the Stryker and Dyonics units were available, analysis and cutter design were limited to those units. Adaptors were designed and developed to allow the same cutters to be used on either motor unit. This effects a cost savings for the system since only one adaptor is needed for all cutters.

ROTATING CUTTER SURFACE DESIGN ANALYSIS

An engineering analysis of the cutting process of the rotating cutters available was undertaken to determine if their efficiency could be increased. It soon became apparent that the surgeon and engineer do not view the word "cut or cutting" in the same light. The surgeon does not have the word "cut" in his vocabulary. The engineer, however, employs the word on a daily basis. We shall use the term "cut" or "shear" in the same context to mean to divide into parts with an edged instrument.[1] This is the engineer's definition.

The rotating shaver cutter action is analogous to cutting tools used in the metalworking industry. Technology from the discipline was found to be readily applicable to these instruments. In the case of a lathe or milling machine, the work (material) is clamped and the cutter rotated against it to remove material. The work may be rotated and the cutter held stationary. In either condition, the cutting action is the same. This is the same condition found in the rotating intra-articular shaver.

Figure 1 shows a vector diagram of the forces present during the cutting action.[2]

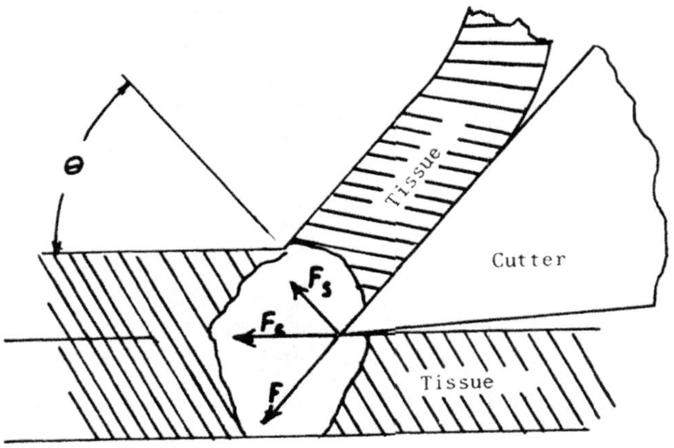

Fig. 1. Vector diagram of cutting action. θ = cutting angle, F_s = shear force, F = frictional force, F_c = cutting force.

---

[1] David B. Guralnik, Editor, Webster's New World Dictionary, (Williams Collins Publishers, Inc., 1979), p. 349.

[2] H. W. Pollack, Tool Design, (Reston Publishing Company, 1976), p. 328.

The cutting force is given by the equation.[3]

(1) $F_c = LtS_s$ Where L = Length of cut edge
t = Thickness of material S_s = Shear strength of material

One can see that the cutting force is directly proportional to the length of the cutting edge in contact with the material. If this length is minimized, a point contact condition would exist. Analysis of the cutting action of scissors reveals that the point-contact condition is closely approximated with only material thickness being the limiting factor. The material simply slides along the edges of the scissors until the frictional force F (Fig. 1) is equal to the cutting force F_c and then cutting begins.

Analysis of available cutters revealed that the cutting edges were for the most part straight. Some attempts had been made to duplicate the action of scissors with little success. In most cases, the angled surfaces increased the stresses at the tip of the instrument causing premature failure.

The task then became that of designing a cutter edge or window that would duplicate the cutting action of scissors and minimize tip stresses. Evaluation revealed that the elliptical window would accomplish both requirements since it duplicates the curved meniscal scissors cutting edge.

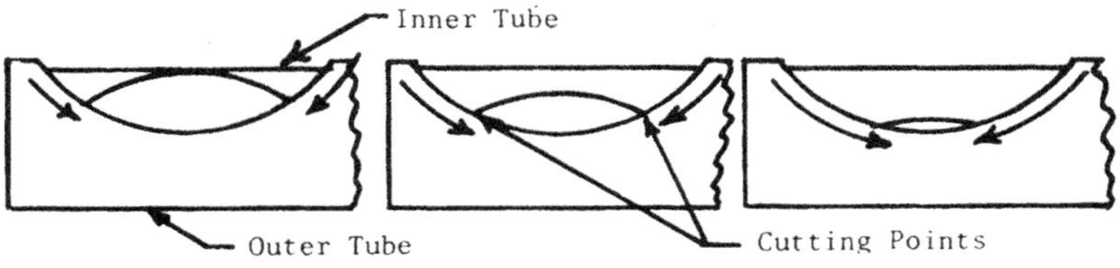

Fig. 2. Cutting action of elliptical window.

As seen in Fig. 2, as the inner tube rotates, the ellipse closes in the stationary outer tube causing point contact cutting to progress from both ends toward the center of the ellipse. This action moved the point of highest stress from the tip itself to the center of the ellipse and minimized the longitudinal stress component.

THE SYSTEMS APPROACH

In the development stage of the project it became evident that one or two cutters would not adequately perform all requirements. The variables encountered for intra-articular sugery were: a) vertical opening knee joint space, b) shear strength of tissue, and c) size of tissue defect. What was needed was an entire surgical instrument system. The elliptical side-window design worked well on soft tissue and chondromalacia. Cutter diameters ranging from 3 mm. to 5 mm. in 0.4 mm. increments served all needs encountered.

End cutters were designed using the elliptical window concept for cutting meniscus because of restrictions of angular approach.

---

[3] H. W. Pollack, Tool Design, (Reston Publishing Company, 1976), p. 413.

ARTHROSCOPIC SURGICAL TECHNIQUES OF ROTARY CUTTERS

During development of the cutters, it became apparent that surgical technique was equally important. Conventional approaches started at the base of the tear or tissue and severed the material at that point. The rotary cutters work just opposite. Cutting must start at the outside edge and proceed toward the base of the tissue or material. Otherwise, the window becomes clogged and the cutter tears or chews at the material.

Proper use of the suction feature greatly enhances powered cutter efficiency. By first sucking the ends of fibrillated material into the window, a smooth clean cut is achieved.

Another misconception uncovered was the idea that since this was a powered cutter it must be very aggressive. The surgeon tended to crowd too much material into the window causing clogging and tearing. This tendency is compounded by the suction action pulling more and more material into the window. When the cutter was allowed to cut and clear the window before more material entered, the cutting action was clean and efficient. The shear strength of the material also had a marked effect on the cutting edge efficiency. Therefore, various tissues have different cutting requirements. This technique was developed over many cases by multidisciplinary collaboration of surgical and engineering personnel.

CONCLUSIONS

The application of cutting tool technology to rotating intra-articular arthroscopic surgical instruments has shown dramatic improvement in cutter efficiency. A system of cutters has been developed which allows the surgeon great latitude in performing the many and varied procedures encountered during arthroscopic surgery.

With the design of drive adaptors the system becomes standardized for the Stryker and Dyonics motors. Other motor adaptors will be forthcoming.

ACKNOWLEDGEMENTS

The authors wish to thank the personnel of St. Joseph Hospital, St. Joseph, Missouri, for their assistance in cutter analysis and development, and to Phyllis Jarman and Paul Wheeler for their help in preparing this paper. Also we wish to thank Dr. Jerold Lancourt and Leon Herring for technical advice and long hours of support and help.

REFERENCES

Eikelaar, H. R. (1975). Arthroscopy of the Knee Royal United Printers. The Netherlands.
Gondring, W. H. (1981). "Arthroscopic Surgery of the Knee - A Preliminary Report". Accepted for publication in Military Medicine.
Johnson, L. L. (1977). Comprehensive Arthroscopic Examination of the Knee. C. V. Mosby Company, St. Louis.
McGinty, J. B. and Berci, G. (1981). Arthroscopy Guidelines Karl Storz Endoscopy. America, Incorporated.
Metcalf, R. W. (1979). Arthroscopic Meniscectomy Seminar Handbook Materials. Self published.

A MICROPROCESSOR BASED AUTOMATED H RESPONSE MONITOR

T.M. SRINIVASAN and K. VINOD KUMAR
Biomedical Engineering Division,
Indian Institute of Technology,
Madras - 600 036, INDIA.

1. INTRODUCTION

The study of the human spinal reflex was first introduced by Hoffman (1918). The reflex described by him was subsequently named the H reflex and is accepted as a spinal reflex. (Hoffman, 1918; Magladery and Mc Dougal, 1950). The H reflex is a sub-threshold monosynaptic discharge and is usually a triphasic response of short duration. Low intensity stimulation of the posterior tibial nerve evokes the classical H reflex in the calf muscle.

The H wave has been used as a measure of monosynaptic reflex activity and hence spinal motoneuron excitablity. With this technique, patients with spasticity and rigidity have been studied and compared with normals in an attempt to understand the mechanisms of altered motor function. (Teasdall et al, 1952; Angel and Hoffman, 1963; Landau and Clare, 1964).

The maximum M response in muscles represent the activity of all motor units in the muscle, while the maximum H reflex recorded with the same position of electrodes represents the total activity of the motor units activated by the IA afferents. The ratio of H/M is an index of motoneuron excitablity, and of what portion of the motoneuron pool can be activated by this reflex. (Angel and Hoffman, 1963; Landau and Clare, 1964).

The spinal motoneuron excitablity in man can be tested by studying the recovery cycle of the H reflex after a conditioning electrical stimulus using the paired shock (H1, H2) technique (Magladery, 1955; Pillard, 1955; Yap, 1967).

This paper describes a dedicated microprocessor (μP) based system for automatic neurological monitoring. The associated instrumentation used includes a stimulator controlled by the μP, EMG/ECG preamplifier, electrodes, and an oscilloscope. The system can be used for obtaining various neurophysiological data with ease in clinical work and also for analysing this required information and processing them to obtain requisite parameters and waves. An extra feature of averaging sensory evoked potentials (typically auditory) is also incorporated in the system.

2. INSTRUMENTATION

The system is built around an Intel 8085 CPU. An 8 bit A/D converter is interfaced to an 8 bit port of the μP system for acquiring analog data from the preamplifier. This is controlled by the serial output data (SOD) line of the μP. Waveforms are displayed on the oscillos cope after conversion by an 8 bit D/A converter interfaced to another 8 bit port. The timer in the 8155 is used to control the stimulator output pulse width and frequency. This is also used to control an analog gate when the system is used for evoked potential monitoring, to gate the auditory stimulus to the subject.

A set of 8 function keys are connected to a third port of the μ P sys tem for the user to select the different available user functions or programs. In addition a key board with 24 keys is interfaced to the μ P using an 8279 kayboard/display interface chip. With these keys the user can program the μ P for any other required function. This IC is used to interface six 7 segment alphanumeric displays.

A block schematic of the system is presented in Fig 1.

3. FUNCTIONS

Different user functions are provided which can be selected through a set of 8 different function keys. The functions include initialis ation, data aquisition and display of wave forms (H and M waves, and evoked potentials), addition of subsequent datasets and display of number of datasets acquired, average the summed amplitudes of the da tasets acquired and display averaged wave form, calculation of H/M ratio and display this value, automatically plot the recovery curve by the paired shock technique on the oscilloscope. In addition, the user can execute any other defined program.

3.1. INITIALISATION : The system is initialised for data acquisiti on and storage after the requisite memory locations and scratch pads are cleared.

3.2. DATA AQUISITION AND DISPLAY : The μP initiates the stimulator to send out a 1 msec pulse to the subject. This stimulus duration is chosen since it is more selective for the afferent axons with lo ng utilisation times, and since a shorter duration would favour the activation of α motor axons (Hugon, 1973). The pulse duration can be varied by reprogramming the 8155 timer. Simultaneously, the μP initiates data conversion and aquisition using the SOD line, and acquires 256 samples to form a dataset which is stored in the memory.

For H reflex and M waves, 256 samples are taken in a 40 msec inter val after stimulation, with an inter-sample time of 156 μsec. This gives a sampling rate of 6400 HZ. For evoked potential monitoring, 256 samples are taken in 1.28 seconds with an inter sample-time of 5 msec. This gives a sampling rate of 200 HZ.

After data aquisition, the μP enters a display loop whereby it out puts the stored dataset waveform on the oscilloscope via the D/A

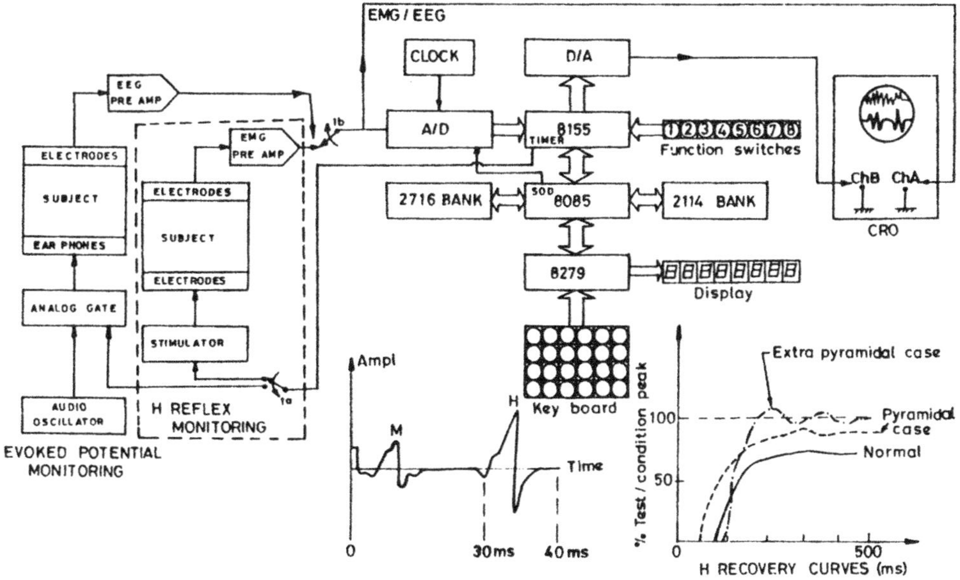
EMG / EEG
CLOCK
D/A
EEG PRE AMP
EMG PRE AMP
A/D
8155
TIMER
Function switches
CRO
ChB
ChA
ELECTRODES
SUBJECT
EAR PHONES
ANALOG GATE
AUDIO OSCILLATOR
EVOKED POTENTIAL MONITORING
ELECTRODES
SUBJECT
ELECTRODES
STIMULATOR
H REFLEX MONITORING
1a
1b
2716 BANK
SOD
8085
2114 BANK
8279
Display
Key board
Ampl
M
H
Time
0
30ms
40ms
Extra pyramidal case
Pyramidal case
Normal
% Test / condition peak
100
50
0
500
H RECOVERY CURVES (ms)

FIG.1

converter. This waveform is displayed continuously until another functi on is initiated by the user.

3.3. H/M RATIO : Data aquisition is carried out as before on the H and M waves elicited. After this, μP automatically detects the peak values of these waveforms and stores them. Then using a divide routine, the ratio of these peak values is calculated in decimal notation to give the H/M ratio with an accuracy of two decimal places. This value is displayed.

3.4. RECOVERY CURVE : The recovery curve is got by using the paired sho ck (H1, H2) technique. Pairs of stimuli (one conditioning and one test stimulus) are given to the subject, each of 1 msec duration. The inter pulse time is increased in steps 25 msec, upto a maximum of 1 second, thus giving 40 pairs of stimuli to the subject. Between each pulse pair, a delay of 5 seconds is introduced by calling a 'Delay Routine'. For ea ch pair, the ratio of the peaks of the H waves elicited by the test and conditioning stimuli is calculated and stored. After corresponding peak ratio values are obtained for all 40 pairs of stimuli, and stored in the display buffer, the recovery curve is displayed on the oscilloscope.

4. DISCUSSION :

A dedicated μP system has been developed and used in our laboratory for neurological monitoring. The system is completely automatic and the user need only select the required function from a set of different fun ction keys.

H reflex studies have been conducted using this system on normals and patients with spasticity and parkinsonism. Systematic monitoring of sub jects during therapeutic exercises have been carried out with ease usi ng this system. The system is also being used for post operative monit oring of spastic patients undergoing cerebellar stimulation.

The signal averaging facility on the system has been used to obtain aud itory evoked potentials from subjects under normal and biofeedback cond ition..

Note: Software developed for this available from the authors on request.

5. <u>REFERENCES</u>:

Angel.R.W. and Hoffman.W.W.(1963). Arch.Neurol.(Chic.),8; 591-596.
Hoffman.P.(1918).Z.Biol.,68: 351-370.
Hugon.M.(1973).New Developments in Electromyography and clinical Neurophysiology, edited by J.E.Desmedt, Vol 3,pp 277-293.
Landau.W.M.and Clare.M.H.(1964) Arch.Neurol.(Chic) 10:128-134.
Magladery.J.W.and Mc Dougal.D.B.Jr.(1950). Bull.Johns. Hopkins. Hosp., 86: 265-290.
Magladery.J.W.(1955).Pflug.Arch.Ges.Physiol., 261:302.
Pillard.J.(1955). Paris, Librarie Arnette.
Teasdall.R.D. et al. (1952) Bull.Johns.Hopkins.Hosp., 91, 245-256.
Yap.C.B. (1967). Brain, vol. 90, Part IV, 1967, pp.887-897.

AUTHOR

INDEX

*Not available at time of publication.

Printed in Dunstable, United Kingdom

91343002R00246